Natural Surfactants

Natural Surfactants

Science, Technology, and Industrial Aspects

Guido Bognolo
Everberg
Belgium

Krister Holmberg
Department of Chemistry and Chemical Engineering
Chalmers University of Technology
Gothenburg, Sweden

WILEY

This edition first published 2026

Registered Office(s)
John Wiley & Sons, Inc., 111 River Street, Hoboken, NJ 07030, USA
John Wiley & Sons Ltd. New Era House, 8 Oldlands Way, Bognor Regis, West Sussex, PO22 9NQ, UK

For details of our global editorial offices, customer services, and more information about Wiley products visit us at www.wiley.com.

The manufacturer's authorized representative according to the EU General Product Safety Regulation is Wiley-VCH GmbH, Boschstr. 12, 69469 Weinheim, Germany, e-mail: Product_Safety@wiley.com.

Wiley also publishes its books in a variety of electronic formats and by print-on-demand. Some content that appears in standard print versions of this book may not be available in other formats.

Library of Congress Cataloging-in-Publication Data Applied for:

Hardback ISBN: 9781394283781
ePDF: 9781394283804
ePub: 9781394283798

Cover Design: Wiley
Cover Image: © FotoKunstenaar/Pixabay

Set in 9.5/12.5pt STIXTwoText by Straive, Chennai, India

Contents

Preface

The 1991 Nobel Prize winner Pierre-Gilles de Gennes is quoted for having once said, "Without surfactants we would be powerless opposite 90% of the industrial problems." This is absolutely true. Almost every single object surrounding us has met surface active agents at one point or another of its life cycle.

The inks of our printers use surface-active agents for dispersing pigments, the cases of our personal computers could not be produced without surface-active auxiliaries and processing aids. The textiles of our clothes have used surface-active agents for fiber spinning, weaving, and dyeing. The food we eat meets surface-active agents either during the growth of the plants, or in its preparation, or in its conditioning, or in its packaging, sometimes in all of these steps. We decorate our homes with paints that use pigment dispersants, wetting, emulsifying, and leveling agents in their formulation. The metal of our cars has been produced using metalworking fluids, often in the form of emulsions. Not to mention the cleansers, emulsifiers, and dispersants used in cosmetics and toiletries, the household and industrial laundry detergents, the hand and machine dishwashing agents, and so forth and so on.

On the back of an ever-increasing demand, surfactants have grown into a global industry, with an output of 15–18 million tons per year, worth billions of USD.

In the past, when the large array of surface-active agents now available did not exist, surfactant effects were obtained using products derived from nature: different varieties of soaps, discovered nearly 5000 years ago, as well as products from plants, such as saponins, or from animals, e.g. phospholipids. The industrial revolution and the evolving socioeconomic environment in the industrialized countries evidenced the limitations of these products already at the onset of the nineteenth century, but it took more than a century before there occurred a dramatic change. During the period from 1930 to 1970, on the back of progress in organic chemistry, the surfactant industry experienced a spectacular growth, both in terms of technology development and market expansion.

It is not exaggerated to say that today surfactants contribute significantly to the quality of the life we enjoy. However, two spots stain this rosy picture.

First, the technological development has slowed down considerably. Few entirely new products have reached the market on a large scale during the last 50 years. With a few exceptions, we use the same types of surfactants today as we did in the 1970s. The advances in recent decades have mainly concerned formulation science rather than development of new surfactant types.

Second, a considerable proportion of the synthetic surfactants relies entirely on nonrenewable raw materials. The balance uses partially or totally sustainable feedstock but remains based on chemical processes that often need noxious chemicals in the synthesis and that may be harmful to the environment.

Environmental concern, particularly related to aquatic toxicity and rate of biodegradation, started to be an issue during the 1970s and has gradually increased in importance. More recently, questions related to carbon footprint have also become important, along with the source of the starting material for the synthetic surfactants.

The producers of surfactants and their raw materials have put forward readily implementable solutions, proposing the bio-based surfactants and the bio-attributable concept that are steps, albeit partial, in this direction. It should be noted, however, that it cannot be automatically assumed that these proposals represent a real environmental benefit. Complete and rigorous life cycle analyses are missing, and it is fair to say that in this area emotion and assumption prevail over scientific evidence.

Against this background, it is not surprising that recent years have witnessed a renewed interest in natural surfactants, that is, those occurring in nature or that are synthesized by microorganisms through naturally occurring processes. However, much of the information reaching the public has been poorly and misleadingly presented, highlighting benefits (sometimes imaginary) and ignoring drawbacks, thereby fueling unrealistic hopes and causing in consequence disillusion and even antagonisms when these remained unfulfilled.

This has hampered the commercial development of biosurfactants (those produced from microorganisms). In addition, there have been inadequate attempts to correct their major drawback, i.e. the cost, which for many years has made them noncompetitive on a cost-effectiveness basis against petrochemical or bio-based surfactants.

Much remains to be done to exploit, to a larger extent, the natural surfactants. This includes the cultivation of surfactant-rich plants in barren or hostile environments unsuitable for agricultural exploitation and the use of the derived surfactants as such or in synergy with conventional surfactants in detergent formulations, which is and will remain the major consumption segment of surface-active agents.

A biosurfactant may not be suitable to substitute a conventional surfactant based on a single effect. It has a better chance to be successful when it offers a combination of effects, for example, cleansing and antimicrobial action. Taking advantage of the multitude of features that many natural surfactants possess will be the way forward when formulating products based on the natural amphiphiles.

Research on biosurfactants still has plenty of challenges ahead. The producing strains must be optimized and so must the fermentation conditions and the process design. There are also challenges related to the formulation of natural surfactants, translating their physicochemical properties into consumer-perceived benefits.

While it is unrealistic to think that natural surfactants will completely replace conventional surfactants, the last few years have seen significant progress in their underlying biochemistry and in understanding their physicochemical behavior. They have entered new applications, and it is likely that they will find more in the coming years. Natural surfactants appear to have entered a virtuous cycle that holds promises for future rewarding and possibly unexpected achievements.

This book attempts to bridge the physicochemical properties of naturally occurring surfactants with their established and potential commercial achievements. The scope of the book is broad, ranging from biochemistry, via biotechnology and surface chemistry, to existing and possible future applications. It also lists producers and distributors of the products in Europe, the United States, China, India, Korea and Japan.

The authors' intention has been to present naturally occurring surfactants in a balanced context for what they really are, dismantling some of the exaggerated and misleading presentations of the past, while highlighting features and potentials that may have gone unnoticed. The authors hope that the book will foster imaginative thinking and be a springboard for a fresh, innovative approach to the realm of the surface-active agents.

Acronyms

Å	Angstrom
Acyl-CoA	acyl coenzyme A
AIST	National Institute of Advanced Industrial Science and Technology – Japan
a.m.	active matter
A/N	author's note
APase	alkaline phosphatase
APG	alkylpolyglucosides
ara	D-arabinose
BCE	before Christian era
BE	Belgium
bw	by weight or body weight, depending on context
ca.	circa
CMC	critical micelle concentration
CSCL	Japan Chemical Substances Control Law
DE	Germany
DHA	docosahexaenoic acid
DLCA	dynamic life cycle assessment
DMSO	dimethyl sulphoxide
EC	European Commission
EC50	concentration that gives half-maximal response
ECHA	European Chemicals Agency
ELINCS	European List of Notified Chemical Substances
EMA	European Medicines Agency
EO	ethylene oxide
EOR	enhanced oil recovery
EPA	Environmental Protection Agency (USA)
EPA	eicosapentaenoic acid
ES	Spain
EU	European Union
FCB	frozen carbonated beverages
FDA	Food and Drug Administration (USA)
FEMA	Flavor and Extract Manufacturers Association (USA)

fru	D-fructose
g	gram
gal	D-galactose
GHz	gigahertz
g/L	grams per liter
glc	D-glucose
glcA	D-glucuronic acid
g/L/h	productivity in grams per liter per hour
G6PD	glucose-6-phosphate dehydrogenase
GMO	genetically modified organisms
GRAS	generally recognized as safe (by the FDA)
GSFA	General Standard for Food Additives
GWP	global warming potential
GYPA	glucose yeast peptone
h	hour
HF	high foaming
HAA	3-(3-hydroxyalkanoyloxy)alkanoic acid
HPLC	high-pressure liquid chromatography
IC50	half-maximal inhibitory concentration
INC	International Cosmetic Ingredient Nomenclature Committee
IP	identity preservation
ISCOM	immunostimulating complex
kg	kilogram
KGI	killing and growth inhibiting
L	liter
LB	Luria-Bertani broth
LC_{Lo}	lowest detected lethal dose
LCA	life cycle analysis
LCC	life cycle costing
LCE	life cycle expenditure
LCSA	life cycle sustainability assessment
LD_{50}	the amount of a material that, given all at once, causes the death of 50% of the individuals in a group
LDL	low-density lipoproteins
M	molar
MAG	monoammonium glycyrrhizinate
MALDI-TOF-MS/MS	matrix-assisted laser desorption ionization time-of-flight tandem mass spectrometry
M-EOR	microbially enhanced oil recovery
mg	milligram
MIC	minimal inhibitory concentration
min	minutes
mM	millimolar
MNNG	N-methyl-N′-nitro-N-nitrosoguanidine
MSDS	material safety data sheet

MTT	3-(4,5-dimethylthiazol-2-yl)-2,5-diphenyltetrazolium bromide
NADPH	nicotinamide adenine dinucleotide phosphate, reduced form
nm	nanometer
µm	micrometer
NOEC	no observed effect concentration
NOEL	no observed adverse effect level
NOS	nitric oxide synthase
NRPS	nonribosomal peptide synthetase
O/W	oil-in-water
PCP	peptidyl carrier protein
PDA	potato dextrose agar
psi	pounds per square inch
q.s.	quantum sufficit
QS	quorum sensing
REACH	Regulation, Evaluation, Authorisation and Restriction of Chemicals
rha	D-rhamnose
RL	rhamnolipids
SANS	small-angle neutron scattering
SAXS	small-angle X-ray scattering
SDS	sodium dodecyl sulfate
SL	sophorolipids
SSF	solid state fermentation
STTP	sodium tripolyphosphate
SUR	substrate utilization rate
SwF	submerged water fermentation
TEM	transmission electron microscopy
TSCA	Toxic Substances Control Act
UK	United Kingdom
U/ml	units per milliliter
USA	United States of America
v/v	volume by volume
v/w	volume by weight
WCO	waste cooking oil
WEO	waste engine oil
WHO	World Health Organization
W/O	water-in-oil
WOC	winterized oil cake
w/v	weight by volume
w/w	weight by weight
XRD	X-ray diffraction
xyl	D-xylose
Yp/x	yield in grams of biosurfactant per gram of biomass

1

General Views on Natural Surfactants

1.1 Introduction

This book has been written with the view of presenting a state-of-the-art bridge between the academic research, the commercial achievements, and the future opportunities of naturally occurring surfactants. The title of the book is natural surfactants, but in the text we sometimes also use the terms naturally occurring surfactants or biosurfactants for the amphiphiles produced by plants, animals, or microorganisms through a naturally occurring biosynthesis. The term "bioamphiphile" has also been proposed in the literature [1].

Biosurfactants are not to be confused with bio-based surfactants that are man-made surfactants produced by chemical reactions from natural building blocks, such as sucrose and sorbitan esters of fatty acids, alkylpolyglucosides, and fatty alcohol ethoxylates built from naturally derived fatty alcohols and bio-based ethylene oxide, or other synthetic surfactants containing different proportions of moieties obtained from nature.

Bio-based surfactants are excluded from this book, because they represent a totally different reality than the biosurfactants. Bio-based surfactants are carbon copies of petrochemical surfactants, made from building blocks of natural origin, but involving chemical reactions, processing conditions, and catalysts that are anything but natural. Because of the raw material origin, they can claim sustainability compared to their petrochemical brothers, but they fall short of being entirely natural. We have dedicated a short section to this category just to describe more completely their features and to position them correctly vis-à-vis the biosurfactants and the petrochemical surfactants.

The term "green surfactants" has become popular and is frequently used by surfactant-producing companies and by formulators, such as the detergent and personal care industries. The term is vague, however, and can mean anything from biosurfactants to man-made surfactants with a reasonable share of the building blocks coming from natural sources. We have avoided the term "green surfactants" in this book.

References to patents are reduced to a minimum as these are not scientific documents, nor do they provide useful information on commercial aspects. The few patent references quoted are because of their conceptual value rather than for scientific or commercial interest.

Regulatory information is immensely complex as it is country and application dependent and in various stages of legal development (and therefore in a constant goalpost moving mode). Only if something is approved or forbidden for a very clear reason, a firmer

Natural Surfactants: Science, Technology, and Industrial Aspects, First Edition. Guido Bognolo and Krister Holmberg.
 Published 2026 by John Wiley & Sons Ltd.

statement on regulatory acceptance/nonacceptance can be made. For this reason, only a few references are reported in the book.

The information given in the different chapters on regulatory matters and commercial issues represents the situation known to the authors at the moment of writing.

1.2 Surfactants from Nature

Nature is a forge of surface-active agents and produces scores of substances with amphiphilic structure with considerable surface and/or interfacial activity, some of which have been exploited for a long time. For sure, the people who inhabited the Blombos caves in South Africa around 100 000 years ago did not know that they were dealing with colloid chemistry when they prepared the color mixtures, of which the remains have been recently discovered. But they did, and they must have used some natural ingredients in the process. Closer to our times, polysaccharides from acacia trees (e.g. *Acacia nilotica*, *Acacia senegal*, and *Vachellia (Acacia) seyal)*, commonly referred to as Gum Arabic were used in the preparation of inks by the ancient Egyptians. By reference to these products and the effects that can be obtained from them, modern researchers have used Gum Arabic to separate agglomerates of carbon nanotubes into individual fibers [2].

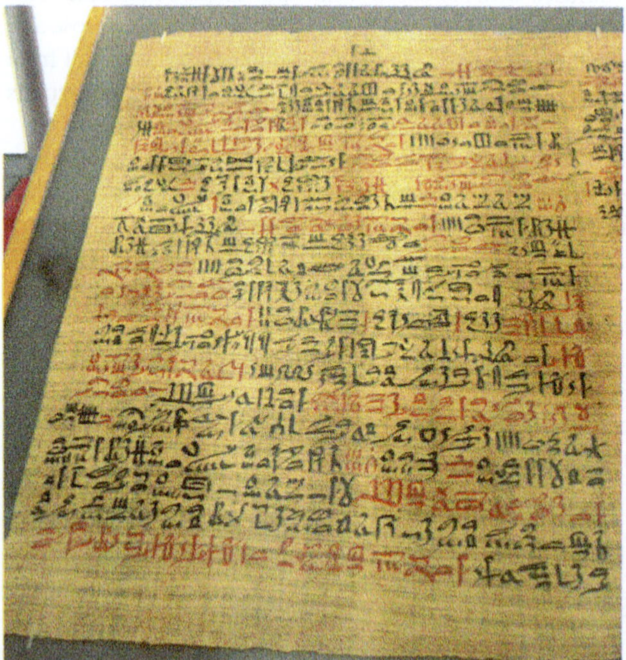

Figure 1.1 *The Ebers Papyrus*, ca. 1550 BCE. (*Source:* Einsamer Schütze / Wikimedia Commons / CC BY-SA 3.0.)

The Ebers Papyrus (ca. 1550 before the Christian era [BCE]), see Figure 1.1, along with the *Kahun Gynaecological Papyrus* (ca. 1800 BCE), the *Edwin Smith Papyrus* (ca. 1600 BCE), the *Brugsch Papyrus* (ca. 1300 BCE), and the *London Medical Papyrus* (ca. 1300 BCE) are among the oldest preserved documents of ancient Egyptian medicine. They contain recipes for the treatment of various diseases that involve the use of both natural and man-made surface-active agents (essentially soap) for their preparation [3].

Surface-active agents are produced by plants and macro- and microorganisms in an amazing variety of molecules and in relatively small quantities for functions that are often not fully understood yet. Several hundred species have been isolated from the two principal vegetal sources, *Quillaja saponaria* and *Yucca schidigera*, and from the other numerous plants producing saponins (e.g. bouncing bet, (*Saponaria officinalis*), a few from animals (e.g. starfish (Asteroidea), and sea cucumber (Holothuroidea)). Phospholipids and bile acids are found in the bodies of animals and humans, where they facilitate pulmonary functions and ensure food absorption, respectively. A large variety of sugar lipids and lipoproteins are synthesized by bacteria, fungi, and yeasts.

With the exception of saponins, which have been used for centuries in many countries around the world to wash clothes, and the phospholipid lecithin, isolated for the first time in 1845 and purposedly used as emulsifier in food since then (but certainly already used long before not as an isolated product but via the main product containing it, i.e. egg yolk), until recent years there was no interest in natural surfactants.

The use of saponins for washing clothes had fallen into oblivion with the advent of broader availability, first of soaps and later of synthetic surfactants. However, saponins are being proposed again by regional formulators because of their performance, natural connotation, and ecological profile. They have found a wide variety of profitable niche markets, ranging from aquaculture and adjuvants for vaccines to foaming agents in nonalcoholic drinks. These are specialty applications where synthetic surfactants are inadequate. An example of a niche segment where the peculiar effects of saponins are exploited is aquaculture. The hemolytic properties of saponins were recognized long ago and saponin extracts were spread on rivers and ponds to daze or kill fish, making their gathering easy. This practice is still common with primitive tribes and, in more sophisticated ways, in industrial aquaculture in developed countries. In fact, saponins are toxic to some aquatic species but not to others, e.g. shrimps and shellfish in general. Not only are they nontoxic but they also have nutritional value. They have become an important ingredient in fisheries, killing predators on one side and contributing to the growth of the valuable breed on the other. The historical source of saponins, *Quillaia saponaria*, was inadequate to respond to the growing demand and new plant sources, such as *Yucca schidigera*, and new cultivation methods have been introduced.

Lecithin is also entering new markets as its phospholipids are essential ingredients to produce the high added value organized structures, liposomes, and ethosomes, capable of encapsulating and controlling the penetration or the release of drugs or cosmetic actives for pharmaceutical and personal care formulations.

While the presence of surface-active species in plants and animals had been recognized long ago, it was only in 1946 that an amphiphilic molecule produced by a microorganism was mentioned [4] and identified a few years later by Jarvis and Johnson as the metabolites of *Pseudomonas* species [5].

In 1961, Gorin et al. described the oil formed during fermentation by a strain of *Torulopsis magnoliae* that mainly consisted of partly acetylated 2-O-β-D-glucopyranosyl-D-glucopyranose units attached β-glycosidically to 17-L-hydroxyoctadecanoic and 17-L-hydroxy-9-octadecenoic acids, the first sophorolipids isolated [6]. In 1968 came the discovery of surfactin, the biologically active secondary metabolite first identified in cultures of *Bacillus subtilis* strains [7].

Surfactants from microorganisms, notably the lipopeptide surfactin, are used in medical and pharmaceutical research. Sophorolipids and rhamnolipids are increasingly included in cosmetic preparations, and have become an ingredient for mass consumers, the former in household cleaners and the latter in hand dishwashing formulations.

Many other surface-active species produced by microorganisms were identified in the years after the initial discoveries, and with them came a better selection of the producing strains and the progressive improvement of the production process, such that there are now on offer microbial surfactants in quantities and at prices that eventually enable their use in commercial formulations.

On the back of these achievements, the chemical industry is beginning to take a real interest in biosurfactants. Many major surfactant producers like BASF, Dow, Sasol, Solvay, and Evonik, as well as formulators like Unilever and Ecover, have biosurfactants in their selling range, in their research programs, or in their mass consumer products. All this means that the volumes of biosurfactants produced are steadily increasing. Early manufacturers such as Holiferm and Evonik increase their capacities, and new ones emerge.

1.3 Natural Surfactants Versus Bio-Based Surfactants

After years of warning messages from the scientific community, the public opinion worldwide has gained consciousness and has started to react to the reality that our planet is warming up, that the climate is changing, and that in the absence of concrete and immediate actions major disruptions ahead are to be expected. And there is a growing consensus on the cause of this phenomenon: the ever-increasing emission of carbon dioxide from the use of fossil fuels, i.e. petroleum, gas, and coal. In many countries, large segments of the population have developed an almost instinctive revulsion against anything with a "petro" connotation and an equally instinctive, sometimes even irrational, attraction for what looks natural, sustainable, renewable, in one word, "green." Something that has eventually coalesced around that magic and captivating word "bio."

The surfactants industry did not have much "bio" in its basket when this movement began to gain momentum and started to impact on consumer choices, reversing the trend of the previous generations. In fact, since the early years of the twentieth century in the industrialized countries, there had been a progressive penetration of synthetic detergents in all the activities of an evolving society, from household cleaning and personal care to the agriculture and the industry in general. All human activities consumed more and more petrochemical products: alkylbenzene, alkylphenol, ethylene oxide, synthetic fatty alcohols, paraffins, and olefins. Only a generation ago, the surfactants industry relied on nonrenewable raw materials for most of the production.

Looking *a posteriori* it is evident that, in the absence of facts, the first response from the surfactant producers was haphazard to say the least. It took the form of highlighting to the consumers the natural/sustainable content of their products. An example of such an approach is the document, shown in Table 1.1, published by Cognis (at that time not yet part of BASF) in 2008 for some of their personal care products, showing the classification of their products based on the content of natural moieties.

Table 1.1 shows that few molecules could have the label "100% natural" and those that did were biological actives rather than surface active agents. A more thought-after response came when the surfactant producers realized that they could create the new label "bio-based" to attach to their surfactants by switching the raw materials from petrochemical to natural sources.

Fatty alcohol ethoxylates became bio-based if made from "a natural fatty alcohol" and "bio-ethylene oxide," regardless of the fact that the "natural" raw materials, fatty alcohol, and ethylene oxide were produced via nonnatural chemical processes involving consumption of nonrenewable resources. To produce a fatty alcohol a triglyceride oil is normally converted to a fatty acid methyl ester in a process called methanolysis, consuming methanol and energy, and the methyl ester is subsequently hydrogenated, using hydrogen that does not exists in nature and has to be petrochemically produced, consuming energy and generating offal, without considering the use of land, fertilizer, pesticides for growing the plants, or the feeding for growing the animals, for the production of the triglycerides, not to mention the energy consumption involved in the harvesting and working-up of the crop.

Bio-ethylene oxide is made by dehydration of ethanol produced from sugar canes or sugar beet (again use of land, etc.) followed by oxidation with a silver catalyst at high temperature and pressure (again energy, offal, and chemical consumption). Few people are probably aware of the fact that 50% of the silver extracted globally every year is immobilized in catalysts used to produce ethylene oxide.

The European Commission for Standardization has defined a bio-based surfactant as follows [8]:

- *Wholly bio-based:* >95% bio-based carbon; surfactants where all raw materials can be considered as bio-based
- *Majority bio-based:* 95–50% bio-based carbon; surfactants where the majority of the raw material is bio-based
- *Minority bio-based:* 50–5% bio-based carbon; surfactants where a minor part of the raw material is bio-based
- *Non-bio-based:* <5% bio-based carbon; surfactants where no raw material is bio-based

The percentage of bio-based carbon is usually calculated based on data from the surfactant supplier. It can also be determined experimentally by isotope analysis. Carbon in nature consists of 99% ^{12}C, 1% ^{13}C, and a tiny amount (1 per 10^{12} carbon atoms) of ^{14}C. The ^{14}C isotope is radioactive with a half-life of 5730 years. This means that it has more or less completely disappeared in fossil petroleum. Thus, an isotope analysis of the carbon in a linear alcohol made by oligomerization of ethylene will show a negligible amount of ^{14}C while the same alcohol obtained from a triglyceride by methanolysis followed by hydrogenation of the methyl ester will give an abundance of ^{14}C of 1 per 10^{12} carbon atoms. The methodology

Table 1.1 Cognis' classification model based on the content of natural ingredients in their personal care products (2008). (*Source:* Cognis release, reprinted from author's archives.)

Category	Description	Product examples	Chemical structure/INCI
Cognis naturals	From 100% natural, renewable feedstocks; physically purified by water, alcohol, or energy treatment.	Coviox® T50	Tocopherol
		Aloveria®	*Aloe barbadensis*
		Generol® R	*Brassica campestris* (rapeseed) Sterols
Cognis natural modified	From 100% natural, renewable feedstocks; chemically processed using catalysts or other reaction aids.	Cetiol® OE	Dicaprylyl ether
		Plantacare® 818 UP	Coco-glucoside
		Glucopon® 600 CSUP	Aqueous dispersion of alkylglucosides based on natural fatty alcohol C10–C16, free of preservatives
Cognis ecohybrids	From a mixture of natural, renewable, and synthetic feedstocks, combined in a chemical process. Higher level of natural C-atoms.	Lamesoft® Care	PEG-4 distearyl ether (and) sodium laureth sulfate (and) distearyl ether (and) dicaprylyl ether
		Cosmedia® DC	Dimethylcarbonate copolymer
		Texapon® N70	Sodium laureth sulfate
Cognis hybrids	From a mixture of natural, renewable, and synthetic feedstocks, combined in a chemical process. Lower level of natural C-atoms.	Plantapon® LC7	Sodium lauryl glucose carboxylate (and) lauryl glucoside (CTFA)
		Dehypon® LS 54	Fatty alcohol C12-C14 with approx. 5 moles EO and approx. 4 moles PO
		Cosmedia® CTH (E)	Polyquaternium-37 (and) propylene glycol Dicaprylate/dicaprate (and) PPG-1 trideceth-6 (CTFA)
Cognis synthetics	From synthetic feedstocks, chemically processed.	Ultragel® 300	Polyquaternium-37 (CTFA)
		Polyquart® Ampho 149	Aqueous solution of an acrylic copolymer
		Dehyquart® E-CA	Hydroxycetyl Hydroxyethyl dimonium chloride (CTFA)

is the same as archaeologists use to determine the age of organic artifacts, where it is often referred to as radiocarbon dating or carbon-14 dating.

In the literature, and in particular in the promotion material from raw materials producing companies, the term "bio-attributed" is sometimes seen. The label "bio-attributed" for a given product indicates that the use of renewable feedstock has been ascribed to that product using the "mass balance approach." The mass balance approach provides a means to calculate and declare that an equivalent amount of biogenic carbon was used compared to nonbiogenic carbon, but gives no indication of the total quantity of biogenic substitution. Thus, "bio-attributed" is not necessarily synonymous with "bio-based."

Abstracting from the chemicals involved in the raw material preparation and in the surfactant synthesis the following surfactant classes are examples of amphiphiles that could claim the label "wholly bio-based" if produced with the appropriate raw materials, such as natural fatty alcohols and bio-ethylene oxide (although the presence of certain groups, such as sulfate, sulfonate, ammonium, and phosphate, may make the "wholly bio-based" attribution questionable):

- Fatty alcohol ethoxylates
- Fatty alcohol sulfates, sodium salts
- Fatty alcohol ether sulfates, sodium salts
- Alkylpolyglucosides
- Sucrose esters
- Natural fatty amine ethoxylates
- Natural fatty amide ethoxylates
- Sorbitan esters
- Polysorbates
- Fatty acid esters of glycerol and of other naturally occurring polyols
- Castor oil ethoxylates
- Fatty amino acid surfactants
- Phosphate ester ethoxylates
- Natural fatty amine oxides
- Fatty acid sulphosuccinates and their ethoxylated derivatives

In general, switching surfactant raw materials, such as a C12–14 alcohol, from a petrochemical to an oleochemical origin is not an issue, as for most applications, there is no noticeable difference in performance. However, given the choice cosmetic formulators will use bio-based products. In special formulations, for example, those used in crop protection or for pharmaceutical purposes, a change in ingredients may require re-registration.

To the best of knowledge, as of autumn 2023 in the United States, only two companies, Croda and Clariant, produce bio-ethylene oxide, and they do so exclusively for in-house consumption. This drive to nature was probably prompted, at least in part, by the prohibitive cost of transporting ethylene oxide from the US Gulf, where it is produced, to the ethoxylation sites of these companies in the Northeast of the country, combined with the risk of disruptions in the transport chain (e.g. stricter regulations or limited availability of rail cars certified for EO transport). India is the only other country in the world where bio-ethylene oxide is produced on an industrial scale. India Glycol has been producing it since 1989 (India Glycol Ltd. Company Summary – India Infoline).

In July 2022, INEOS (London, UK) launched bio-attributed ethylene oxide which is claimed to have 100% substitution of fossil feedstock and give 100% greenhouse gas saving compared to ethylene oxide produced in the conventional way, as certified by both RSB (Roundtable on Sustainable Biomaterials) and ISCC+ (International Sustainability & Carbon Certification Plus) (INEOS press release, July 27, 2022).

During recent years, there has been an interest in making bio-based surfactants from a wide range of natural raw materials, some with quite a complex structure, which have been synthesized. The driving force for such investigations, apart from scientific curiosity, has been to find new and green building blocks for surface-active molecules, in particular amino acids and peptides. This has resulted in scientific publications of high standard with respect to preparative organic chemistry. However, these amphiphiles are also man-made in the laboratory by organic – or sometimes bioorganic – synthesis and are therefore examples of bio-based surfactants, not natural surfactants. Thus, they are outside the scope of this book.

1.4 Advantages and Disadvantages of Naturally Occurring Surfactants

Much literature has been produced portraying natural and biosurfactants as the Holy Grail that will enable to correct all the wrongdoings of the synthetic surfactants, perhaps on the back of the negative connotation of anything that has to do with chemistry and in particular petrochemistry, or maybe by ignorance of the facts. Whatever the reason, the reality is that it has contributed to a distorted, false image of the benefits of biosurfactants.

A few observations are appropriate and may hopefully contribute to a more realistic picture.

First, biosurfactants will never be able to completely replace the synthetic surfactants. The best evidence is that synthetic surfactants were developed and encountered a rapid and widespread acceptance because of the shortcomings of the sustainable surfactants previously used, in particular, soaps.

Another reason is that while the synthetic surfactants comprise all categories, i.e. anionic, nonionic, cationic, and zwitterionic, the majority of the naturally occurring surfactants are essentially nonionic with a weak anionic character conferred to some of them by a carboxylate moiety. And in surfactancy terms, a carboxylate group does not have the power of a sulfate or sulfonate group.

It should also be recognized that a key feature of nature, with consequences on its products, is variability. The chemistry and structure of its produce depend, for example, on the soil composition, the meteorological conditions, and the identity of the producing organisms. Sometimes, small differences in congener molecular structure can result in significant differences in functionality. For example, sophorolipid congeners can possess the same length and chemical structure of the fatty acid chain but can exist either in the lactonic or the acidic form. The lactonic form has strong antimicrobial properties that are missing in the acidic form.

A feature of some of the synthetic surfactants is the suitability of one single product for a large range of different applications. Alkylbenzene sulfonates or alkylphenol ethoxylates can cover, if not all, at least a great number of applications and application conditions from the point of view of the physical chemical performance. The fact that in practice other surfactants or combinations of surfactants are often used instead is a matter of toxicology, environmental impact, cost effectiveness, consumer habits, regulatory, etc. Before their use was restricted because of environmental and human safety concerns, one could find alkylphenol ethoxylates in household detergents and cosmetic products, as well as in crop protection, textiles, paints, paper, and metalworking formulations. Their safer replacements can also be used in a broad, though considerably reduced, range of applications. None of the naturally occurring surfactants discovered so far has shown such versatility.

The main advantage of naturally occurring surfactants is the sustainability, in the sense that they are the result of the natural transformation of renewable substrates. There is the widespread perception that since they are a produce of nature, they must be good, overlooking that some of the most potent poisons, such as curare and strychnine, are naturally occurring, respectively from *Chondrodendron tomentosum* and *Strychnos nux-vomica,* not to mention the taxine alkaloids, that can be found in garden's hayes, such as those of *Taxus baccata* (English yew), *Taxus cuspidata*, *Taxus canadensis*, *Taxus floridana*, and *Taxus brevifolia* (Pacific yew or Western yew).

It is generally believed that natural surfactants must be readily biodegradable and safe to humans and the environment, but this lacks scientific demonstrations, as discussed later.

Biosurfactants are natural products produced by microorganisms colonizing a variety of environmental niches; some can maintain their effectiveness even under much adverse conditions. Certain biosurfactants have been shown to maintain their physical chemical activity at high temperatures and within a pH range of 3–12 and salt concentrations up to 10%. Most synthetic anionic surfactants are deactivated by high concentrations of divalent cations such as calcium and magnesium, while ethoxylated nonionic surfactants are sensitive to monovalent cations.

For sure, natural surfactants do not need for their production hazardous and harmful chemicals such as SO_3, HCl, HBr, HF, oleum, H_2SO_4, H_3PO_3, ethylene oxide, epichlorohydrin, just to mention a few. They also do not require organic solvents for their synthesis. But it should not be overlooked that high biosurfactant yields sometimes require pathogenic species (e.g. *Pseudomonas aeruginosa* for producing rhamnolipids or *Bacillus subtilis* for surfactin) or genetic modification (again for the production of rhamnolipids). This is just to say that many of the generally accepted beneficial features of biosurfactants may originate from debatable premises.

There are much less toxicological data reported for the natural surfactants than for the synthetic surfactants, for which there is also a wealth of information from in-use conditions. The knowledge available suggests in general, a favorable toxicological profile for the natural surfactants, save for the strong hemolytic effect and the sternutatory effect of the powder saponins, but the field is still in its infancy. Also, the regulatory for most biosurfactants ranges from ill-defined to nonexistent.

Table 1.2 summarizes the most relevant strengths and weaknesses of naturally occurring surfactants compared with synthetic surfactants.

1.5 Production Process of Natural Surfactants

Natural surfactants of plant and animal origin can be isolated with traditional technologies involving solvent (generally water) extraction, filtration to separate impurities, and possibly concentration and drying. Biosurfactants from microorganisms involve a more

Table 1.2 Strengths and weaknesses of natural surfactants.

Criterion	Strength	Weakness
Environmental impact	x	—
Social acceptability	x	—
Sustainability	x	—
Availability	—	x
Cost	—	x
Heterogenicity of raw materials	x	—
Product consistency	—	x
Competitivity for food production	—	x
Process yield	—	x
Biodegradability	x	—
Performance	—	x
Global surfactancy	—	x
Versatility	—	x
Tolerance to external environment	x	—
Physical chemical characterization	—	x
Formulation expertise	—	x
Toxicity	Ill defined	Ill defined
Regulatory	Ill defined	Ill defined

complex process with the intervention of several variables, as discussed below. Most of these biosurfactants are exolipids, although in some cases they can be cell-bound. The excretion occurs in certain phases of the fermentation cycle: the late exponential growth or the stationary phase.

The surfactant-producing microorganisms can be divided into three groups with respect to hydrocarbon utilization and the synthesis of extracellular lipids:

- Microorganisms that produce biosurfactants exclusively from hydrocarbons, e.g. some strains of *Arthrobacter* sp., *Corynebacterium* sp., and *Nocardia* sp.
- Microorganisms that produce biosurfactants from both hydrocarbon and water-soluble substrates. This is by far the largest category, the best-known example being the rhamnolipids-producing *Pseudomonas aeruginosa*.
- Microorganisms that produce biosurfactants exclusively from water-soluble substrates. Surfactin-producing *Bacillus subtilis* is one example.

Figure 1.2 shows the microorganisms that produce the most promising biosurfactants: *Candida albicans* for the sophorolipids, *Pseudomonas aeruginosa* for the rhamnolipids, and *Bacillus subtilis* for the lipopeptide surfactin.

The microorganisms feeding on hydrocarbons attracted much attention in the early days of the biosurfactants development, in the expectation that they could assist in oil recovery and/or in soil bioremediation, which has happened only to a moderate extent. Today, the

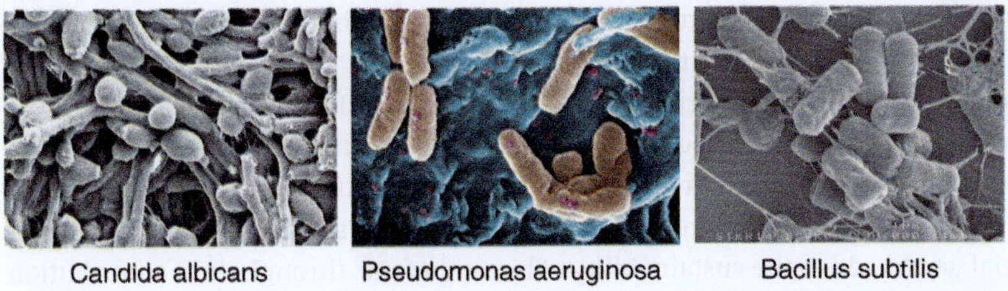

Figure 1.2 The three microorganisms producing the most promising biosurfactants visualized using scanning electron microscopy. (The left image is from Vader1941 / Wikimedia Commons / CC BY-SA 4.0; the middle image is from Janice Carr / Pixnio / Public domain; and the right image is from Reproduced with permission from Zweers et al. (2008) / Springer Nature / CC BY 2.0.)

commercially most successful biosurfactant, a rhamnolipid, is produced by genetically modified *Pseudomonas putida*, grown on corn glucose.

There is a distinction between the biological processes leading to the formation of biosurfactants in nature and their laboratory or industrial production. This section will focus on the latter, while the biological pathways are discussed separately in the chapters on the different classes of biosurfactants. The natural biosynthesis generally yields biosurfactants with specific properties in concentrations of a few grams/liter or less, which makes their costs unacceptable except for a few extremely high added value applications. However, the production yield, the productivity, the chemical structure, the composition, and the physical chemical properties of the biosurfactants can be modified to a certain extent by genetic, biological, or chemical manipulation.

1.5.1 Selection of the Microorganism

The selection of the microorganism is perhaps the most critical step for the production of a biosurfactant, as it will be decisive for the yield, with only limited (if any) information to guide the choice [10]. So, the first step in the process is the choice of microorganism, using criteria of interest to the researcher. Strains of interest need to be harvested from a suitable source and then isolated using various culture techniques. Strains are usually selected that efficiently feed on the raw materials chosen, preferentially as the sole energy source. These can be refined raw materials, such as glucose, but they can also be crude products if the objective is to reduce the cost of the feed, while retaining an acceptable yield. The strains that give the highest cell density at a certain optical density are retained for further optimization. Several tests can be performed at this point, e.g. Gram-staining analyses, motility, shape (e.g. rod-shaped, coccoid, bacilli), catalase (or other enzyme) sensitivity, and genetic identification using 16S rDNA gene sequence analysis. A phylogenetic tree is useful for detecting evolutionary relationships.

1.5.2 Medium Composition

The main parameters to consider are:

Carbon source: The carbon source is the primary variable to consider in the production of biosurfactants, as it directly influences the growth of the microorganism, as well as the structure and yield of the target biosurfactant molecule. A variety of carbon sources have been studied, ranging from crude oil to refined hydrocarbons of different molecular weight, glycerol, glucose, and their mixtures [11, 12], as well as carbon sources derived from waste products from industrial processes [13]. While in the case of more or less pure raw material, the prime objective is to maximize yield and productivity, the drive for using waste raw material is to reduce production costs, both directly and because of the added value of reducing or eliminating disposal problems and costs. The use of waste raw material would add to the sustainability of biosurfactants through their contribution to a circular economy.

The company Evonik reports that glucose from corn, nongenetically modified, is the carbon source for the industrial production of their rhamnolipid (Evonik management, personal communication).

Nitrogen source: The nitrogen source can be of inorganic origin, such as ammonium nitrate and ammonium sulfate, or organic, such as urea, amino acids, and yeast extract. The choice of nitrogen source depends on the composition of the medium and the producing microorganism [14]. An example of where the nitrogen source affects biosurfactant production has been identified in *Bacillus* species. Increased surface activity or emulsifying efficiency of *Bacillus* biosurfactants is linked to the use of organic or inorganic nitrogen sources, respectively, during growth [15]. When both organic (yeast extract) and inorganic (ammonium nitrate) sources are added simultaneously to the medium, there is synergism between the two sources in the production of biosurfactants [16]. Nurfarahin et al. have reviewed various nutrients and elements considered in the formulation of a production medium describing, among others, the effect of organic and inorganic nitrogen sources in the production of biosurfactants [17].

C/N ratio: The ratio between the carbon source and the nitrogen source is an important parameter for the production yield. The C/N ratio determines the endpoint of the exponential microorganism growth and the onset of the metabolic process. However, the optimal C/N ratio varies widely depending on the strain, as well as on the carbon and nitrogen source, and needs to be established experimentally for each individual biosynthesis.

Fermentation medium: A typical medium for the production of biosurfactants will consist of a source of carbon and energy (ideally glucose, but hydrocarbons, oils, and degraded products from other industrial processes may also be employed) and a source of nitrogen. It further contains potassium dihydrogen phosphate and/or dipotassium hydrogen phosphate to control the pH and serve as a source of phosphorous. The medium may also contain microelements like Ca^{2+}, Mg^{2+}, Fe^{2+}, Fe^{3+}, Mo^{2+}, and Mn^{2+}. Although not critical, the choice and level of the microelements must be optimized in each specific case.

Biosurfactants are produced mostly in the stationary stage of the fermentation process when the feed is no longer used for producing cells but is metabolized into surface-active molecules. Ideally, the process should be performed under conditions such that cells are produced as fast as possible and in volumes as high as possible. From this phase of exponential growth, the system should move to a stationary stage and remain in that producing stage for as long as possible.

The possibility of recovering the cells and using them for the exclusive production of metabolites has been investigated and holds promise.

Subsanguan et al. studied the possibility of maintaining the biosurfactant-producing activity of a strain of a GRAS ("generally recognized as safe" by the US Food and Drug Administration) lactic acid bacterium, *Weissella cibaria* PN3. The bacterium was immobilized on a commercial porous carrier. For biosurfactant production, 2% soybean oil was used as the carbon source. After 72 hours, the immobilized cells were reused by replacing the production medium. The extracellular and cell-bound biosurfactants were extracted from the resulting cell-free broth and cell pellets, respectively. To maintain biosurfactant production activity, the immobilized cells were reactivated every three production cycles by incubating the washed immobilizing carriers in LB medium (LB stands for Luria-Bertani broth) for 48 hours. The maximum yields of purified extracellular and cell-bound biosurfactants were 1.46 g/L and 1.99 g/L, respectively, and were achieved in the fourth production cycle. The repeated biosurfactant production of nine cycles was completed within one month, while only 2 g of immobilized cells/L were applied. The extracellular and cell-bound biosurfactants had comparable surface tension values (31–33 mN/m); however, their CMC values were different, 1.6 and 3.2 g/L, respectively [18]. Although the total quantity of biosurfactants produced was not large (24.67 g/L for the sum of extracellular and cell-bound biosurfactants), the process is relatively cheap, given that two types of biosurfactants were produced simultaneously, and no new bacterial inoculum was required. Furthermore, the yields were higher than observed in other studies [19, 20].

However, there are some disadvantages of cell immobilization for long-term usage, such as cell inactivation during the process, mass transfer limitations, and accumulation of toxic metabolites or inhibitor products inside the carrier [21].

1.5.3 Feeding Strategies

The feed can be supplied in one (batch feeding) or sequential installments (fed-batch feeding) according to one of the following strategies.

1.5.3.1 Pulse Feed

The nutrient solution is added in separate installments at fixed time intervals. The advantage is the simplicity of the feed addition; the disadvantage is that the optimal feed volume and the feed interval times have to be optimized by trial and error and may vary with time. In addition, the yield and productivity are lower compared with other strategies.

1.5.3.2 Constant Feed Rate

A constant feed rate strategy involves adding a nutrient solution at a fixed rate throughout the process. The advantage of this strategy is that it is easy to implement and monitor, and it can prevent substrate accumulation and inhibition. The disadvantage is that it does not account for the changes in the biomass concentration, the specific growth rate, and the product formation rate during the process. Therefore, it may not be optimal for maximizing the productivity or the product quality.

1.5.3.3 Exponential Feed Rate

An exponential feed rate strategy is based on the principle of maintaining a constant specific growth rate of the microorganisms or cells. It involves adding a nutrient solution at a rate that is proportional to the biomass concentration. The advantage of this strategy is that it can achieve a high biomass yield and a high product concentration, as well as reduce the risk of overflow metabolism and by-product formation. The disadvantage is that it requires accurate measurement and control of the biomass concentration, which can be challenging at a large scale. Moreover, it may not be suitable for processes that require a low or variable specific growth rate.

1.5.3.4 Feedback Control Feed Rate

A feedback control feed rate strategy is a more advanced and flexible feeding method for fed-batch bioprocesses. It involves adding a nutrient solution at a rate that is adjusted according to the feedback from one or more process variables, such as substrate concentration, dissolved oxygen, pH, or product concentration. The advantage of this strategy is that it can adapt to the dynamic behavior of the bioprocess and optimize the feeding rate according to the desired objectives. The disadvantage is that it requires a complex and robust control system, as well as reliable and accurate sensors and actuators.

As an example, Bouassida et al. compared batch and fed-batch strategies for biosurfactant production by *Bacillus subtilis* SPB_1. Experiments of fed-batch fermentations were carried out through three different glucose feeding strategies, namely the pulsed, the constant feed rate, and the exponential feeding. The comparison between different fermentation processes revealed that the fed-batch process is a more efficient cultivation strategy than the batch process in terms of cell biomass and biosurfactant production and productivity. Among the three different feeding strategies, the exponential feeding process achieved the highest biosurfactant concentration, an almost twofold increase compared to batch fermentation [22].

Industrially, biosurfactant production is mostly carried out under submerged fermentation (SmF), which implies cultivating microorganisms in a liquid nutrient medium. This is because of history: SmF was used in the early days and therefore more knowledge has been accumulated. In the case of SmF, there is also a possibility of stricter and simpler control over various parameters, whereas there is a knowledge gap for the solid-state fermentation process (SSF). SSF is performed on a solid substrate with a low moisture content, with the advantage that relatively low energy is consumed. The required water content in SSF can be provided by the feeding medium and is absorbed by the substrate in a solid matrix. This offers advantages for the transfer of oxygen to support the growth of microorganisms. Several agricultural wastes, such as rice straw, sugarcane bagasse, wheat straw, rice hulls, and corn cobs, are used as substrate for SSF. One possible approach to produce biosurfactant is to utilize SSF with substrates derived from other industrial processes. For example, in two days of fermentation using a simple downstream process with 5.58 g of soy flour and 3.67 g of rice straw, 50.01 mg/g of dry lipopeptide was produced by *Bacillus amyloliquefaciens* XZ-173 in SSF [23].

In addition, through SSF and using molasses as substrate, a strain of *Bacillus subtilis* producing 74 mg/g of biosurfactant has been reported [24]. Eras-Muñoz et al. have investigated the sustainable production of sophorolipids through an SSF process, which seems to be a promising alternative for re-valorizing industrial or food waste that cannot be exploited by SmF. Industrial wastes like sunflower oil cakes, waste-water from the cleaning of industrial equipment used in the production of cosmetics and household cleaners, and organic fractions of municipal solid wastes were used as feeding substrates. A production of 0.033 g/L/h was achieved under optimum conditions, with a yield of 0.047 g/g for the diacetylated lactonic C18:1 and of 0.141 g/g for the sophorolipid crude extract [25]. Bouassida et al. conducted a comparative study of SmF and SSF for the production of lipopeptides using Aleppo pine waste and confectionery effluents as cheap substrates. The SmF gave a yield of 0.0172 g/g versus 0.0276 g/g for the SSF, thus showing the superiority of SSF under the test conditions [26].

The above cases demonstrate the feasibility of using SSF to produce lipopeptides and glycolipids, saving operational costs related to SmF. However, more work is needed before SSF is ready for large-scale processing, notably in terms of reaction times (longer than with SmF) and heat transfer. Bioreactor design and sustainability are some critical limiting factors for SSF, which need to be tailored based on the existing advancements in SSF techniques.

1.5.4 Biosurfactants Transport

One of the key contributors to reaching high productivity titers is the transport of biosurfactants outside the microbial cell environment where they are synthesized. Apart from the obvious economic advantage of an increased efficiency of the biosynthesis, a high productivity also lowers downstream processing costs and reduces the risks of product toxicity.

The biosurfactants transfer from the interior of the cells to the exterior in yeasts (mannosylerythritol lipids, cellobiose lipids, sophorolipids) and bacteria (trehalose lipids and rhamnolipids) fermentation is mediated by transport proteins and has been addressed in several works. Yet, in contrast to the yeast glycolipids, the genetic basis of transporters of bacterial glycolipids is largely unknown.

Apart from moving biosurfactant molecules to the extracellular environment, transport proteins also intervene in many intracellular steps along the biosynthetic pathway. In fact, the eukaryotic producers of glycolipids present the compartmentalization of building blocks within the cell, hence the importance of intramolecular transport of feedstuff and of synthesized moieties.

The direct and indirect involvement of transport proteins in the production of biosurfactants has been recently reviewed by Claus et al. [27].

1.5.5 Recovery and Purification

Gravity is generally sufficient to separate the liquid phase containing the biosurfactants from the biomass.

The supernatant liquid can be used as such for applications that do not demand high purity, but it must be noted that the impurities are likely to affect the physical and chemical properties of the isolate. If high purity is demanded, ultrafiltration or solvent extraction has to be used, which negatively impacts the production costs. The solvent to be used depends upon the biosurfactant and may be hydrocarbons, such as *n*-heptane or *n*-octane, alcohols such as 1-octanol, ethyl acetate, acetone, and chloroform/methanol blends.

Ultrafiltration was used in a one-step process to purify and concentrate the lipopeptide surfactin and rhamnolipids from the culture supernatant fluids. Surfactants form micelles at concentrations above the critical micelle concentration and the micelles can be retained by relatively high molecular weight cut-off membranes. Lower molecular weight impurities, such as salts, free amino acids, and small peptides, are then easily removed [28].

The performance of a laboratory-scale fixed-bed adsorption column for surfactin isolation from the culture medium over commercial active carbon was assessed by Montastruc et al. Using methanol elution in up-flow mode in the column at 25 °C gave a surfactin recovery of 94% [29].

1.6 Raw Materials in Natural Surfactants Production

For the following discussion, the term "raw materials" must be viewed in a broad sense and encompass not only the classical chemicals necessary for plants, animals, or microorganisms to grow but also all other materials consumed in their life cycle and the physical space needed to accommodate their growth. Saponins and soy lecithins require acreage for growing the plants, yolk lecithin needs poultry to lay eggs, the most performing fermentation processes for microbial surfactants rely on feedstock like glucose or corn syrup, again the produce of agriculture.

Time ago, there was much concern and debate about the use of agricultural resources, and acreage in particular, to produce natural fatty alcohols, one of the major feedstocks for synthetic nonionic and anionic surfactants. So, in all likelihood with the advent of natural surfactants the issue will only be shifted, but it will remain: is it ethical to use soil and agricultural resources, for example corn crops (see Figure 1.3) to produce surfactants and their raw materials rather than to feed people?

Figure 1.3 Corn, one important raw material source for biosurfactants. (*Source:* Silverije / Wikimedia Commons / CC BY-SA 3.0.)

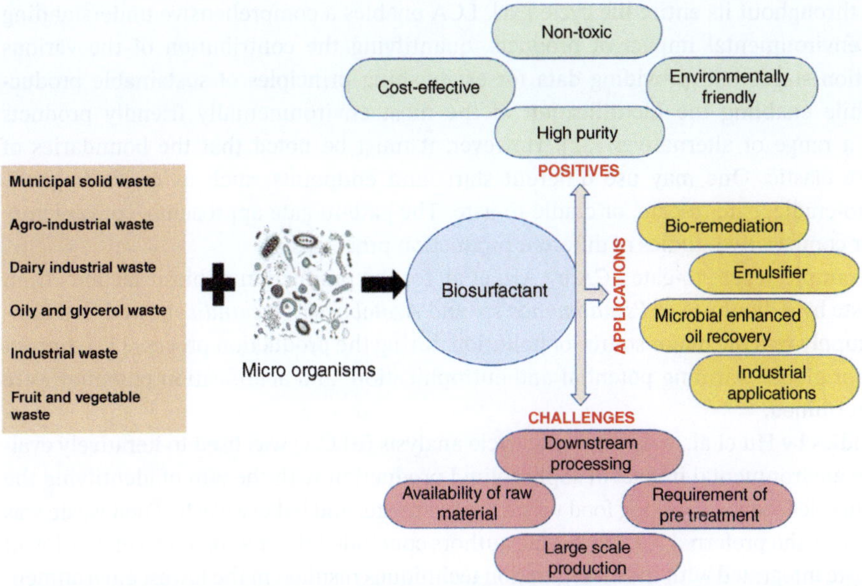

Figure 1.4 Biosurfactants from microorganisms using different wastes as raw material. (From Mohanty et al. [30]. Mohanty, S.S., Koul, Y., Varjani, S., et al. (2021) / A critical review on various feedstocks as sustainable substrates for biosurfactants production / Springer Nature / Public domain.)

For the microbial production of biosurfactants, the conundrum has been approached, along with the context of reducing production costs, by exploring the use of wastes from domestic or industrial uses or from agricultural products processing.

Figure 1.4 is an illustration of the concept of using wastes of different types for biosurfactant production, but it is fair to anticipate that so far, the efforts have not been very successful. It has been demonstrated that biosurfactants can be produced from wastes, but for the time being, not economically. The yields are insufficient and the production costs for commercial volumes always exceed the savings from the reduced feedstock costs.

Different types of wastes have been explored for different types of natural surfactants. This is described and discussed in some detail in the chapters devoted to the individual biosurfactants.

1.7 Life Cycle Analysis of Natural Surfactants

It is an almost omnipresent claim associated to naturally occurring surfactants that they are environmentally friendly, based on the imaginative consideration that they are a product from nature, built on renewable resources, easily biodegradable, and that they return to nature as nonnoxious catabolites. Yet, for years, the supporting factual evidence has been lacking. It is only in the last few years that scientific principles have begun to be implemented in the assessment of the environmental impact of naturally occurring surfactants.

Life cycle analysis (LCA) methodologies provide a scientifically valid technique for evaluating the inputs, outputs, and potential environmental consequences of a product system throughout its entire life cycle [31]. LCA enables a comprehensive understanding of the environmental impact of products, quantifying the contribution of the various production stages and providing data for establishing principles of sustainable production while enabling the identification of the most environmentally friendly products among a range of alternatives [32]. However, it must be noted that the boundaries of LCA are elastic. One may use different start- and endpoints, such as cradle-to-grave, cradle-to-cradle, gate-to-gate, or cradle-to-gate. The gate-to-gate approach is conveniently used for comparative studies of different production processes.

For example, a gate-to-gate LCA by Aru et al. for the production of biosurfactants from oily waste by a diculture of *Pseudomonas* sp. and *Azotobacter vinelandii* showed that electricity supply was the major source of pollution during the production process [33]. Impact values for global warming potential and eutrophication, and acidification potential were also determined.

In studies by Hu et al., a dynamic life cycle analysis (dLCA) was used to iteratively evaluate the environmental impact of sophorolipid production with the aim of identifying the most suitable feedstock among food waste, textile waste, and bakery waste. Food waste was identified as the preferred feedstock. The authors concluded that fed-batch fermentation of food waste integrated with in situ separation techniques resulted in the lowest environmental impact, mainly due to the relatively low cumulative energy demand and global warming potential [34, 35]. In addition, energy consumption was identified as accounting for over 80% of the total impact in sophorolipids production.

Kopsahelis et al. conducted a gate-to-gate LCA of sophorolipids and rhamnolipids. The authors found that biogenic emissions (CO_2) and thermal energy and electricity needs during the synthesis of the biosurfactants were the major contributors to the environmental effects. The environmental impact of sophorolipids synthesis was 22.7% higher than that of rhamnolipids. The reason for this is that production of sophorolipids is more time consuming than that of rhamnolipids and, as a result, it requires more thermal energy in order to maintain the culture medium at a steady temperature [36].

Schonhoff et al. studied biosurfactants produced from two waste products from the sugar industry, molasses and sugar beet pulp. They found that production of mannosylerythritol lipids had a lower environmental impact than that of rhamnolipids due to differences in microbial properties and process designs. The choice of substrate did not play an essential role [37].

The literature on LCA is still modest. Most of the recent works cover the production of single biosurfactants, comparing the effect of different feedstocks, or of different products (generally two) from the same feedstock. Comparative studies of naturally occurring surfactants versus synthetic surfactants are missing, and therefore, the widespread claim of the superiority of naturally occurring surfactants over synthetic surfactants as far as environmental impact still rests on immaterial considerations.

1.8 Production Costs of Naturally Occurring Surfactants

Only general principles are given here; figures for specific natural surfactants are reported, when available, in the respective chapters.

The production costs of naturally occurring surfactants vary considerably depending upon their origin, whether animal/vegetal or from microorganisms because in the former the process is extractive, while in the latter it is fermentative, generally followed by an isolation step.

The cost components include:

- plant capital (land, equipment, construction)
- extraction (for animal/plant-derived surfactants) or fermentation (for microbial surfactants)
- isolation and purification (homogenization, precipitation, filtration, centrifugation, solvent recovery, drying, etc.)

and the parameters involved are feedstock, yield, energy, chemicals, labor, and sundries.

Various cost models have been developed, which however are largely based on extrapolations, as the actual figures for industrial production either do not exist, or in the few instances when they do, they are proprietary intellectual property. Schonhoff et al. have produced a comprehensive study of the industrial production of rhamnolipids and mannosylerythritol lipids by extrapolating laboratory and pilot plant data. This is further discussed in the chapter on mannosylerythritol lipids [38].

In the production of biosurfactants, electricity is by far the major contributor to the variable costs, approaching in some instances 80% of the total [39]. However, simulation of different scenarios with aeration units providing higher oxygen transfer rates and adjusting power input to match oxygen uptake can significantly decrease the electricity consumption [40].

For many years, the costs of biosurfactants have been prohibitively high, making them totally noncompetitive against synthetic surfactants, even in a cost/effectiveness perspective. However in the case of rhamnolipids, advances in the production technology have shifted the balance. Although still remaining on the edge of cost/effectiveness, the progress has opened their use in mass consumer brands. Here there are still opportunities unexploited for further reducing production costs. This may involve identification of new producing strains as well as genetic engineering. Advances in formulation technology will also be of importance. Together, this development is likely to result in broadly recognizable cost-effectiveness progress, hence a deeper industrial penetration.

1.9 Natural Surfactants are Chiral

There is currently a considerable interest in chiral surfactants, i.e. amphiphiles with one or more chiral centers. For self-assembling species such as natural surfactants, there is chirality in the individual molecules but also in the self-assemblies. Highly organized chiral supramolecular nanostructures, such as fibers, ribbons, tubes, and helices, have been described [41–43]. Applications that have been proposed for such chiral surfactant aggregates include enantiomeric catalysis and chiral separation [44, 45].

Synthesis of chiral surfactants is cumbersome and difficult to perform in large scale. Since natural surfactants are chiral molecules, they are attractive alternatives to the synthetic chiral amphiphiles. Extraction and purification of biosurfactants is usually less demanding than organic or bioorganic synthesis of chiral surfactants. This is discussed and exemplified in a recent review on chiral surfactants [46].

1.10 A Glimpse into the Future

The commercial opportunities for naturally occurring surfactants are evolving under several influences, such as public opinion perceptions, environmental concern, and regulatory constraints.

While naturally occurring surfactants will never replace in full the synthetic surfactants, they could achieve a much broader market penetration if two conditions are realized:

- The availability of even larger volumes at even reduced prices. Cost-effectiveness and availability are achieved in certain circumstances, but for the time being, this is the exception, not the rule.
- Substantial progress in formulation technology, resulting from both a better knowledge and understanding of the physical and chemical properties of natural surfactants and an extended testing under real-use conditions. More specifically, the possible synergies between natural and synthetic surfactants have been only marginally studied until now and published information on the topic is virtually nonexisting.

For naturally occurring surfactants to become real winners they will have to make substantial entries in the major application, the household detergents, over and above what already achieved in hand dishwashing products (see Figure 1.5). Although there is no published information, there are reasons to believe that certain natural surfactants and notably

Figure 1.5 Biosurfactants are already used in hand dishwashing liquids and in household cleaners. (Courtesy of Joakim Holmberg.)

Figure 1.6 The cosmetic industry has recorded the first commercial entry for biosurfactants. (*Source:* Wikimedia Commons / CC BY-SA 3.0.)

the saponins, may have the potential, probably in combination with synthetic surfactants, to solve today's two major challenges for domestic laundry detergents:

- stains removal
- energy conservation (shorter washing cycles at lower temperature)

These are discussed in the section on potential applications in the chapter on saponins.

It is obvious that a massive entry into the household detergents will require volumes far in excess of what can be supplied today or can be envisaged in a foreseeable future, and this is a serious matter that needs attentive consideration, as discussed in section 1.6.

Outside the household laundry detergents, there are other markets open to natural surfactants. The cosmetic industry has already responded favorably (see Figure 1.6). It has become an established consumer of saponins, sophorolipids, and rhamnolipids and is eying with interest to the mannosylerithritol lipids. But it is fair to recognize that what has happened so far has been just scratching the surface. There are many more opportunities ahead, driven both by objective benefits and the emotional appeal that the natural/sustainable/renewable connotation elicits.

In the authors' opinion, agriculture/crop protection has received less attention than it deserves. While there is no evidence that biosurfactants on their own can show pesticidal and fungicidal activity superior to that of existing synthetic actives, they could perform additively or synergistically with plant protection actives, while enabling formulations in convenient delivery forms, for example for aerial spray(see Figure 1.7).

Soil remediation is a realistic and environmentally beneficial application target if it can be demonstrated beyond doubt that biosurfactants not only leach soil contaminants but

Figure 1.7 An aerial crop duster applying a fine mist. (Image credit: Stephan Krause / Wikimedia Commons / Public domain.)

Figure 1.8 Remediated soil can be returned to agriculture. (*Source:* SURASIT / Adobe Stock Photos.)

also enhance their ultimate degradation into nonnoxious catabolites, thus enabling the re-start of agricultural exploitation (see Figure 1.8). Unless this is happening, any surfactant, no matter if bio, bio-based, or petrochemical, will only move the issue from one place to another, from soil to aquifers. A confirmation that this enhanced biodegradation is possible comes, among others, from a recent study by Patowary et al. [47].

Figure 1.9 Microbial enhanced oil recovery, where biosurfactants are generated in situ, may have a future. (Image credit: Ficelloguy / Wikimedia Commons / CC BY-SA 3.0.)

Microbially enhanced oil recovery, often written M-EOR (or MEOR), is the technique of injecting a surfactant-producing microorganism into the reservoir to generate in-situ production of biosurfactants. The crude oil is normally the sole carbon source necessary, but a nitrogen source and other nutrients need to be injected together with the microorganism. The microorganism is supposed to generate amphiphiles that will reduce the interfacial tension between the crude and the injection water to very low levels, of the order of 10^{-3} mN/m or lower, which is key for successful mobilization of the oil (see Figure 1.9). The microorganism should also decompose the oil into shorter hydrocarbons, which will facilitate the recovery due to lower viscosity of the oil. In successful cases, the decomposition is so powerful that a substantial amount of gas is also generated. M-EOR field trials of this type are currently being performed in large scale in China and the results so far are promising. Injecting ex-situ generated biosurfactants into the reservoir does not hold promise, however. That would fall short from cost and maybe also from technical reasons.

It is an undeniable fact that natural surfactants are today (2025) playing a real, sustained role in colloid chemistry and its industrial realizations, a step change from the situation of just a few years ago. The renewed interest of the industry, the public opinion, and the regulatory drive for sustainability, the improved knowledge gained in understanding their physical and chemical properties and in-use performance, the progress in formulation technology, and the positive impact of process improvement for both cost reduction and availability are positive steps forward for a promising future.

As a final remark, natural surfactants may be the shot in the arm that the surfactant industry is in need of. The volumes of surfactants on the market are impressive and growing but from a technological point of view, the surfactant industry is in many ways a sunset industry. Most of the important types of surfactants that we use today appeared on the market during the half-century 1920–1970. The important processes involved in surfactant production, such as alkylation of aromatics, development of processes to make synthetic alcohols, sulfonation, sulfation, alkoxylation, etc., were developed and commercialized during this period. After 1970, only a few new products have reached the market on a large scale, the paraffin sulfonates and the ester quats being the most important exceptions. The advent of these two surfactant classes also dates long back. They were both produced already in the 1930s, although on a relatively small scale.

There have also been substantial innovations during the last half-century related to silicone surfactants, acetylenic surfactants, and fluorinated surfactants, but these are small volume, specialty products. In addition, silicone surfactants and fluorinated surfactants will probably experience problems in the future because they face problems in meeting the increasingly strict environmental regulations.

In short, there is time for something new in the surfactant arena. Natural surfactants may be the answer in a longer-term perspective.

References

1 Bacille, N. (2023). Are microbial *biosurfactants* actually only *surfactants*? *Current Opinion in Colloid and Interface Science* 68: 101747.

2 Bognolo, G. (2023). *Surface Active Agents – Historical Perspectives and Future Development*. Boca Raton, FL: CRC Press.

3 Guerini, V. (1909). *A History of Dentistry: From the Most Ancient Times Until the End of the Eighteenth Century*. Philadelphia, PA: Lea & Febiger.

4 Bergström, S., Theorell, H., and Davide, H. (1946). Pyolipic acid, a metabolic product of *Pseudomonas pyocyanea*, active against *Mycobacterium tuberculosis*. *Archives of Biochemistry* 10: 165–166.

5 Jarvis, F.G. and Johnson, M.J. (1949). A glycol-lipide produced by *Pseudomonas aeruginosa*. *Journal of the American Chemical Society* 71: 4124–4126.

6 Gorin, P., Spencer, J., and Tulloch, A. (1961). Hydroxy fatty acid glycosides of sophorose from Torulopsis magnolia. *Canadian Journal of Chemistry* 39: 846–855.

7 Arima, K., Kakinuma, A., and Tamura, G. (1968). Surfactin, a crystalline peptidelipid surfactant produced by *Bacillus subtilis*: isolation, characterization and its inhibition of fibrin clot formation. *Biochemical and Biophysical Research Communications* 31: 488–494.

8 Tropsch, J. (2017). A journey to standardization of bio-based surfactants in Europe. *Inform* 28: 20–22.

9 Zweers, J.C., Barak, I., Becker, D. et al. (2008). Toward the development of *Bacillus subtilis* as a cell factory for membrane proteins and protein complexes. *Microbial Cell Factories* 7: 10.

10 Jaime, J.B., Romero, L.G., and Leon, J.C. (2023). Selection of biosurfactant-producing microorganisms. In: *From Biorefineries Production to Versatile Applications* (ed. P.R.F. Marcelino, S.S. da Silva, and A.O. Lopez) Chapter 2. Chichester, UK: Wiley.

11 Lee, D.W., Lee, H., Kwon, B.-O. et al. (2018). Biosurfactant-assisted bioremediation of crude oil by indigenous bacteria isolated from Taean beach sediment. *Environmental Pollution* 241: 254–264.

12 Patowary, K., Patowary, R., Kalita, M.C., and Deka, S. (2017). Characterization of biosurfactant produced during degradation of hydrocarbons using crude oil as sole source of carbon. *Frontiers in Microbiology* 8: 1–14.

13 Liepins, J., Balina, K., Soloha, R. et al. (2021). Glycolipid biosurfactant production from waste cooking oils by yeast: review of substrates, producers and products. *Fermentation* 7: 136.

14 Jimoh, A.A. and Lin, J. (2019). Enhancement of *Paenibacillus* sp. D9 lipopeptide biosurfactant production through the optimization of medium composition and its application for biodegradation of hydrophobic pollutants. *Applied Biochemistry and Biotechnology* 187: 724–743.

15 Liang, X., Shi, R., Radosevich, M. et al. (2017). Anaerobic lipopeptide biosurfactant production by an engineered bacterial strain for *in situ* microbial enhanced oil recovery. *RSC Advances* 7: 20667–20676.

16 Abbasi, H., Hamedi, M.M., Lotfabad, T.B. et al. (2012). Biosurfactant producing bacterium, *Pseudomonas aeruginosa* MA01 isolated from spoiled apples: physicochemical and structural characteristics of isolated biosurfactant. *Journal of Bioscience and Bioengineering* 113: 211–219.

17 Nurfarahin, A.H., Mohamed, M.S., and Phang, L.Y. (2018). Culture medium development for microbial-derived surfactants production – an overview. *Molecules* 23: 1049.

18 Subsanguan, T., Khondee, N., Nawavimarn, P. et al. (2020). Reuse of immobilized *Weissella cibaria* PN3 for long-term production of both extracellular and cell-bound glycolipid biosurfactants. *Frontiers in Bioengineering and Biotechnology* 8: 751.

19 Joy, S., Rahman, P.K.S.M., Khare, S.K. et al. (2019). Statistical and sequential (fill-and-draw) approach to enhance rhamnolipid production using industrial lignocellulosic hydrolysate C_6 stream from *Achromobacter* sp. (PS1). *Bioresource Technology* 288: 121494.

20 Bustos, G., Arcos, U., Vecino, X. et al. (2018). Recycled *Lactobacillus pentosus* biomass can regenerate biosurfactants after various fermentative and extractive cycles. *Biochemical Engineering Journal* 132: 191–195.

21 Partovinia, A. and Rasekh, B. (2018). Review of the immobilized microbial cell systems for bioremediation of petroleum hydrocarbons polluted environments. *Critical Reviews in Environmental Science and Technology* 48: 1–38.

22 Bouassida, M., Mnif, I., and Ghribi, D. (2023). Enhanced biosurfactant production by Bacillus subtilis SPB_1 using developed fed-batch fermentation: effects of glucose levels and feeding systems. *Bioprocess and Biosystems Engineering* 46: 555–563.

23 Zhu, Z., Zhang, F., Wei, Z. et al. (2013). The usage of rice straw as a major substrate for the production of surfactin by *Bacillus amyloliquefaciens* XZ-173 in solid-state fermentation. *Journal of Environmental Management* 27: 96–102.

24 Al-Dhabi, N.A., Esmail, G.A., and Arasu, M.V. (2020). Enhanced production of biosurfactant from *Bacillus subtilis* strain Al-Dhabi-130 under solid-state fermentation using date molasses from Saudi Arabia for bioremediation of crude-oil-contaminated soil. *International Journal of Environmental Research and Public Health* 17: 8446.

25 Eras-Muñoz, E., Gea, T., and Font, X. (2023). Carbon and nitrogen optimization in solid-state fermentation for sustainable sophorolipid production using industrial waste. *Frontiers in Bioengineering and Biotechnology* 11: 1252733.

26 Bouassida, M., Mnif, I., Hammami, I. et al. (2023). *Bacillus subtilis* SPB_1 lipopeptide biosurfactant: antibacterial efficiency against the phytopathogenic bacteria *Agrobacterium tumefaciens* and compared production in submerged and solid state fermentation systems. *Food Science and Biotechnology* 32: 1–15.

27 Claus, S., Sánchez, L.J., and Van Bogaert, I.N.A. (2021). The role of transport proteins in the production of microbial glycolipid biosurfactants. *Applied Microbiology and Biotechnology* 105: 1779–1793.

28 Mulligan, C.N. and Gibbs, B.F. (1990). Recovery of biosurfactants by ultrafiltration. *Journal of Chemical Technology and Biotechnology* 47: 23–29.

29 Montastruc, L., Liu, T., Nikov, I. et al. (2011). Integrated process for production of surfactin (III). Modeling of adsorption column. *Chinese Journal of Chemical Engineering* 19: 357–364.

30 Mohanty, S.S., Koul, Y., Varjani, S. et al. (2021). A critical review on various feedstocks as sustainable substrates for biosurfactants production. *Microbial Cell Factories* 20: 120.

31 Sala, S., Amadei, A.M., Beylot, A., and Ardente, F. (2021). The evolution of life cycle assessment in European policies over three decades. *The International Journal of Life Cycle Assessment* 26: 2295–2314.

32 Valdivia, S., Backes, J.G., Traverso, M. et al. (2021). Principles for the application of life cycle sustainability assessment. *The International Journal of Life Cycle Assessment* 26: 1900–1905.

33 Aru, O.O. and NE, O.I. (2018). Life cycle assessment of the environmental impact of biosurfactant production from oil waste by a diculture of *Azotobacter vinelandii* and *Pseudomonas* sp. *Journal of Bioremediation & Biodegradation* 9: 435.

34 Hu, X., Subramanian, K., Wang, H. et al. (2021). Guiding environmental sustainability of emerging bioconversion technology for waste-derived sophorolipid production by adopting a dynamic life cycle assessment (dLCA) approach. *Environmental Pollution* 269: 116101.

35 Hu, X., Subramanian, K., Wang, H. et al. (2021). Bioconversion of food waste to produce industrial-scale sophorolipid syrup and crystals: dynamic life cycle assessment (dLCA) of emerging biotechnologies. *Bioresource Technology* 337: 125474.

36 Kopsahelis, A., Kourmentza, C., Zafiri, C., and Kornaros, M. (2018). Gate-to-gate life cycle assessment of biosurfactants and bioplasticizers production via biotechnological exploitation of fats and waste oils. *Journal of Chemical Technology and Biotechnology* 93: 2833–2841.

37 Schonhoff, A., Ihling, N., Schreiber, A., and Zapp, P. (2022). Environmental impacts of biosurfactant production based on substrates from the sugar industry. *ACS Sustainable Chemistry & Engineering* 10: 9345–9358.

38 Schonhoff, A., Stöckigt, G., Wulf, C. et al. (2023). Biosurfactants production with substrates from sugar industry. Environmental, cost, market and social aspects. *RSC Sustainability* 1: 1798–1813.

39 Moutinho, L.F., Moura, F.R., Silvestre, R.C., and Romão-Dumaresq, A.S. (2021). Microbial biosurfactants: a broad analysis of properties, applications, biosynthesis, and techno-economical assessment of rhamnolipid production. *Biotechnology Progress* 37: e3093.

40 Oraby, A., Rupp, S., and Zibek, S. (2022). Techno-economic analysis as a driver for optimisation of cellobiose lipid fermentation and purification. *Frontiers in Bioengineering and Biotechnology* 10: 913351.

41 Pi-Boleda, B., Bouzas, M., Gaztelumendi, N. et al. (2020). Chiral pH-sensitive cyclobutane β-amino acid-based cationic amphiphiles: possible candidates for use in gene therapy. *Journal of Molecular Liquids* 297: 11.

42 Berthier, D., Buffeteau, T., Léger, J.-M. et al. (2002). From chiral counterions to twisted membranes. *Journal of the American Chemical Society* 124: 13486–13494.

43 Koc, M.H., Ciftci, G.C., Baday, S. et al. (2017). Hierarchical self-assembly of histidine-functionalized peptide amphiphiles into supramolecular chiral nanostructures. *Langmuir* 33: 7947–7956.

44 Liang, X., Gui, Y., Li, K. et al. (2020). A novel chiral surfactant-type metallomicellar catalyst for asymmetric Michael addition in water. *Chemical Communications* 56: 11118–11121.

45 Liu, L., Sun, D., Li, F. et al. (2019). Enantioselective liquid-liquid extraction of valine enantiomers in the aqueous two-phase system formed by the cholinium amino acid ionic liquid copper complexes and salt. *Journal of Molecular Liquids* 294: 111599.

46 Xiong, Z., Zhou, L., Wang, J. et al. (2024). Chiral surfactants: design, aggregation behaviors and applications. *Current Opinion in Colloid & Interface Science* 73: 101842.

47 Patowary, R., Jain, P., Malakar, C., and Devi, A. (2023). Biodegradation of carbofuran by *Pseudomonas aeruginosa* SO_7: biosurfactant production, plant growth promotion, and metal tolerance. *Environmental Science and Pollution Research* 30: 115185–115198.

2

Saponins

2.1 Introduction

There are 250 000–500 000 species of plants on earth with the ability to synthesize an almost limitless array of chemicals. Among the different compounds derived from plants, the saponins deserve a special mention. These chemicals are believed to be part of the plants' natural defense systems and are included in a large group of protective molecules named "phytoanticipins" or "phytoprotectants" [1]. They are present in the plant even before the attack by pathogens. Saponins are amphiphilic, i.e. surface active, and like all such molecules, they assemble at interfaces, where they can act as emulsifier or foaming agent. In this, they behave like a soap and the name is derived from "sapo," latin for soap.

Saponins are mainly found in angiosperms, i.e. flowering plants, and they occur in many different plant parts, such as roots, shoots, bark, flowers, leaves, and even the stem. When a saponin-rich part is ground and shaken in water, a soapy lather is generated. Such a test was used for identification of saponins in plants and even for a semi-quantitative assessment. In many parts of the world, such saponin-rich plants have been used as a soap substitute.

A list of some saponin-rich plants that have traditionally been utilized as a source of natural surfactants is presented in Table 2.1.

As can be seen from Table 2.1, the saponin content is high in some plants. This is, of course, important from an industrial point of view. The content may differ between regions of the world, however, and may also show a season-specific variation. The composition and concentration will also vary with the state of development of the plant.

The use of saponins as cleaning agents probably pre-dates recorded history. Since antiquity, *Sapindus mukorossi* has been used in Eastern Asia and the Himalayas as a detergent for shawls and silk and by jewelers for cleaning silver.

Saponaria officinalis, which Linnaeus named "Bouncing bet - the Soap Weed" in his *Species Plantarum* from 1753, was known to the Biblical Magi as "Chalyriton," to the Egyptians as "Oeno," and to the Africans as "Syris" [4]. The Greeks identified the plant with the popular name, "Catharsis," which meant, and still means, "a Cleanser or Purifier," perhaps referring to its use as soap, as well as to its purgative effect when used medicinally [5]. The Romans called it "Radix" (i.e. root) or "Herba lanaria" (i.e. wool herb) and already 2000 years ago, clothiers used decoctions of *Saponaria officinalis* to clean wool before weaving it into cloth [6]. Its use was common in France and Germany already in the tenth

Natural Surfactants: Science, Technology, and Industrial Aspects, First Edition. Guido Bognolo and Krister Holmberg.

Table 2.1 Saponin content in plants.

Latin name	Common name	Saponin content (%)
Aesculus hippocastanum	Horse-chestnut	3
Avena sativa	Oat	0.10–0.13
Beta vulgaris	Sugar beet (leaves)	5–6
Chenopodium quinoa	Quinoa	0.14–2.3
Cicer arietinum	Chickpea	0.20–0.30
Crocus sativus	Saffron crocus	1.2–3.4
Glycine max	Soybean	0.22–0.49
Glycyrrhiza glabra	Licorice (root)	22–32
Hedera helix	Ivy	5
Medicago sativa	Alfalfa	0.14–1.71
Panax notoginseng	Chinese ginseng	2–3
Panax quinquefolius	American ginseng	1.4–5.6
Pisum sativum	Grean pea	0.18–4.2
Polygala spp.	Milkwort	8–10
Primula spp.	Primula	5–10
Quillaia saponaria	Quillaia bark	9–10
Saponaria officinalis	Soapwort	2–5
Smilax officinalis	Sarsaparilla	1.8–2.4
Trigonella foenum-graecum	Fenugreek	4–6
Yucca schidigera	Yucca	10

(*Sources:* Güclü- Üstündağ and Mazza [2]; Mir et al. [3].)

century (where it remained in use until the nineteenth century) [7] and presumably across other countries in Europe in the late Middle Ages. When the industrial revolution enabled large-scale commercial manufacture of cloths, textile industrialists cultivated fields of the "weed" as an inexpensive detergent [8].

Nowadays in certain industrialized countries saponin-based cleaning and laundry washing products begin to re-surface, on the back of their cleaning power and the natural connotation and have also found applications in a plethora of consumer products ranging from flavoring and sweeteners for American-type tobacco, chewing gums, candies, bakeries, and ice creams, via ingredients in pharmaceutical and veterinary products and in vaccines to industrial applications, such as photographic emulsions, sustainable anti-mist agents in acid baths, emulsifiers of colorants and flavors in food and drinks, feed additives, agriculture adjuvants, soil remediation, waste water treatment, foaming agents in fire extinguishers, and mild cleanser and foaming agents in shampoos. Some of these applications are discussed further down in this chapter.

In a number of nonindustrialized countries, saponins continue to be used for washing clothes, cleaning delicate ribbons, garments and wool, washing hair, removing greasy spots,

Figure 2.1 *Quillaia saponaria* in a Chilean medical plant postage stamp series. (*Source:* Stamp Boy / flickr / CC BY-SA 2.0.)

and disinfecting wounds. Saponins prevent moth attack on clothes, an indication of an interesting natural insecticide property.

Since prehistoric times, cultures throughout the world have used fish-killing plants, often containing saponins, for fishing, a practice still widely used, for example by indigenous tribes in Guyana and by the Gondi people in the Indian subcontinent [9]. In China (and presumably also in some Southeast Asian countries), tea saponins are extensively used nowadays in aquaculture.

Chemical synthesis of saponins is an alternative to the extraction of ground saponin-containing material Figure 2.1; however, the procedures are cumbersome and involve many steps [10]. Semisynthetic procedures, i.e. postmodifications of natural saponins, may be a more realistic way forward.

2.2 Chemical Structures and Properties

Saponins are glycosides with one or two (or occasionally three) oligosaccharide chains connected to the aglycone. The aglycone, which is sometimes called sapogenin, is usually a triterpene but may also be a steroid and saponins are commonly classified as either triterpenoid or steroid saponins. As can be seen from Figure 2.2, the triterpenoid and the steroid backbones are very similar, both having five condensed rings with a hydroxyl group in 3-position. The triterpenoid skeleton is slightly bigger than the steroid skeleton, with 30 and 27 carbon atoms, respectively. Figure 2.2 shows the structures of the skeletons of the two types of products.

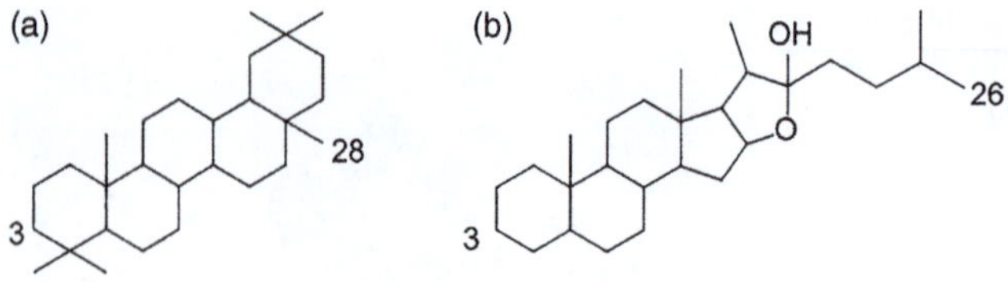

Figure 2.2 Structure of (a) a triterpenoid and (b) a steroid skeleton with numbers given for the carbon atoms where oligosaccharide chains are attached. (From Kregiel, D., Berlowska, J., Witonska, I.A., et al. (2017) / Saponin-based, biological-active surfactants from plants / IntechOpen / Public domain.)

When the saponin contains one oligosaccharide chain, it is usually connected with a glycosidic bond at the 3-OH of the aglycone. When there are two oligosaccharide chains, one is connected with a glycosidic linkage at the 3-OH and the other is usually attached by an ester bond in position 28 for triterpenoid saponins and by a glycosidic bond at position 26 for steroid saponins.

The oligosaccharide chains are short, typically three or four units, and may be branched or linear. The most common monosaccharides are D-glucose (glc), D-galactose (gal), L-rhamnose (rha), L-arabinose (ara), D-xylose (xyl), D-fructose (fuc), and D-glucuronic acid (glcA). Figure 2.3 shows representative examples of saponins with one and two oligosaccharide chains.

Steroidal glycoalkaloids are related compounds that contain a nitrogen atom in the skeleton. They are sometimes included in the saponin family. Figure 2.3 shows an example of a steroidal glycoalkaloid. It is important to bear in mind, however, that saponins are seldom well-defined single molecules. They are usually a complex mixture of glucosides of the same or of different sapogenins. However, such heterogeneity is not unique to saponins or to natural surfactants in general; synthetic surfactants are also often heterogeneous and commercial products can be regarded as a mixture of many substances.

Typical synthetic surfactants contain one hydrophilic part, the polar headgroup, which can be charged or uncharged, and one hydrophobic part, often called the tail, which can be a linear or branched alkyl or an alkylaryl chain. Saponins with one oligosaccharide chain, for example the one shown in Figure 2.3, *top left*, would consequently fall into the category of an anionic surfactant (if the carboxyl group in the oligosaccharide chain is deprotonated) and the one in Figure 2.3, *bottom left*, would be classified as a nonionic surfactant. However, the hydrophobic part is unusual. Steroid-based synthetic surfactants do exist on the market. They are typically sterols with the 3-OH group transformed into a polar headgroup, which may be charged or uncharged [12]. Cholesterol ethoxylate is the most well-known example and is included in Figure 2.3 as a comparison. Cholesterol ethoxylates are used in cosmetic formulations, but the volumes are small.

The saponin skeleton with several fused rings is very rigid compared to the hydrophobic tail of most synthetic surfactants. This influences the packing of the molecules, in bulk as well as at interfaces. The saponins with one oligosaccharide chain form micelles like almost all synthetic surfactants but the self-assembling process is not as smooth as for surfactants with flexible tails and the critical micelle concentration (CMC) is therefore higher than for most nonionic surfactants [13, 14]. The shape of the micelles varies from spherical via elongated to thread-like depending on the size of the oligosaccharide chain and

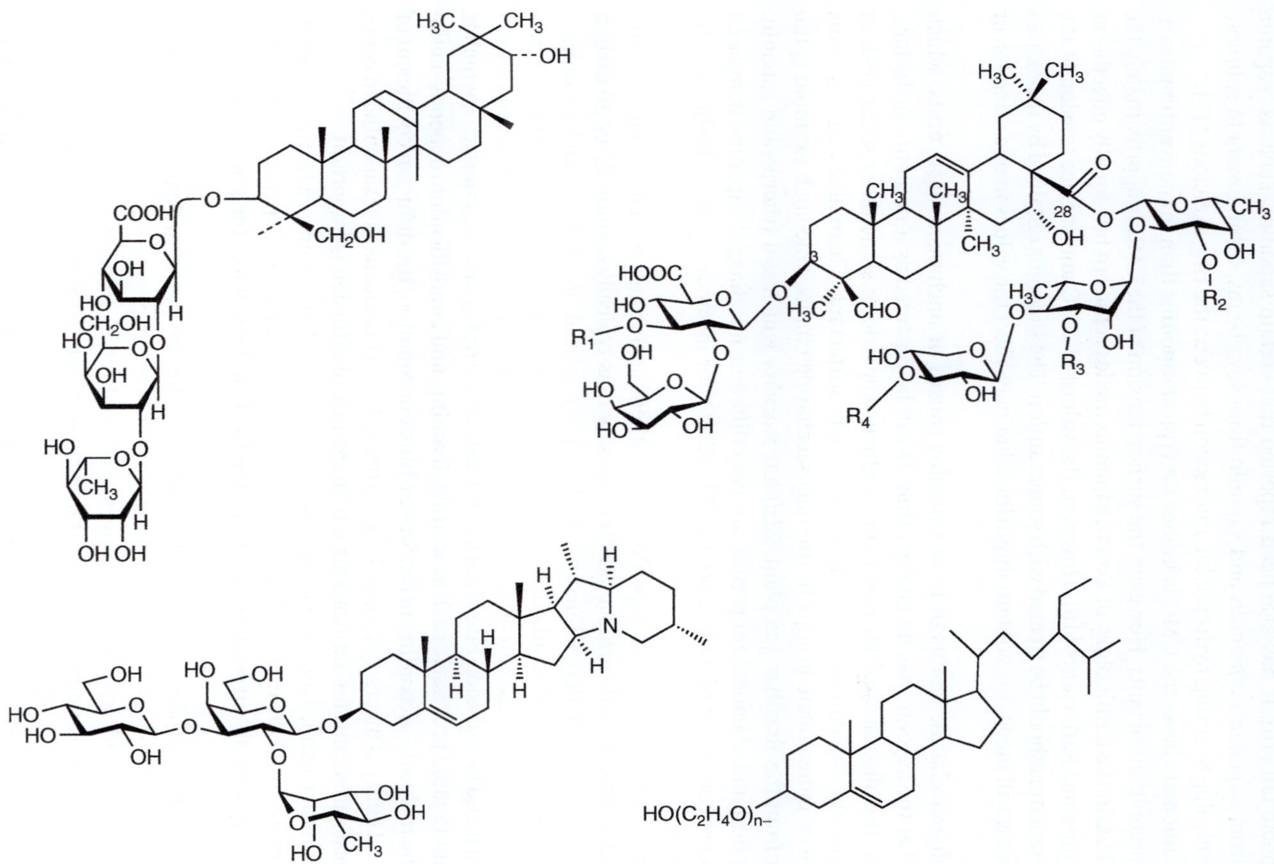

Figure 2.3 Examples of saponins with the synthetic surfactant cholesterol ethoxylate included as a comparison. *Top left:* a triterpenoid saponin with an oligosaccharide chain attached at C-3; *top right:* a triterpenoid saponin with oligosaccharide chains attached at C-3 and C-28; *bottom left:* a steroidal glycoalkaloid with an oligosaccharide chain attached at C-3; *bottom right:* cholesterol ethoxylate.

the same behavior has been reported for phytosterol ethoxylates with varying length of the polyoxyethylene chain [15]. This behavior is according to expectations; the shape of surfactant micelles is governed by the area per headgroup and the length and volume of the hydrophobic tail [16]. It has also been reported that certain saponins, such as "saponin white" from *Saponaria officinalis* and saponin from soya beans, form dimers in solution, presumably due to strong hydrophobic interaction between the ring skeletons [17].

As mentioned above, the CMC is higher for typical saponins than for conventional synthetic nonionic surfactants. However, the surface tension at the CMC is approximately the same. Thus, in the terminology of Rosen, saponins are less efficient but equally effective as regular nonionic surfactants [18]. However, the values of CMC and surface tension at the CMC of saponins should be treated with some caution. The saponins obtained by the extraction process will inevitably contain impurities that may affect the values in one direction or another.

The oligosaccharide chains of most saponins consist of uncharged sugar units, which means that the saponin has no charge; thus, it can be regarded as a nonionic surfactant. However, the oligosaccharides may contain glucuronic acid or some other sugar acid, in which case the saponin can be seen as an anionic surfactant. There also exist saponins with one or more sulfate groups in the oligosaccharide chains. One such example is the saponin from the Brazilian juá plant, *Ziziphus joazeiro*, which is a triterpenoid saponin. The physical and chemical properties are very different for charged saponins compared to the more common nonionic saponins. The CMC, for instance, is much higher for the charged ones.

Another difference between saponins and synthetic surfactants is that the saponins are larger than conventional surfactants. Whereas saponins typically contain 27 or 30 carbon atoms in the skeleton, depending on whether it is a steroid or a triterpenoid saponin, a synthetic surfactant rarely contains more than 18 carbon atoms in the hydrophobic tail, and many have tails as short as 10 or 12 carbon atoms. A higher molecular weight can be both good and bad. The diffusion rate of surfactant unimers to newly created interfaces, which is important in rapid processes, such as wetting, foaming, and emulsification, depends on the size of the molecule. According to the Stokes-Einstein equation, the diffusion coefficient of a spherical entity is inversely proportional to the radius. This means that saponins may not be the best surfactants for applications where the rate of diffusion is important.

On the other hand, large surfactants are preferred over small ones when it comes to properties like solubilization, stabilization of emulsions and foams, dispersion of particles, etc. Thus, as with all surfactants, saponins are likely to be suitable for some, but not all, applications.

Saponins with two oligosaccharide chains, such as the one shown in Figure 2.3, *top right*, are very different from those with one chain when it comes to physical and chemical behaviors. Surfactants with hydrophilic groups at both ends of a sufficiently long hydrophobic chain are called bola amphiphiles and saponins with two oligosaccharide chains can be regarded as examples of this surfactant class. At the oil–water or the air–water interface, a bolaamphiphile will fold to orient its polar parts into the water and the hydrophobic chain into the oil or the air. The saponins with two hydrophilic chains are likely to try to

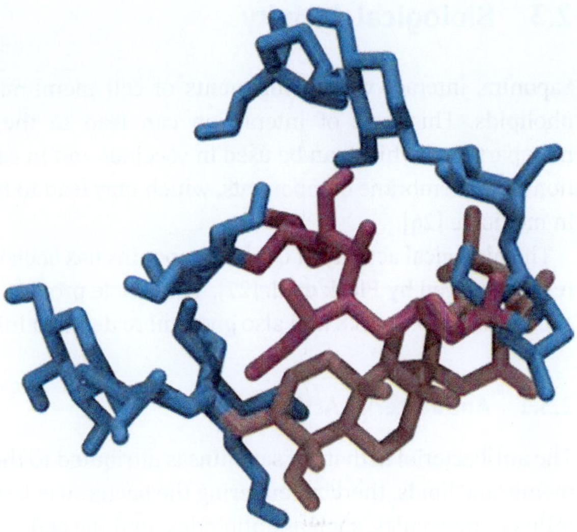

Figure 2.4 A drawing based on molecular dynamics simulations of QS-21, a saponin extracted from *Quillaia saponaria*. The brown and purple parts represent the hydrophobic region and the blue parts the oligosaccharide chains. (Adapted from Pedebos, C., Pol-Fachin, L., Pons, R., et al. (2014) and Liao, Y., Li, Z., Zhou, Q., et al. (2021).)

orient themselves in such a manner but the lack of flexibility of the skeleton will make this configuration difficult. Figure 2.4 shows a computer-assisted drawing of such a saponin, where the stiffness of the condensed ring system (in brown) is clearly seen. The area that a saponin molecule occupies at the air–water interface has been determined from adsorption isotherms and found to be around 1 nm^2 for saponins with two oligosaccharide chains, while for saponins with one hydrophilic chain, the value is around 0.5 nm^2, a sign of how difficult it is for the double-chain saponins to align tightly at an interface [21].

Saponins from various sources have been evaluated as emulsifiers, both for macroemulsions and for nanoemulsions. Most attention has been put on the saponin from *Quillaia* bark. This is a triterpenoid saponin with a glucuronic acid moiety in one of the two oligosaccharide chains. Thus, at neutral and higher pH, it behaves as a nonionic surfactant with an additional anionic character. Emulsions with small droplet sizes (diameters below 200 nm) with high thermal stability (up to 90 °C) and insensitivity to NaCl variations in the range 0–500 mM have been documented [22, 23]. *Yucca* saponins have also been subject to several studies related to their use as emulsifiers. These are steroid saponins without any charged moieties in the oligosaccharide chain. The *Yucca* saponins have proven to be efficient emulsifiers giving emulsions with small droplet size. However, the emulsions seem not to be as resistant to changes in temperature and electrolyte concentration as the emulsions made with *Quillaia* saponin as emulsifier [23, 24].

The oligosaccharide chains of saponins are highly hydrated in aqueous solution. A neutron reflectivity study on *Quillaia* bark saponin, which has two such chains, indicates that each saponin carries around 60 water molecules [25].

Saponins can be regarded as chemically stable amphiphiles and they are also stable to heat. The glycosidic linkage between the oligosaccharide chain and the aglycone, as well as between the sugar units, is stable in alkali but labile at low pH. Saponins that contain an ester bond, such as the one shown in Figure 2.3, *top right*, are also labile in alkali, however.

2.3 Biological Activity

Saponins interact with components of cell membranes, such as cholesterol and phospholipids. This type of interaction can lead to the formation of aggregates, such as nanoparticles, which can be used in vaccines and in cancer therapy. However, the interaction with membrane components, which may lead to hemolysis, limits the use of saponins in medicine [26].

The biological activity of *Quillaia* saponins has been widely studied and a comprehensive review is given by Fleck et al. [27]. The article provide reference to the effects described in subsections 2.3.1–2.3.8 and also gives more detailed information.

2.3.1 Antibacterial Activity

The antibacterial activity of saponins is attributed to their detergent properties that destroy membrane lipids, thereby rendering the bacterial cell membranes permeable, allowing the influx of molecules, such as antibiotics, into the cell.

2.3.2 Antiviral Activity

A strong antiviral activity has been demonstrated both *in vitro* and *in vivo*. The action of saponins is believed to be due to strong affinity of the aglycone part to membrane cholesterol. Without properly functioning cholesterol, the fluidity of the cell membrane increases, causing it to be beyond the control of enzymatic activities. Thus, saponins may act by preventing the specific binding of viral receptors to cellular proteins.

2.3.3 Antifungal Activity

Membranes of regenerated cellulose nanofibers supplemented with *Quillaia saponaria* saponins have been shown to have antifungal activity as shown in experiments on several fungi, including *Penicillium roquefortii* and *Aspergillus ochraceus*.

2.3.4 Antiparasitic Activity

The saponins from *Quillaia saponaria* bark showed antiparasitic activity, *in vitro*, in a ruminal fermentation system. The antiparasitic effect may be due to the detergent action of the saponins from the genus *Quillaia* on cholesterol in the protozoa membranes, resulting in cell lysis. However, the activity is not seen in all protozoan species. Some species may not be susceptible to the toxic effects of saponins, or the parasite may develop resistance to the saponins.

2.3.5 Antiinflammatory Activity

Tea saponins have been found to have both antiinflammatory and antioxidant properties. Saponins isolated from root extracts were found to inhibit carrageenan-induced paw oedema in rats. The antioxidant activity of these compounds was evaluated using the xanthine-xanthine oxidase system [28].

2.3.6 Antitumor Activity

Fractions rich in *Quillaia* saponins are described as potential therapeutic agents for treating and preventing cancerous diseases. One of the first studies to demonstrate the saponins' antitumor potential was conducted with Quil-A®, which prolonged the survival time of mice with leukemia. However, since the saponins interact with the cholesterol of cell membranes and, consequently, cause membrane damage, cell lysis, and necrosis, they show too high toxicity for clinical practice. These toxic effects have been reversed by incorporating soluble *Quillaia saponaria* into nanoparticles, where they interact with cholesterol. The *Quillaia saponaria* fraction 21 (QS-21; a more hydrophobic fraction with an acyl segment in the oligosaccharide chain attached at C-28) formulated in "killing and growth inhibiting" (KGI) particles selectively killed nine out of ten tumor cell lines at a concentration 30 times lower than the concentration required to kill normal monocytes. In addition, the lytic effect on red blood cells was inhibited by the formulation with KGI particles. The use of QS-21 in vaccines is discussed in Section 3.8.

2.3.7 Hepatoprotective Activity

A study demonstrated that bark saponins from *Quillaia saponaria* are interesting hepatoprotective agents. The hepatoprotective effect is probably due to the amelioration of oxidative stress and suppression of nitric oxide synthase (NOS) expression, maintaining hepatocyte integrity and function.

2.3.8 Immunoadjuvant Activity

An adjuvant in a vaccine is an ingredient that enhances the immune system's response to that vaccine. Adjuvants help the immune system to better recognize what is in a vaccine and remember it longer, thus increasing the length that a vaccine may offer protection. One remarkable biological activity of saponins is their ability to modulate immunity toward a more cell-mediated response or to enhance antibody production, such that they act as immunoadjuvants.

The strong immunoadjuvant activity reported for Quil-A, a crude extract of *Quillaia saponaria* bark, and its derivatives, makes them ideal substances for use in vaccine formulations

[29, 30]. Since Quil-A is a heterogeneous mixture of saponins, its various components may exhibit different biological properties, including toxicological effects.

An improvement in reducing saponin toxicity and enhancing immunoadjuvant activity has been reported with the development of immunostimulating complexes (ISCOMs). ISCOMs are cage-like particles, approximately 40 nm in size, composed of saponins derived from *Quillaia saponaria*, phospholipids, cholesterol, and the targeted antigen. Recently, a purified fraction of *Quillaia brasiliensis* saponin was also tested as ISCOMs, and the typical cage-like structures were observed.

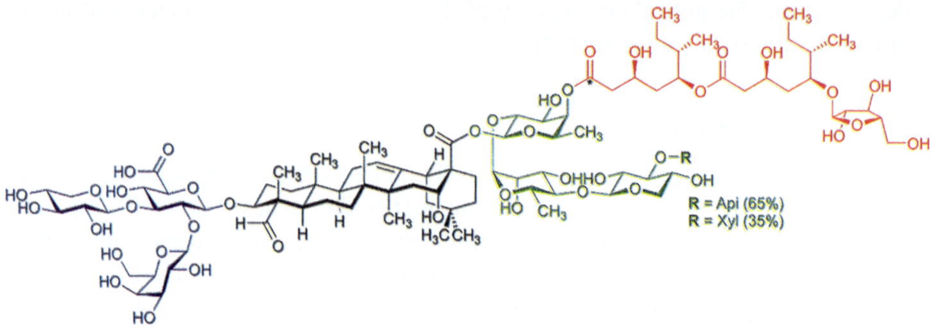

Figure 2.5 Structure of QS-21. The adjuvant is a 65:35 mixture of the apiose- and the xylose-substituted variants of the above molecule. (From Ragupathi et al. [31]. Wikimedia Commons / CC BY 1.0.)

The majority of adjuvants on the market today, including alum, mainly activate the humoral immune response. *Quillaia* saponins, on the other hand, are well known to induce strong activation of both the cell-mediated (Th1) and the humoral arms (Th2) of the immune system. Moreover, *Quillaia*-based adjuvants stimulate cytotoxic T cell production against exogenous antigens, thus providing an enhanced immune response.

Quillaia-based adjuvants are currently found in many commercially available veterinary vaccines from the main vaccine manufacturers in the world and are being used or tested in over 15 human vaccine candidates ranging from oncology to infectious diseases. A recent example is the use of the *Quillaia* adjuvant QS-21 in vaccines against COVID-19. QS-21 is a triterpenoid saponin with the structure shown in Figure 2.5. As can be seen, it has a branched trisaccharide chain in 3-position (shown in blue) and a more complex oligosaccharide chain attached at C-28 (shown in green), which is the normal pattern for triterpenoid saponins with two oligosaccharide chains, as was discussed above in Section 2.2. The unique structure of QS-21 is the acyl chain connected to a 6-membered sugar ring by an ester linkage and ending with a 5-membered sugar ring (shown in red). The ester linkage is hydrolytically labile.

In the vaccine against COVID-19, QS-21 is mixed with phospholipids and cholesterol to give the adjuvant, which is subsequently mixed with the antigen to give the final vaccine product.

2.4 Sources of Commercial Saponins

Quillaia saponaria and *Yucca schidigera* are the two principal sources of commercial saponins for medical, pharmaceutical, personal care, food, and general industrial preparations [13], while liquorice extracts are used by the food industry. *Quillaia saponaria*, the soap bark tree or soapbark, is an evergreen tree in the family Quillaiaceae, native to warm temperate heights of central Chile (populations are found up to 2000 m) in the forests of La Campana National Park and Cerro La Campana, between 32° and 40° south latitude approximately. A picture of the tree is given in Figure 2.6. The variant *Quillaia brasiliensis*

Figure 2.6 A *Quillaia saponaria* tree. (Photo by Matt Ritter, Cal Poly, San Luis Obispo.)

grows between 38° and 45° in southern Brazil in the states of Rio Grande do Sul, Santa Catarina, and Paraná, in northern Uruguay, northeastern Argentina, and eastern Paraguay. *Quillaia* contains more than 100 different triterpenoid saponins.

Saponin was originally extracted from the inner bark, which was a destructive practice leading to progressive deforestation. The later discovery that also other parts of the plant, like leaves and stems, contain appreciable quantities of saponin caused a change in the habits and today the pruning of *Quillaia* plantations provides the volumes of biomass demanded by the market.

Yucca is a genus of perennials and trees in the family Asparagaceae, subfamily Agavoideae. An image of the plant is shown in Figure 2.7. Its 40–50 species are native to the Americas and the Caribbean in a wide range of habitats, from humid rainforests and wet subtropical ecosystems to the hot and dry deserts and savannah regions. A total of 108 saponins from eight species of the genus *Yucca* have been described [32].

Yucca schidigera is the most commercially exploited plant among the yuccas and also the one that has been subject to the largest number of studies [33, 34]. The plant is common

Figure 2.7 *Yucca schidigera* in flower. (Stan Shebs / Wikimedia Commons / CC BY-SA 3.0.)

in the Mexican province of California, where it is domesticated and grown in controlled plantations. The content of saponins is high, around 10%.

Licorice *(or liquorice*; *Glycyrrhiza glabra)* is a plant belonging to the bean family Fabaceae, An image of the plant is shown in Figure 2.8. It is an herbaceous perennial legume native to western Asia, north Africa, and southern Europe. The roots are particularly rich in saponins, up to 20–25% have been reported.

The use of licorice has a long history in traditional medicines and folk remedies. The first documents can be traced back to ancient Assyrian, Egyptian, Chinese, and Indian cultures. Different geographical areas and periods show different uses [35]. Due to their sweet taste, licorice extracts are used as flavorings and sweeteners for American-type tobacco, chewing gum, candies, bakeries, soft drinks, dietary supplements, and ice creams. They are reported to have depigmentation properties and as such they are incorporated in personal care preparations.

China is not home to either *Quillaia* or *Yucca*; instead, other sources of saponins have been exploited, sometimes directed toward different uses than in Western countries, for example aquaculture. Tea saponins are extracted from the *Camellia oleifera*

Figure 2.8 *Licorice* in flower. (*Source:* Stamp Boy / flickr / CC BY-SA 2.0. "Pharaoh han / Wikimedia Commons / CC BY-SA 3.0.")

(see image of the plant in Figure 2.9) seed meal and from the leaves and flowers of *Camellia sinensis*. Water extraction, which may be ultrasound-assisted, combined with acetone precipitation, is recommended for the industrial extraction and purification of tea saponins.

Other sources of saponins in China include extracts of *Tribulus terrestris* (goat's head, devil's eyelashes, bullhead), *Trigonella foenum-graecum* (fenugreek), *Panax ginseng* (Chinese ginseng), *Hedera helix* (ivy), and *Glycine max* (soya bean). Many studies have focused on Chinese traditional herbal medicine plants, e.g. *Rhizoma paridis*, *Aralia chinensis*, *Clematis chinensis*, *Radix astragali*, *Mussaenda pubescens*, and *Clinopodium chinensis*.

Table 2.2 gives examples of saponins available in China, their source, the manufacturers, the approximate prices, and the recommended uses.

2.5 Extraction and Purification

To obtain saponins from plant material, different extraction methods may be employed, using solvents such as water, methanol, ethanol, or mixtures of these in Soxhlet extractors

Figure 2.9 *Camellia oleifera* in flower. (Photo courtesy of USDA Agricultural Research Service.)

or in orbital shakers. Other solvents, such as glycerol and aqueous or alcoholic surfactant solutions, have also been reported. Novel procedures use lower amounts of solvent but additional physical/chemical treatment: multistage extraction, pressure, microwaves, ultrasound, or supercritical fluid extraction. Such methods can lead to an increase in the process efficiency. However, it should be considered that under harsher conditions (higher temperature in particular) saponins may be hydrolyzed and degraded, which is detrimental to the quality [14, 36–39]. Table 2.3 summarizes various methods for the extraction of saponins.

Quillaia saponin extracts can be classified according to their saponin content into nonrefined, semirefined, and highly refined *Quillaia* saponin products. Simple, nonrefined *Quillaia* saponins are produced by coarsely shredding the biomass and using water as an extraction solvent [41]. Semirefined products are obtained from the nonrefined ones by removing low molecular weight organic compounds and salts by ultrafiltration or affinity chromatography [42]. Commercial nonrefined extracts typically contain 190–200 g saponins per kg solids, while semirefined extracts contain 750–800 g saponins per kg solids. The extracts can be further refined to fractions or individual saponins to yield a range of *Quillaia* extracts that vary in composition (and thus exhibit different functionalities) by column chromatography on silica gel or other stationary phases, using either normal or

Table 2.2 Sources, producers, approximate prices, and recommended uses of saponins in China.

Source	Description	Producer	Price (USD/kg)	Application
Camellia sinensis	Tea saponin, 70%	Promax Safety Co., Ltd.	3–4	Shrimp feed
Camellia sinensis	Tea saponin, 85%, powder	Changsha Xian Shan Yuan Agricult. and Technol. Co., Ltd.	3–4	Aquaculture – predator fish killing
Camellia sinensis	Crushed tea seed pellets	Changsha Richfield Agri Tech Co., Ltd.	0.2–0.5	Plant growth
Camellia sinensis	Chinese herbal medicine saponin	Qingdao Qingmei Biotech Co., Ltd.	18–20	Heart disease prevention
Camellia sinensis	Bulk alfalfa extract powder saponin: 5%, 20%, 40%	Qingdao Hilda Jingyi Trading Co., Ltd.	1–30	Food, pharmacy, health care
Camellia sinensis	Food, pharmacy, health care	Shaanxi Qingya Biotechnology Co., Ltd.	20–150	
Camellia oleifera	Camellia seed powder saponin	Anhui Yaliya Biotechnology Co., Ltd.	3–4	Aquaculture – shrimp feed
Camellia oleifera	High protein content camellia seed, saponin extract	Purelife Biotech Co., Ltd.	10–25	Food
Eleutherococcus senticosus	Acanthopanax saponin root powder	Wellgreen Technology Co., Ltd.	35–55	Injectables, suppositories, lotions
Hedera helix	*H. helix* extract	Shaanxi Science and Biotech Co., Ltd.		
Momordica charantia	Bitter melon extract, powder, 1% active	Hunan Nutramax Inc.	10–60	Personal care, food, pharmacy
Panax notoginseng	Natural source radix, 5–80% active	Herb Green Health Biotech Co., Ltd.	68–245	Pharmacy
Sapindus mukorossi	*Sapindus* extract saponin	Kingherbs Co., Ltd.	25–28	Pharmacy, food
Tribulus terrestris	*T. terrestris*, 98% active	Xian Ruierli Biological Engineering Co., Ltd.		
Tribulus terrestris		Shaanxi Saiyang Food Co., Ltd.	10–11 (wholesale)	

Table 2.2 (Continued)

Source	Description	Producer	Price (USD/kg)	Application
Tribulus terrestris	*T. terrestris* extract	Shaanxi KBC Biotech. Co., Ltd.	8–15	Health care
Trigonella foenum-graecum	Fenugreek natural herbal plant extract	Nanjing Muzhi Biotechnology Co., Ltd.	1–20	Food, pharmacy
Yucca schidigera	*Yucca* saponin extract, 30%, 60%, 80%, powder	Hunan E.K. Herb Co., Ltd.	1–30	Food, pharmacy, health care

(*Source:* Alibaba, consulted January 2024.)

reverse phase chromatography [43–46]. Use of preparative high pressure liquid chromatography (HPLC) as a chromatographic method to separate individual fractions has also been reported in the literature [47, 48].

2.6 Toxicology

Saponins vary widely in the kind and intensity of their biological activity. Some of the more important effects include hemolysis (strong *in vitro*, much weaker *in vivo*), local irritation, inflammation (intestine), antiinflammatory and antimicrobial activity, cytotoxicity, and antihypercholesterolemia in laboratory animals. After large oral doses, severe toxic effects were reported, such as liver damage, respiratory failure, gastric pain, diarrhea, hemolysis of erythrocytes, convulsions, and coma [49].

2.6.1 Acute Toxicity

Tested in mice, saponins extracted from the soapbark tree were less acutely toxic when administered orally (LD_{50} of 1600 mg/kg bw) than when administered subcutaneously (LD_{50} of 650 mg/kg bw), intraperitoneally (LD_{50} of 280 mg/kg bw), or intravenously (LD_{50} of 280 mg/kg bw) [50].

2.6.2 Long-Term Toxicity in Mice

Groups of 48 male and 48 female TO strain mice were fed diets containing 0, 0.1, 0.5, or 1.5% *Quillaia* extract for 84 weeks. The mice were observed regularly for abnormal condition or behavior, and some males were weighed at intervals. Hematological parameters were measured at weeks 24, 56, and 84. No compound-related effects were reported. At the highest dose, male mice showed decreased weight gain. *Quillaia* at these concentrations had no adverse effect on condition, behavior, or death rate. A detailed autopsy and histopathological examination of tissues and organs at the end of the study showed no compound-related effects. No carcinogenic effects were seen. The no observed adverse effect level (NOEL) was 0.5%, equivalent to 700 mg/kg bw per day [51].

Table 2.3 Methods for extraction of saponins.

Extraction method	Extraction principle	Solvent	Temperature	Time	Procedure	Advantage/disadvantage
Maceration	Solid–liquid interface extraction based on the solubility and diffusion of the solute into the solvent	Ethanol or methanol, either pure or as mixture with water	From room temperature to the boiling point of the solvent	Varies from few hours to weeks	Defatted plant material is soaked in solvent, sometimes assisted by mechanical stirring	Long extraction time, high amount of extraction solvent, low saponin yield
Soxhlet extraction and reflux extraction	Extraction by continuous distillation	Mostly 50–98% aqueous ethanol	Heated to the boiling point of the extracting solvent	1–3 d for Soxhlet extraction; 1–4 h for reflux extraction	The process involves heating the solvent to boiling, then returning the condensed vapor to the flask containing plant material. In case of Soxhlet extraction, plant material is separately placed in the thimble	Degradation of thermally labile components
Ultrasound-assisted extraction	Ultrasound radiation disrupts the cell structure and facilitates release of intracellular content	Ethanol or methanol, either pure or as mixture with water		10–20 min	Plant material dissolved in solvent is subject to ultrasound irradiation	High saponin yield. Relatively short extraction time. Minimum extraction solvent
Microwave-assisted extraction	Absorption of microwave radiation by the water in the plant material disrupts the cell structure, releasing desired components	Ethanol or methanol, either pure or as mixture with water		10–20 min	Plant material dissolved in solvent is subject to microwave irradiation (0.3–300 GHz)	High saponin yield. Relatively short extraction time. Minimum extraction solvent
Pressurized solvent extraction	Automated pressurized solvent extraction	Water or methanol	Usually 100 °C at 1500 psi	15–25 min	Extraction solvent is continuously pumped through the sample vessel applying high pressure	High saponin yield. Relatively short extraction time. Minimum extraction solvent

(*Source:* Rai et al. [40] / MDPI / CC BY 4.0.)

2.6.3 Short-Term Toxicity in Rats

Groups of 15 male and 15 female weanling rats were housed five per cage and fed diets containing 0 (control), 0.6, 2, or 4% *Quillaia* extract for 13 weeks. Groups of five male and five female rats from the same lot of animals were fed diets containing 0 (control), 2%, or 4% *Quillaia* extract for two or six weeks. Body weights and food intake were measured at the beginning and weekly throughout the study. Urine was analyzed during the final week of the study. At sacrifice, the absolute and relative organ weights were determined, and tissues and organs from the group given 4% diet were analyzed histologically. Hematological and serum chemical parameters were tested in all groups. No abnormalities of behavior or condition were seen in the rats receiving *Quillaia* extract. The NOEL was 0.6% in the diet, equivalent to 400 mg/kg bw per day [52].

2.6.4 Long-Term Toxicity in Rats

Groups of 48 male and 48 female Wistar-derived rats from a specific pathogen-free breeding colony were housed in groups of four and fed diets containing 0 (control), 0.3, 1, or 3% *Quillaia* extract for 108 weeks. Hematological examinations were made at weeks 15, 25, 52, and 108, and urine was analyzed at weeks 13, 24, and 78. At the end of the study, a complete autopsy was carried out, including histological examination of tissues and organs.

Male rats at the highest dietary concentration had lower body weights than controls throughout the experiment, the differences being statistically significant between 10 and 22 months. Females at the lowest dietary concentration had significantly higher body weights than controls during the first six months of the study. The lower body weights and reduced incidence of glomerulonephrosis in male rats fed 3% *Quillaia* were believed to be due to reduced food consumption. Hematological and urinary parameters were within the normal ranges.

In general, the incidence of histopathological lesions was similar in treated and control animals, as there was no dose–response relationship and no similar occurrence in males. The only tumor for which the incidence was statistically significantly different from that in controls was thyroid adenoma, which occurred more frequently in females fed 1% *quillaia* in the diet. This finding was not considered to be related to the treatment, as the incidence did not increase with the dietary concentration of *Quillaia* extract, and the total incidence in rats fed 1% *Quillaia* extract was not statistically significantly different from the total control incidence [53].

2.7 Producers and Vendors

Sigma-Aldrich (Saint Louis, MO, USA) supplies a wide range of *Quillaia* saponins for laboratory work: analytical reference, molecular biology, nonionic surfactants, permeabilization of membrane, as well as products complying with Pharmacopeia standards.

American Extracts (Strathmore, CA, USA) produces *Yucca* extracts for agriculture, wastewater management, and animal feed.

- Products in the ThermX range are concentrated plant meals containing steroid saponins. They protect crops against heat and water stress, soil compaction, poor water penetration, salinity, and alkali. The saponin improves water penetration into the soil, and water and nutrient uptake by wetting the root surfaces. The products are registered with California Department of Agriculture and with Organic Materials Review Institute (Eugene, OR, USA).

- BREAKDOWN70 is a concentrated plant extract containing steroid saponins. When placed in the growing environment of the microorganisms that digest organic wastes, the steroid saponin acts as an antistress agent, i.e. it dampens the harmful effects of the fluctuations from overloading, toxic input, pH, and temperature and it accelerates the decomposition process.

American Extracts has developed products for animal nutrition starting from the first recorded feeding trials that involved the feeding of steroid saponins from *Yucca schidigera* to lamb and cattle [54]. Very significant improvements were seen in weight gain and feed efficiency. Later publications confirmed similar results in other animals.

BAJA Yucca (Scottsdale, AZ, USA) commercializes a range of *Yucca schidigera* and *Quillaia saponaria*.

- *Whole yucca powder:* a natural dry concentrate of 100% natural *Yucca schidigera* produced from *Yucca* that has not been pressed or de-juiced. The powder contains 30% *Yucca* solids and a minimum of 10–12% saponins. It is nonhygroscopic and produced without the use of preservatives, additives, stabilizers, or artificial chemicals. Not for human consumption.
- *Yucca 50 liquid:* a concentrated natural cold-pressed extract of *Yucca schidigera*. Extraction is by cold-pressed processing to ensure retention of all natural state attributes. Approved for use in food and beverages under FDA Federal Register Paragraph 21 CFR 172.510.
- *Yucca 30 natural powder:* a natural dry concentrate of 100% natural *Yucca schidigera*. It contains about 30% of *Yucca* solids and a minimum of 10–12% saponins. It is nonhygroscopic and is produced without the use of preservatives, additives, stabilizers, or artificial chemicals. Not for human consumption.
- *Yucca 30 spray dry powder:* a premium, soluble, cold-pressed spray-dried extract of pure *Yucca schidigera* containing 30% *Yucca* solids on a maltodextrin base. Approved for use in food and beverages under FDA Federal Register Paragraph 21 CFR 172.510.
- *Yucca 15 natural powder:* a natural, low-temperature dried press cake of 100% natural *Yucca schidigera* that has been pressed or de-juiced. It contains about 15% of *Yucca* solids and a minimum of 5% saponins. It is nonhygroscopic and is produced without the use of preservatives, additives, stabilizers, or artificial chemicals. Not for human consumption.
- *QuillajaChoice®:* 100% pulverized natural *Quillaia saponaria* powder, reduces undesirable emissions, such as ammonia and odors in livestock, swine, poultry, and pet feces, while improving gut health. It also improves performance as an alternative growth promoter and nutrient absorbent, gives an immune response and productive parameters in livestock and pets. It is less expensive than *Yucca schidigera*.

Desert King International (DKI) (San Diego, CA, USA), with subsidiaries in Chile and Mexico, is dedicated to finding scientific solutions using natural components from *Quillaia*, *Yucca*, and *Jojoba*. For the past 35 years, DKI has been the world's largest producer of food-grade saponin extracts and for over a decade, it has produced and supplied the majority of pure saponin extracts, rich in QS-21, used by the pharmaceutical industry.

With its *Quillaia saponaria* forestry operations and production facility in Chile, the company has established a robust, integrated supply chain that ensures production of the highest purity extracts. Desert King Mexico is the manufacturing facility located in Tijuana (Baja California, Mexico) for all the *Yucca schidigera* products, following the rules and regulations of the Mexican Forestry Department to ensure that the *Yucca* production rate does not exceed the plant's natural recovery rate.

The company also offers *Quillaia* Saponin Fractions for Pre-Clinical Research or ISCOM preparations (QS-7 fraction, QS-17 fraction, QS-18 fraction, fraction A, fraction B, fraction C, quillaic acid) and hemolytic agents for diagnostic analyses (e.g. QA21).

Nanochemazone (Leduc, Alberta, Canada) is a high-tech enterprise focusing on research and development of nanomaterials and nanotechnology. They offer *Quillaja* 300 powder, 99% purity, which is presented as an additive for advanced ceramic materials.

Naturex (part of Givaudan) is headquartered in Avignon (France) and has production and sales offices in over 16 countries. It produces plant chemicals for the food and beverage, nutrition and health, and personal care industries. The two main products are Uptaia™, a *Quillaia* extract enhancing foam structure and stability in soft drinks, and Sapnov™, a purified extract of *Quillaia* used as an emulsifier for flavors and colors in food and beverages. They also supply *Yucca* saponins and extracts from American ginseng roots, licorice, and green tea leaves.

In 2014, Naturex acquired Chile Botanicas and their production facilities in Linares (Chile), among others for the extraction of the *Quillaia* saponins from the barks collected in the country.

It has commercialized Bucovia™, a toothpaste that contains an extract of *Solidago virgaurea*, a European plant containing saponins that is particularly effective in controlling *Candida albicans* in the oral cavity.

The *Quillaia* extract Sapnov LS™ was found in initial studies to be effective as a stabilizer in emulsion vaccines (see Section 2.10, Potential Applications).

Nor-Add (Bröndby Strand, Denmark) and its sister companies *Nor-Guard, Nor-Vet, Nor-Feed, Feed-Add, and Dumas* develop, manufacture, and market botanical, natural additives for inclusion in animals' feed and drinking water. The products are primarily based on utilization of active constituents of plants, primarily in the form of special polyphenols and saponins. Along with the traditional sources of saponins, *Quillaia* and *Yucca*, they are experimenting with other plants, such as *Trigonella foenum-graecum* (Fenugreek).

- Quiponin SR™ is a natural antioxidant.
- Quiponin™ S is a natural feed utilization improver, an ammonia controller, and a natural feed intake improver. It contains active elements from *Quillaia saponaria* and *Camellia oleifera*.
- Quiponin BS is a 100% natural soil improver manufactured on the basis of elements from *Quillaia saponaria* and *Camellia oleifera*. The product is available in two variants: Quiponin BS Powder and Quiponin BS Liquid. The powder version further contains

sepiolite as a carrier. The high content of *Quillaia*, a natural fungicide, makes the products effective natural alternatives to synthetic chemicals that are often used as solutions to fungus and nematode-related problems in plant production or as growth promoters.

2.8 Products and Prices

Because of the fragmentation of the market, it is impossible to give more than an approximate estimate of the volume of saponins produced, all grades combined. It has been estimated that in 2020, between 40 and 50 ktons of *Quillaia* biomass were sufficient to respond to the global demand for saponins from this source. Assuming an average 9% yield in the extraction process, this gives a figure of between 3600 and 4500 tons expressed as 100% active. Table 2.4 gives approximate prices of the saponin products.

There are also distributors offering saponins of different structures and from different sources. Below are a few examples.

Founded in 1879, **Carl Roth GmbH + Co. KG** (Karlsruhe, Germany) is a wholesaler of labware, life science products, and chemicals (some produced by the company). They offer Saponin DAB from Quillaia (Euro 805/kg), Saponin Pure (Euro 492/kg), and Saponin Tannin-free (Euro 555/kg).

Otto Chemie Pvt. Ltd. (Mumbai, India) is a distributor of different purity grades of saponin, CAS Nr. 8047-15-2, recommended as dietary supplements and nutraceuticals, an adjuvant in the production of vaccines, food additive, and personal care. The promotional leaflet also says that Otto saponins are used in liquid soaps, jewelry polish, pet shampoo, as detergent and wetting agent, and in animal feed. 100 g sells at USD 25, 500 g at USD 101.

2.9 Existing Applications

Quillaia extracts may be used as a foaming agent in foods and beverages, for example soft drinks, such as ginger beer, root beer, and cream soda, in cocktail mixes, and as an emulsifier in other foods, such as baked goods, candies, frozen dairy products, gelatine, and puddings. The major food use is in soft drinks. *Quillaia* extracts are currently proposed for use in the Codex draft General Standard for Food Additives (GSFA) 14.1.4 "Water-based flavored drinks," including "sport" or "electrolyte" drinks and particulate drinks (WHO Food reports at 500 mg/kg in the food group Additives Series 48, Safety Evaluation of Certain Food Additives and Contaminants – Quillaja Extracts). Traditionally, the quality and price of the extracts are determined using simple foam tests. This may not be adequate since similar foam levels can be obtained by blending *Quillaia* extracts with other low-cost saponin sources. To overcome these problems and standardize the quality of commercial extracts, the use of reverse-phase HPLC techniques has been suggested [42].

Somewhat surprisingly, highly purified *Quillaia* saponin products were found to have reduced functional properties, including the ability to stabilize foams. It has been suggested that this is caused by the removal during the extraction and fractionation of other compounds that act synergistically. These may be polyphenols and/or proteins [42].

Table 2.4 Prices of saponins in the United States and Europe.

Manufacturer	Product description	Packaging	Price (USD)
AK Scientific	Saponin	100 g	80
Alfa Aesar	Saponin	25 g	44
		100 g	92
Apex Biotechnology	Sapogenin glycosides	10 mg	266
		50 mg	756
		100 mg	1120
Biosynth/Carbosynth	Saponin	500 g	210
	Saponin	1 kg	345
	Saponin	2 kg	650
Chem-Impex	Saponin from *Quillaia* bark, spray dry powder, saponin content >20%	25 g	20
	Saponin from *Quillaia* bark, spray dry powder, saponin content >20%	100 g	47
	Saponin from *Quillaia* bark, spray dry powder, saponin content >20%	250 g	80
	Saponin from *Quillaia* bark, spray dry powder, saponin content >20%	1 kg	280
	Saponin from *Quillaia* bark, spray dry powder, saponin content >20%	5 kg	1612
Frontier Specialty Chemicals	Saponin	25 g	36
	Saponin >98%, standard reference grade	20 mg	280
	Saponin, sapogenin content 10–20%, from *Camellia sinensis* (L.) O. Kuntze	25 g	36
Sigma-Aldrich	Saponin from *Quillaia* bark, sapogenin content >10%	25 g	51
	Saponin from *Quillaia* bark, sapogenin content 20–35%	10 g	59
	Saponin CAS 8047-15-2, a glycoside used to	100 g	91

Table 2.4 (Continued)

Manufacturer	Product description	Packaging	Price (USD)
	permeabilize cellular membranes, exhibits hemolytic activity		
	Saponin used as nonionic surfactant	10 g	101
	Saponin for molecular biology, used as nonionic surfactant	50 g	108
	Saponin from *Quillaia* bark, sapogenin content 20–35%	25 g	118
	Saponin from *Quillaia* bark, sapogenin content >10%	100 g	154
	Saponin for molecular biology, used as nonionic surfactant	250 g	441
	Saponin used as nonionic surfactant	500 g	472
TCI America	Saponin	25 g	51
	Saponin	250 g	306
LGC	TRC saponin (technical grade)	25 g	45
US Biological	Saponin	25 g	170
	Saponin	250 g	509

(*Source:* "authors" / reproduced from Chemical Book.)

For a good part of the last century, the use of saponin as a detergent fell into a general oblivion and was reduced to a niche phenomenon limited to hobbyist circles and to a small number of local distributors of traditional cleaning products and practices. With the rising of the "eco" message, a number of formulators emerged, selling locally or regionally or offering their products via Amazon or eBay.

The products being offered are all based on soapwort (*Saponaria officinalis,* see Figure 2.10) that can be grown in local fields, harvested, and easily processed. It is interesting to note that when the industrial revolution enabled the large-scale commercial manufacture of clothes, textile industrialists cultivated fields of the "weed" as an inexpensive detergent [8].

Along with these professionals, there are scores of amateur producers of "Bio" and "Nature" who post on the net recipes to extract saponins from soapwort, soapberries (*Sapindus saponaria, Sapindus mukorossi*), chestnut (*Castanea sativa*), horse chestnut (*Aesculus hippocastanum*), and many other plants.

Figure 2.11 shows an example of a commercial cleaning product formulated with chestnut saponin and the associated claims.

Figure 2.10 Soapwort in flower. (*Source:* Unknown author / Wikimedia Commons / CC BY-SA 3.0.)

2.10 Potential Applications

Beyond the established applications of saponins, it is quite possible that there is more concealed in the wide research conducted over the years. If not yet in these works, it is

Figure 2.11 A commercial detergent formulated with chestnut saponin.

the intrinsic behavior of the saponins and the vast number of structures that may unveil unexpected and yet unidentified effects. The relatively recent creation of a market for saponins in aquaculture is an example of the potential achievable with this amazing class of molecules. What today may appear as a curiosity, or as a valueless laboratory experience, may become tomorrow a large, profitable business. It is conceivable that the research so far has just scratched the surface. Below are reported studies that may hold potential for future developments [33].

It has been reported that saponins can act in synergy with synthetic surfactants to reduce the total surfactant level required for proper cleaning performance of laundry detergents. It was also demonstrated that such mixtures were particularly effective at low temperatures. This is interesting because there is a drive toward low-temperature cleaning in home care laundry. Saponins currently have limited commercial availability, being sourced from natural plants. This supply is insufficient in scale, is of limited structural specificity, and is too expensive for home laundry applications. However, there is a huge potential market if a source and a process could be found that enables saponins to be produced in large volumes at a cost that is competitive with that of synthetic surfactants.

Yucca extract contains two groups of useful substances. One group is based on steroid saponins, which may improve the quality of food products, such as porosity, density, and hardness, and the second one is antioxidants that are able to reduce fat oxidation and extend food quality during storage [55].

It has been known since a long time that extracts from *Yucca* and *Quillaia* inhibit fermentation caused by gut microbes, particularly *Streptococcus bovis, Butyrivibrio fibrisolvens, Escherichia coli,* and rumen protozoa [56]. *Yucca* extracts may be used as natural, nontoxic deodorizers. Powdered so-called biopreparations, consisting of mixtures of microorganisms and enzymes, are widely used to control the odor from animal manure. The gaseous pollutants generated from large animal farms, not least poultry farms, are a severe hygienic problem and considerable efforts are being made to combat the problem. It has been found that combining the biopreparations with extracts from *Yucca schidigera* considerably reduced the amount of volatile odorous compounds [57]. Another study on the emission of pollutants to air from horse studs reported that saponin extracts added to a corn silage basal diet reduced the emission of ammonia by 30% while no effect could be seen on emissions of methane and nitrous oxide [58]. YuccaChoice® is a commercial product added to pet food to reduce fecal odor.

Quillaia saponin is a natural surfactant with properties that most amphiphilic substances possess. It is, for instance, widely used as a foaming agent and for cleaning purposes, as has been discussed above. A less common application where its amphiphilic character is taken advantage of is as stabilizer of suspensions. A mixture of *Quillaia* saponins possessing immune-adjuvant properties has been suggested as a suspension stabilizer for pharmaceutical applications [59].

Food products sometimes need to be heat resistant, which may cause problems of various kinds. The products may, for instance, contain more than one surfactant and such surfactant mixtures may change character during heating. It has now been demonstrated that binary mixtures of *Quillaia* saponins and lecithin are relatively unaffected by heating from 25 to 75 °C. This is practically important information because lecithin is a common ingredient in food formulations. It may also be relevant for pharmaceutical applications [60].

Q-Naturale® is a food-grade surfactant isolated from the bark of the *Quillaia saponaria* Molina tree. Its ability to act as an emulsifier for forming and stabilizing typical food emulsions was tested and compared with that of Tween 80, a synthetic surfactant that is widely used in the food industry [61]. Q-Naturale® was capable of forming relatively small droplets ($d < 200$ nm) at low surfactant-to-oil ratios but the drops were not as small as those prepared with Tween 80 under similar conditions ($d < 150$ nm). The droplets coated with Q-Naturale® were stable to coalescence over the pH range 2–8 and over a broad range of electrolyte concentration and temperature, 0–500 mM NaCl and 20–90 °C, respectively.

Orange oil is obtained as a byproduct of orange juice production. It contains more than 90% D-limonene and is often used as a substitute for limonene in foods and beverages. Nanoemulsions of orange oil, of particular interest for beverages, can be made by microfluidization using *Quillaia* saponins as emulsifier. The nanoemulsions were found to be very long-lived if an ester gum or a medium-chain triglyceride was incorporated in the dispersed phase [62]. Incorporation into the dispersed phase of a small amount of a substance that is insoluble in the continuous phase is known to prevent Ostwald ripening. In a work by Zhang et al., ester gum was incorporated in the oil phase as an Ostwald ripening inhibitor and *Quillaia* saponins were compared as emulsifier/stabilizer with starch that had been modified with octenyl succinate anhydride to transform it into a hydrophobically modified carbohydrate. Nanoemulsions of olive oil made using microfluidization and saponins were found to be superior to the modified starch, producing nanoemulsions, with smaller droplets and longer shelf-life [63].

Many food ingredients, e.g. colorants and fat-soluble vitamins, need to be incorporated through solubilization into micelles of amphiphilic components in the food formulation. Saponins from *Quillaia saponaria* Molina were found to be useful for solubilizing lutein ester extract. The size of the micelles increased from 7.5 nm without lutein ester to 130 nm with maximum loading of the carotenoid, which is indicative of a very efficient solubilization [64].

A high level of cholesterol in the blood is known to increase the risk of coronary heart disease. This is well-known to the consumers and there are today many low-cholesterol products on the market. The ability of saponins to associate with cholesterol has been used to make low-cholesterol dairy products. The use of saponin-impregnated diatomaceous earth to remove cholesterol from butterfat was reported as early as 1967 [65]. Later reports have demonstrated efficient removal of cholesterol from homogenized milk. Particularly good results have been obtained with saponins from the bark of *Quillaia saponaria* with diatomaceous earth (celite) added to adsorb the cholesterol-saponin complex [66].

The literature contains many examples where saponins are used as emulsifiers to make oil-in-water nanoemulsions that are used as carriers for vitamins [67], ω-3 fatty acids [68], and other nutrients [23]. Such nanoemulsions have also been explored for the delivery of antioxidants such as astaxanthin [69].

Subunit vaccines consisting of highly purified antigens require the presence of adjuvants to create effective and long-lasting protective immunity. An oil-in-water emulsion adjuvant composed of mineral oil and with gum Arabic as an emulsifier has previously been found useful as an adjuvant in swine subunit vaccines. Addition of a food-grade *Quillaia* saponin extract, named Sapnov™ LS, to the emulsion was later found to give extended physical stability even at elevated temperature (37 °C) [70].

Tea saponins, found in *Camellia* plants, hold promises for soil remediation, an area where traditional surfactants have largely failed. They have been used for leaching remediation, phytoremediation, and microbial remediation and have been found effective in desorbing heavy metals from contaminated soils and in enhancing their bioavailability. Tea saponins also improve the degradation of organic pollutants by microorganisms [71].

2.11 Regulatory

Saponin-based adjuvants are now approved both by the European Medicines Agency (EMA) and by the Food and Drug Administration (FDA) [72].

2.11.1 Saponins in Pharmacopoeias

Because of its biological properties, *Quillaia saponaria* has been included in the European and Brazilian pharmacopoeias. The raw material is recommended as a cough reliever and expectorant, antiinflammatory, hypocholesterolemic, and hemolytic agent. Topically, it is indicated to treat scalp diseases (dandruff and hair loss) and other dermatological disorders. In the African Pharmacopoeia, *Quillaia* bark is reported only as an emulsifying agent.

2.11.2 Saponins in Food and Cosmetics Regulatory

Nowadays, saponins are approved for use as food additive in 187 signatory countries of the Codex Alimentarius, including the European Union, the United Kingdom, the United States, China, and Japan [73]. The European Commission of Cosmetic Ingredients database has listed some *Quillaia* products, such as bark, bark extract, root extract, and wood extract as cosmetic ingredients. Depending on the material used, different functions are attributed, such as anti-dandruff, cleansing, emulsifying, foaming, masking, moisturizing, skin conditioning, and surfactant [74].

Yucca extract of the whole plant of *Yucca arborescens* or *Yucca schidigera* is reported in the Japanese List of Existing Food Additives. Because the steroid saponins in *Yucca* exhibit antifungal activities, *Yucca* extract is added to food as a "shelf-life extender" in the Japanese market [75].

Japan also allows the use of saponins in cosmetics [76].

Quillaia extracts are classified by the FDA as Generally Recognized as Safe (GRAS) and are also part of the human diet when used as an emulsifier in baked goods, candies, frozen dairy products, gelatin, and puddings (WHO, 2002) [77]. *Yucca schidigera* Roezl. Extracts are also commercialized and approved by the FDA as GRAS [78].

Quillaia and *Yucca* saponins have been classified as GRAS by the Flavor and Extract Manufacturers' Association of the United States (FEMA) with FEMA number 2973 [2].

Licorice extract, the portion of the licorice root extracted by boiling water after maceration, contains glycyrrhizin, a triterpenoid saponin. It is recognized as GRAS. It can be further purified by filtration and treatment with acid and ethanol. Monoammonium glycyrrhizinate (MAG), prepared from ammoniated glycyrrhizin by solvent extraction and separation techniques, is also allowed [79]. MAG is used as a natural sweetener and considered to be a functional ingredient used to enhance a product's sweetness.

Nowadays, there is an increasing interest in using saponins as natural emulsifiers for food additives, alone or blended with other natural ingredients, e.g. lecithin. Numerous studies have reported that *Quillaia* saponins are good emulsifiers for O/W emulsions [76, 80, 81] and for nanoemulsions [81]. A food ingredient containing *Quillaia* bark saponins (Q-Naturale) has been approved by the FDA as an emulsifier for beverages [73, 81].

According to the Codex Alimentarius Commission, extracts from *Quillaia saponaria* may be used as a foaming agent in "water-based flavored drinks," including "sport" or "electrolyte" drinks, ginger beer, root beer, cream soda, in cocktail mixes, and in particulate drinks (as General Standard for Food Additives (GSFA) category 14.1.4, "Water-based flavored drinks" 500 mg/kg maximum use level; WHO FOOD ADDITIVES SERIES: 48, Safety Evaluation of Certain Food Additives and Contaminants – Quillaja Extracts). In soft drinks, unpurified *Quillaia* extracts may be used at doses up to 200 mg/kg. However, in syrups intended for dispensable frozen carbonated beverages (FCBs) or frozen lemonades, *Quillaia* extracts may be used up to 500 mg/kg on dry solid basis [82]. In the European Union, *Quillaia* extract is classified as a foaming agent for use in water-based, flavored nonalcoholic drinks and labeled as E 999 (200 mg/L calculated as anhydrous extract) [82].

Quillaia extracts are also used as an emulsifier for foods, such as baked goods, candies, frozen dairy products, gelatine, and puddings.

References

1 Francis, G., Kerem, Z., Makkar, H., and Becker, K. (2002). The biological action of saponins in animal systems: a review. *The British Journal of Nutrition* 88: 587–605.

2 Güclü- Üstündağ, Ö. and Mazza, G. (2007). Saponins: properties, applications and processing. *Critical Reviews in Food Science and Nutrition* 47: 231–258.

3 Mir, M.A., Parihar, K., Tabasum, U. et al. (2016). Estimation of alkaloid, saponin and flavonoid content in various extracts of *Crocus sativa*. *Journal of Medicinal Plants Studies* 4: 171–174.

4 Gunther, R.T. (1933). *The Greek Herbal of Dioscorides*. New York, N.Y.: Hafner.

5 Jaeger, E.C. (1944). *A Source-Book of Biological Names and Terms*. Springfield, IL: Charles E. Thomas.

6 Bognolo, G. (2023). *Surface Active Agents: Historical Perspectives and Future Developments*, 25–27. Boca Raton, FL: CRC Press.

7 Gassdorf, D., Fischer, C., and Brandt, D. (1996). *It all Started with Water Glass: 100 Years of Revolutionary Research at Henkel. 38–39*. Düsseldorf, Germany: Henkel KGaA.

8 Georgia, A.E. (2007). *A Manual of Weeds*. Whitefish, MT, USA: Kessinger Publishing.

9 Chintala, S.R., Murthy, E.N., Pattanaik, C. et al. (2010). Piscicidal plants used by Gond tribe of kaval wildlife sanctuary, Andhra Pradesh, India. *Indian Journal of Natural Products and Resources* 1: 97–101.

10 Xu, P. and Yu, B. (2021). Chemical synthesis of saponins: an update. *Advances in Carbohydrate Chemistry and Biochemistry* 79: 1–62.

11 Kregiel, D., Berlowska, J., Witonska, I.A. et al. (2017). Saponin-based, biological-active surfactants from plants. In: *Application and Characterization of Surfactants* (ed. R. Najjar) Chapter 6. London, UK: IntechOpen.

12 Folmer, B. (2003). Sterol surfactants: from synthesis to applications. *Advances in Colloid and Interface Science* 103: 99–119.

13 Oleszek, W., Hamed, A., Kjellin, M., and Johansson, I. (2010). Saponin-based surfactants. In: *Surfactants from Renewable Resources* (ed. M. Kjellin and I. Johansson), 239–250. Chichester, UK: Wiley.

14 Ribeiro, B.D., Alviano, D.S., Barreto, D.W. et al. (2013). Functional properties of saponins from sisal (*Agave sisalana*) and juá (*Ziziphus joazeiro*): critical micelle concentration, antioxidant and antimicrobial activity. *Colloids and Surfaces A* 436: 736–743.

15 Folmer, B.M., Svensson, M., Holmberg, K., and Brown, W. (1999). The physicochemical behavior of phytosterol ethoxylates. *Journal of Colloid and Interface Science* 213: 112–120.

16 Kronberg, B., Holmberg, K., and Lindman, B. (2014). *Surface Chemistry of Surfactants and Polymers*, 92–93. Chichester, UK: Wiley.

17 Oakenful, D. (1986). Aggregation of saponins and bile acids in aqueous solution. *Australian Journal of Chemistry* 39: 1671–1683.

18 Rosen, M.J. and Kunjappu, J.T. (2012). *Surfactants and Interfacial Phenomena*, 4the, 53–99. Hoboken, NJ: Wiley.

19 Pedebos, C., Pol-Fachin, L., Pons, R. et al. (2014). Atomic model and micelle dynamics of QS-21 saponin. *Molecules* 19: 3744–3760.

20 Liao, Y., Li, Z., Zhou, Q. et al. (2021). Saponin surfactants used in drug delivery systems: a new application for natural medicine components. *International Journal of Pharmaceutics* 603: 120709.

21 Pagureva, N., Tcholakova, S., Golemanov, K. et al. (2016). Surface properties of adsorption layers formed from triterpenoid and steroid saponins. *Colloids and Surfaces A: Physicochemical and Engineering Aspects* 491: 18–28.

22 Zhu, Z., Wen, Y., Yi, J. et al. (2019). Comparison of natural and synthetic surfactants at forming and stabilizing nanoemulsions: Tea saponin, *Quillaja* saponin, and Tween 80. *Journal of Colloid and Interface Science* 536: 80–87.

23 Schreiner, T.B., Dias, M.M., Barreiro, M.F., and Pinho, S.P. (2022). Saponins as natural emulsifiers for nanoemulsions. *Journal of Agricultural and Food Chemistry* 70: 6573–6590.

24 Ralla, T., Salminen, H., Tuosto, J. et al. (2018). Formation and stability of emulsions stabilized by *Yucca* saponin extract. *International Journal of Food Science and Technology* 53: 1381–1388.

25 Wojciechowski, K., Orczyk, M., Marcinkowski, K. et al. (2014). Effect of hydration of sugar groups on adsorption of *Quillaja* bark saponin at air/water and Si/water interfaces. *Colloids and Surfaces B* 117: 60–67.

26 Lorent, J.H., Quetin-Leclercq, J., and Mingeot-Leclercq, M.-P. (2014). The amphiphilic nature of saponins and their effects on artificial and biological membranes and potential consequences for red blood and cancer cells. *Organic and Biomolecular Chemistry* 12: 8803–8822.

27 Fleck, J.D., Betti, A.H., da Silva, F.P. et al. (2019). Saponins from *Quillaja saponaria* and *Quillaja brasiliensis*: particular chemical characteristics and biological activities. *Molecules* 24: 171.

28 Sur, P., Chaudhuri, T., Vedasiromoni, J.R. et al. (2001). Antiinflammatory and antioxidant property of saponins of tea (Camellia sinensis (L) O. Kuntze) root extract. *Phytotherapy Research* 15: 174–176.

29 Sun, H.-X., Xie, Y., and Ye, Y.-P. (2009). Advances in saponin-based adjuvants. *Vaccine* 27: 1787–1796.

30 Wang, P. (2021). Natural and synthetic saponins as vaccine adjuvants. *Vaccines* 9: 222.

31 Ragupathi, G., Gardner, J.R., Livingston, P.O. et al. (2011). Natural and synthetic saponin adjuvant QS-21 for vaccines against cancer. *Expert Review of Vaccines* 10: 463–470.

32 Jiménez, G.G., Durán, A.G., Macias, F.A. et al. (2021). Structure, bioactivity and analytical methods for the determination of *Yucca* saponins. *Molecules* 26: 5251.

33 Piacente, S., Pizza, C., and Oleszek, W. (2005). Saponins and phenolics of *Yucca schidigera* Roezl: chemistry and bioactivity. *Phytochemistry Reviews* 4: 177–190.

34 Oleszek, W., Sitek, M., Stochmal, A. et al. (2001). Steroidal saponins of *Yucca schidigera* Roezl. *Journal of Agricultural and Food Chemistry* 49: 4392–4396.

35 Armanini, D., Fiore, C., Mattarello, M.J. et al. (2002). History of the endocrine effects of licorice. *Experimental and Clinical Endocrinology & Diabetes* 110: 257–261.

36 Doughari, J.H. (2012). Phytochemicals: extraction methods, basic structures and mode of action as potential chemotherapeutic agents. In: *Phytochemicals – A Global Perspective of their Role in Nutrition and Health* (ed. V. Rao), 1–32. Rijeka, Croatia: IntechOpen.

37 Bitencourt, R.G., Queiroga, C.L., Montanari Junior, I., and Cabral, F.A. (2014). Fractionated extraction of saponins from Brazilian ginseng by sequential process using supercritical CO_2, ethanol and water. *Journal of Supercritical Fluids* 92: 272–281.

38 Cheok, C.Y., Salman, H.A.K., and Sulaiman, R. (2014). Extraction and quantification of saponins: a review. *Food Research International* 59: 16–40.

39 Moghimipour, E. and Handali, S. (2015). Saponin: properties, methods of evaluation and applications. *Annual Research & Review in Biology* 5: 207–220.

40 Rai, S., Acharya-Siwakoti, E., Kafle, A. et al. (2021). Plant-derived saponins. A review of their surfactant properties and applications. *The Scientist* 4: 44.

41 Schlotterbeck, T., Castillo-Ruiz, M., Cañon-Jones, H. et al. (2015). The use of leaves from young trees of *Quillaja saponaria* (Molina) plantations as a new source of saponins. *Economic Botany* 69: 262–272.

42 San Martín, R. and Briones, R. (2000). Quality control of commercial *Quillaja* (*Quillaja saponaria* Molina) extracts by reverse phase HPLC. *Journal of the Science of Food and Agriculture* 80: 2063–2068.

43 Guo, S., Kenne, L., Lundgren, L.N. et al. (1998). Triterpenoid saponins from Quillaja saponaria. *Phytochemistry* 48: 175–180.

44 Higuchi, R. and Komori, T. (1987). Structures of compounds derived from the acyl moieties of *Quillaja* saponin. *Phytochemistry* 26: 2357–2360.

45 Higuchi, R., Tokimitsu, Y., Fujioka, T. et al. (1986). Structure of desacylsaponins obtained from the bark of *Quillaja saponaria*. *Phytochemistry* 26: 229–235.

46 Nyberg, N.T., Kenne, L., Rönnberg, B., and Sundquist, B.G. (2000). Separation and structural analysis of some saponins from *Quillaja saponaria* Molina. *Carbohydrate Research* 323: 87–97.

47 Maier, C., Conrad, J., Carle, R. et al. (2015). Phenolic constituents in commercial aqueous *Quillaja* (*Quillaja saponaria* Molina) wood extracts. *Journal of Agricultural and Food Chemistry* 63: 1756–1762.

48 Nord, L.I. and Kenne, L. (1999). Separation and structural analysis of saponins in a bark extract from *Quillaja saponaria* Molina. *Carbohydrate Research* 320: 70–81.

49 Khan, I.A. and Abourashed, E.A. (2010). *Leung's Encyclopedia of Common Natural Ingredients Used in Foods, Drugs, and Cosmetics*, 3rde. New York, N.J.: Wiley.

50 Efimova, T.G., Pivnenko, G.P., Koutsevich, V.A. et al. (1966). The action of soapbark tree Quillaia Saponaria saponins on blood pressure and cholesterol content in animals. *Pharmaceutical Chemistry Journal (Farm. Zh.)* 21: 45–49. (in Russian).

51 Philips, J.E., Butterworth, K.R., Gaunt, I.F. et al. (1979). Long-term toxicity study of quillaia extract in mice. *Food and Cosmetics Toxicology* 17: 23–27.

52 Gaunt, I.F., Grasso, P., and Gangolli, S.D. (1974). Short-term toxicity of quillaia extract in rats. *Food and Cosmetics Toxicology* 12: 641–650.

53 Drake, J.J.-P., Butterworth, K.R., Gaunt, I.F. et al. (1982). Long-term toxicity study of quillaia extract in rats. *Food and Chemical Toxicology* 20: 15–23.

54 Hale, W.H., Sherman, W.C., Reynolds, W.M., and Luther, H.G. (1961). The value of certain steroidal sapogenins in rations of fattening lambs and cattle. *Proceedings of the Society for Experimental Biology and Medicine* 106: 486–489.

55 Sucharzewska, D., Stochmal, A., and Oleszek, W. (2003). The effect of *Yucca schidigera* extract on the physical structure and on the oxidative stability of sugar-candy foam products. *LWT- Food Science and Technology* 36: 347–351.

56 Sen, S., Makkar, H.P.S., Muetzel, S., and Becker, K. (1998). Effect of Quillaja saponaria saponins and Yucca schidigera plant extract on growth of Escherichia coli. *Letters in Applied Microbiology* 27: 35–38.

57 Matusiak, K., Oleksy, M., Borowski, S. et al. (2016). The use of *Yucca schidigera* and microbial preparation for poultry manure deodorization and hygienization. *Journal of Environmental Management* 170: 50–59.

58 Li, W. and Powers, W. (2012). Effects of saponin extracts on air emission from steers. *Journal of Animal Science* 90: 4001–4013.

59 Kensil, C.A. (2010). Compositions comprising immunoreactive reagents and saponins, and methods of use thereof. US Patent 8,808,692 B2, filed 18 November 2010 and issued 19 August 2014.

60 Reichert, C.L., Salminen, H., Leuenberger, B.H., and Weiss, J. (2016). Influence of heat on miscibility of *Quillaja* saponins in mixtures with a co-surfactant. *Food Research International* 88: 16–23.

61 Yang, Y., Leser, M.E., Sher, A.A., and McClements, D.J. (2013). Formation and stability of emulsions using a natural small molecule surfactant: *Quillaja* saponin 8Q-Naturale®. *Food Hydrocolloids* 30: 589–596.

62 Zhang, J., Bing, L., and Reineccius, G.A. (2015). Formation, optical property and stability of orange oil nanoemulsions stabilized by *Quillaja* saponins. *LWT – Food Science and Technology* 64: 1063–1070.

63 Zhang, J., Bing, L., and Reineccius, G.A. (2016). Comparison of modified starch and Quillaja saponins in the formation and stabilization of flavor nanoemulsions. *Food Chemistry* 192: 53–59.

64 Tippel, J., Lehmann, M., von Klitzing, R., and Drusch, S. (2016). Interfacial properties of Quillaja saponins and its use for micellisation of lutein esters. *Food Chemistry* 212: 35–42.

65 Schwartz, D.P., Brewington, C.R., and Burgwald, L.H. (1967). Rapid quantitative procedure for removing cholesterol from butterfat. *Journal of Lipid Research* 8: 54–55.

66 Chang, E.J., Oh, H.I., and Kwak, H.S. (2001). Optimization of cholesterol removal conditions from homogenized milk by treatment with saponin. *Asian-Australian Journal of Animal Science* 14: 844–849.

67 Lv, S., Zhang, Y., Tan, H. et al. (2019). Vitamin E encapsulation within oil-in-water emulsions: impact of emulsifier type on physicochemical stability and bioaccessibility. *Journal of Agricultural and Food Chemistry* 67: 1521–1529.

68 Uluata, S., McClements, D.J., and Decker, E.A. (2015). Physical stability, autoxidation, and photosensitized oxidation of ω-3 oils in nanoemulsions prepared with natural and synthetic surfactants. *Journal of Agricultural and Food Chemistry* 63: 9333–9340.

69 Chen, Z., Shu, G., Taarji, N. et al. (2018). Gypenosides as natural emulsifiers for oil-in-water nanoemulsions loaded with Astaxanthin: insights of formulation, stability and release properties. *Food Chemistry* 261: 322–328.

70 Burakova, Y., Madera, R., Wang, L. et al. (2018). Food-grade saponin extract as an emulsifier and Immunostimulant in emulsion-based subunit vaccine for pigs. *Journal of Immunology Research.* 2018: 8979838.

71 Yu, X.-L. and He, Y. (2018). Tea saponins: effective natural surfactants beneficial for soil remediation, from preparation to application. *RSC Advances* 8: 24312–24321.

72 Pulendran, B., Arunachalam, P.S., and O'Hagan, D.T. (2021). Emerging concepts in the science of vaccine adjuvants. *Nature Reviews Drug Discovery* 20: 454–475.

73 (2018). *Codex Alimentarius International Food Standards*, vol. 49. Rome, Italy: Food and Agriculture Organization of United Nations World Health Organization.

74 Cosmetic Ingredient Database. http://ec.europa.eu/growth/sectors/cosmetics/cosing_en.

75 JETRO (2011). Specifications and Standards for Foods, Food Additives, etc. Under the Food Sanitation Act (2010). https://www.jetro.go.jp/ext_images/en/reports/regulations/pdf/foodext2010e.pdf.

76 Wojciechowski, K. (2013). Surface activity of saponin from *Quillaja* bark at the air/water and oil/water interfaces. *Colloids and Surfaces B* 108: 95–102.

77 Federal Register "Quillaja Saponaria Extract; Exemption from the Requirements of a Tolerance", a rule by the EPA from 01 August 2007.

78 Durán, A.G., Calle, J.M., Butrón, D. et al. (2022). Steroidal saponins with plant growth stimulation effects; *Yucca schidigera* as a commercial source. *Plants* 11: 3378.

79 Code of Federal Regulations, Title 21, Vol. 3, 21 CFR 184.1408.

80 Ozturk, B. and McClements, D.J. (2016). Progress in natural emulsifiers for utilization in food emulsions. *Current Opinion in Food Science* 7: 1–6.

81 de Faria, J.T., de Oliveira, E.B., Minim, V.P.R., and Minim, L.A. (2017). Performance of *Quillaja* bark saponin and β-lactoglobulin mixtures on emulsion formation and stability. *Food Hydrocolloids* 67: 178–188.

82 FAO (2004). Quillaja Extracts. Type 1 and type 2. Chemical and technical assessment. http://www.fao.org/fileadmin/templates/agns/pdf/jecfa/cta/61/QUILLAIA.pdf.

3
Phospholipids

3.1 Introduction

Phospholipids are amphiphilic molecules consisting of a polar headgroup, which can be anionic or zwitterionic, and a hydrophobic part, which is normally twin-tailed. The hydrophobic part is large, typically consisting of some 40 carbon atoms, which means that phospholipids can be regarded as very hydrophobic surfactants.

Figure 3.1 shows the structure of different types of phospholipids. A common feature for almost all phospholipids is that they are triesters of glycerol. The secondary and one of the primary hydroxyl groups of glycerol is esterified with a long-chain fatty acid. The remaining primary hydroxyl group is esterified with phosphorous acid, which is also esterified with an alcohol containing an additional functional group such as quaternary ammonium, amino (which may become protonated), diol, sugar, or a serine residue.

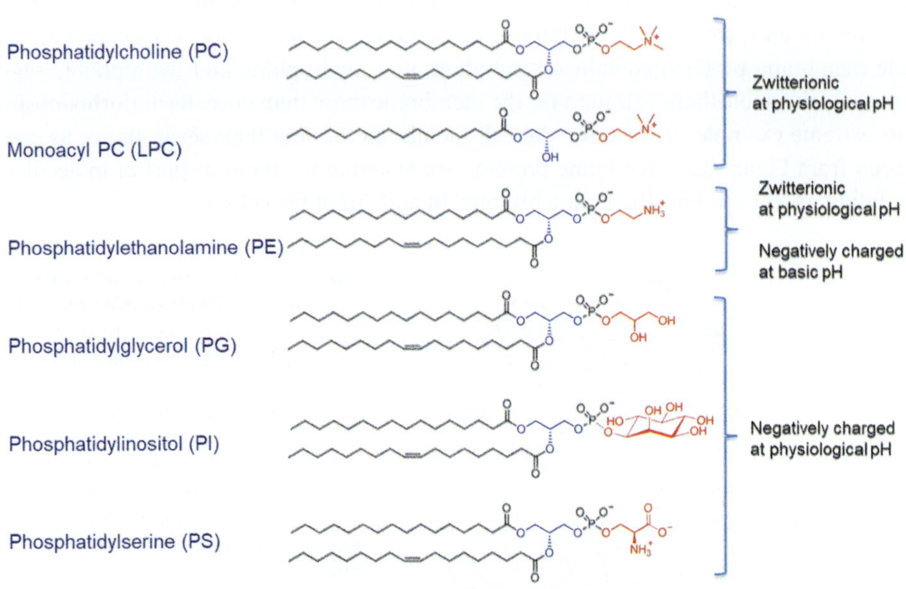

Figure 3.1 Common phospholipids. (Adapted from van Hoogevest et al. [1].)

One of the phospholipids in Figure 3.1, monoacyl PC (LPC), has only one long acyl chain. Such lipids are called lysophospholipids, the name being derived from the fact that such amphiphiles were found to be hemolytic. Lysophospholipids are usually the result of the enzyme phospholipase A catalyzing the hydrolysis of regular phospholipids. From a surfactant viewpoint, lysophospholipids constitute a more balanced amphiphile than the phospholipids. This will be further discussed below.

As shown in Figure 3.1, the acyl chains of phospholipids can be saturated or unsaturated. Marine phospholipids typically have acyl chains from the ω-3 fatty acids eicosapentaenoic acid (EPA) and docosahexaenoic acid (DHA) incorporated in the structure. The geometry of the double-tailed phospholipids promotes self-assembly into bilayers with the hydrophobic tails lining up and with the hydrophilic heads facing the aqueous medium on both sides. The curvature of such bilayers will be governed by the composition since the geometry of the phospholipids varies and the bilayers are made up of mixtures of different species. Cell membranes are built up of phospholipid bilayers together with other biomolecules, such as cholesterol and proteins.

The structure and function of the cell membrane were explained by the "fluid-mosaic model" in 1972 by Singer and Nicolson [2]. The model describes the structure of the cell membrane as a dynamic, flexible arrangement of different components, principally phospholipids and proteins with cholesterol contributing to the membrane fluidity by preventing a too regular packing of the phospholipid tails. There are two types of proteins in the cell membranes, "integral membrane proteins" and "peripheral membrane proteins." The integral membrane proteins are amphiphilic, and the polar groups protrude from the membrane into the aqueous environments inside and outside of the cell while the nonpolar part is largely buried in the hydrophobic interior of the membrane. The peripheral membrane proteins do not extend across the membrane but are transiently associated with the cell membrane. The peripheral membrane proteins are normally glycosylated.

The membrane proteins are very interesting examples of amphiphilic block copolymers. Some membrane proteins contain several alternating hydrophilic and hydrophobic segments, which enable them to transverse the membrane more than once. Bacteriorhodopsin is an extreme example. It traverses the cell membrane no less than seven times, as can be seen from Figure 3.2. Membrane proteins are essential for the transport of molecules (including water) and ions (including protons) in and out of the cells.

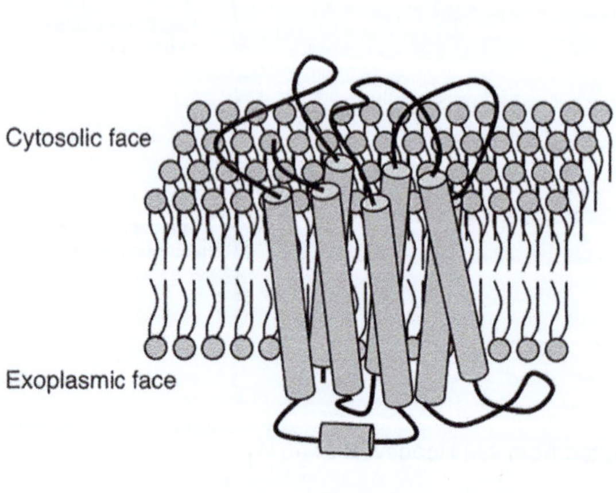

Figure 3.2 The membrane protein bacteriorhodopsin traversing the cell membrane seven times. The membrane is made up of phospholipids oriented in a bilayer with the polar headgroups protruding into the aqueous environments inside and outside of the cell. The cytosolic face is the part of the bilayer facing inward toward the cytoplasm and the exoplasmic face is the part facing outward.

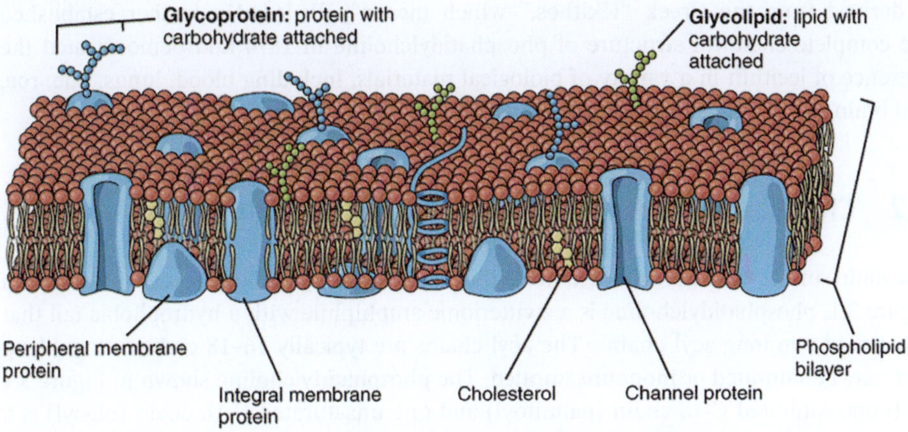

Figure 3.3 Schematic of a cell membrane. (*Source:* Wikipedia Commons.)

Singer and Nicolson's original model has later been somewhat modified. For example, it has been suggested that proteins and glycoproteins do not exist as single elements as originally proposed but rather as diffusing complexes in the cell membrane structure [3]. However, the basic feature of the model still holds. Figure 3.3 illustrates the structure of the biological membrane with the phospholipids coexisting with cholesterol and proteins.

Pulmonary surfactants are phospholipid-protein complexes that cover the respiratory surfaces. They play a fundamental role in facilitating inflation/deflation during breathing and they are essential to prevent collapse of the lung at the end of expiration. In fact, breathing and proper gas exchange would simply not be possible without pulmonary surfactants.

The basic task of pulmonary surfactants is to drastically reduce the surface tension but, in addition to this critical feature, an effective pulmonary surfactant will adsorb rapidly onto the air–liquid interface and spread over it during inspiration. In doing so, it reduces the work of breathing.

The reduction in surface tension – primarily mediated by the major phospholipid component, phosphatidylcholine, especially in its fully saturated form dipalmitoylphosphatidylcholine – facilitates lung inflation and deflation while preventing small airway collapse [4]. It has been demonstrated that both components of the pulmonary surfactant, the protein as well as the phospholipid, are essential. Membranes reconstituted from whole pulmonary surfactant organic extract, containing all the lipids plus the hydrophobic proteins, give very rapid oxygen diffusion, substantially faster than achieved by membranes prepared from the phospholipid fraction only [5].

Lecithin is a generic term that is used to designate a crude mixture of phospholipids extracted from a vegetable or animal source. The composition differs between different sources and between different procedures for the work-up, but phosphatidylcholine is usually the most abundant phospholipid. A sample of crude lecithin usually also contains nonnegligible amounts of triglycerides, fatty acids, and sterols. Refined lecithin is a commercial product that contains essentially only phospholipids. Lecithin is one of the most important surfactants produced by macroorganisms.

Lecithin was first isolated in 1845 by the French chemist and pharmacist Théodore Gobley. The raw material was fish eggs and Gobley named the extract "lécithine," which is derived from the Greek "lékithos," which means yolk [6]. He further established the complete chemical structure of phosphatidylcholine in 1874 and demonstrated the presence of lecithin in a variety of biological materials, including blood, lungs, bile, roe, and brain of humans, sheep, and chickens [7].

3.2 Chemical Structure and Properties of Lecithin

The main surface-active component in lecithin is phosphatidylcholine. As can be seen from Figure 3.1, phosphatidylcholine is a zwitterionic amphiphile with a hydrophobic tail that consists of two long acyl chains. The acyl chains are typically 16–18 carbon atoms long and may be saturated or monounsaturated. The phosphatidylcholine shown in Figure 3.1 with one saturated C-16 chain (palmitoyl) and one unsaturated C-18 chain (oleoyl) is a representative example. However, it is not uncommon that both chains are fully saturated. The pulmonary surfactants discussed above are an important example of such structures.

Lecithin is virtually insoluble in water; thus, it does not form micelles in water and values of critical micelle concentration (CMC) do not exist. Lecithin is soluble in ethanol, however. Lysolecithin, with only one long acyl chain, is water soluble, but the reported CMC values differ considerably depending on the length of the remaining chain and on the purity. CMC values of 2.9–8.4 μM and 0.25–0.4 μM for the palmitoyl and the stearoyl species, respectively, have been reported [8].

Pure phosphatidylcholine is positively charged at very low pH, when the normally deprotonated oxygen of the phosphate group becomes protonated. The pK_a value is 1.0. Above pH 3 phosphatidylcholine is a perfect zwitterion with a net zero charge. This behavior is not necessarily representative of commercial lecithin products, however, because other phospholipids exhibit a different net charge versus pH pattern. Phosphatidylserine is positively charged at low pH and negatively charged at neutral and high pH with an isoelectric point of 4.0. Phosphatidylethanolamine is positively charged below pH 3.5, zero net charged between 3.5 and 8.0, and negatively charged above pH 8 with a pK_a for the protonated amine of 9.8. Phosphatidylinositol and phosphatidylglycerol are both negatively charged over the entire pH range with pK_a values of 2.5 and 3.0, respectively [9]. The behavior may also change with time due to hydrolysis of the ester bonds, generating fatty acids.

As mentioned above, lecithin has very low solubility in water, but it swells with it. This behavior is the same as for synthetic hydrophobic double-tailed surfactants, for instance the so-called "quats" and "ester quats," used as fabric softeners [10]. The phase diagram for the system soybean-lecithin is shown in Figure 3.4. In the range 7–35% water, an extended liquid crystalline phase, which is very temperature stable, forms. This phase has a lamellar structure. Further addition of water gives a two-phase system in which the lamellar liquid crystalline phase coexists with water [12].

Agitating a lamellar liquid crystalline phase can give rise to liposomes. Liposomes were first described by Bangham in the early 1960s and have since been a very active area of research. Lecithin liposomes are readily obtained and by choosing the right procedure, either unilamellar or multilamellar liposomes can be produced. Unilamellar liposomes are composed of one phospholipid bilayer while multilamellar liposomes contain many layers. The size of the liposomes is also governed by the processing method used. The liposomes

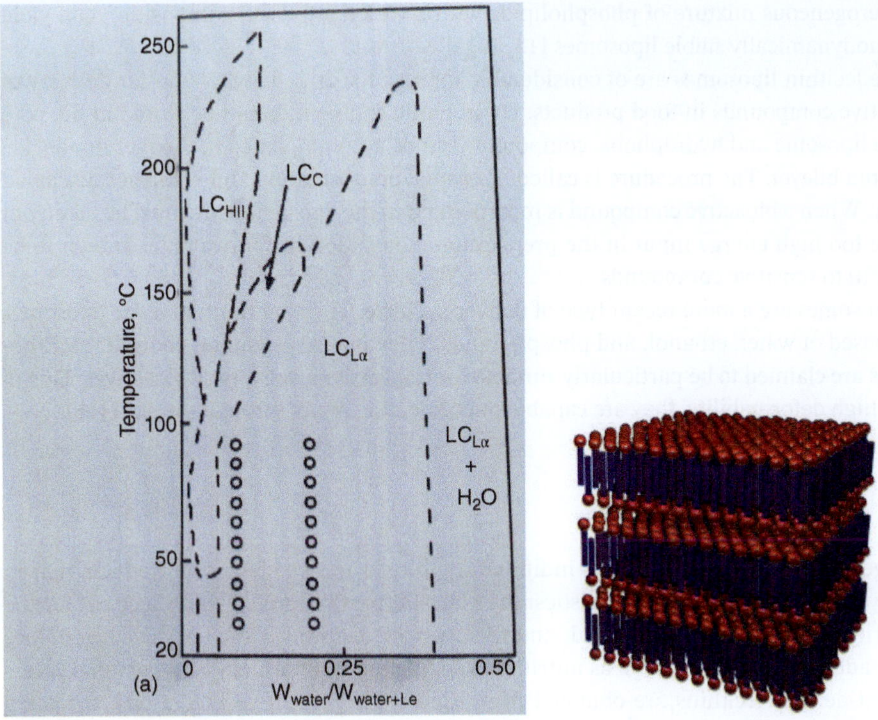

Figure 3.4 Left: Phase diagram of the system soybean lecithin-water. $LC_{L\alpha}$, LC_{HII}, and LC_C stand for liquid crystalline phase with lamellar structure, liquid crystalline phase with reversed hexagonal structure, and liquid crystalline phase with cubic structure, respectively. (From da Cruz Francisco et al. [11]. da Cruz Francisco, J., Topgaard, D., Sivik, B., et al. (2004) / with permission.) Right: Structure of a lamellar liquid crystalline phase.

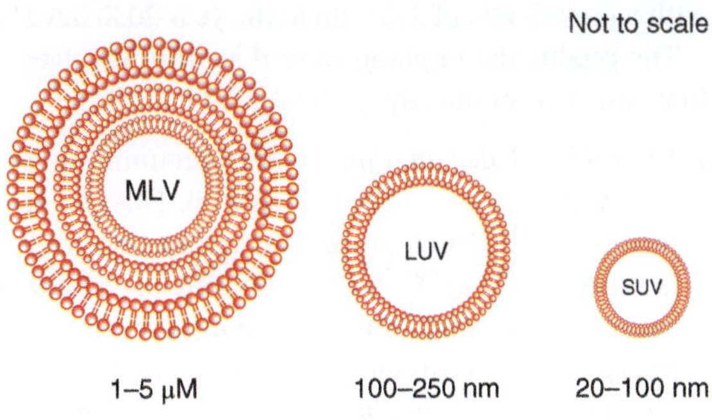

Figure 3.5 From left to right: A multilamellar vesicle (MLV), a large unilamellar vesicle (LUV), and a small unilamellar vesicle (SUV).

are sometimes referred to as vesicles; however, in a strict meaning of the word, vesicles are natural biological entities while liposomes are man-made objects. Thus, a liposome can be said to be an artificial vesicle. Figure 3.5 illustrates vesicles with different number of bilayers and with different sizes.

An open question ever since liposomes were first produced is whether they can be thermodynamically stable. This issue is still not settled but there are strong indications that a heterogeneous mixture of phospholipids with a varied fatty acid composition can yield thermodynamically stable liposomes [13, 14].

The lecithin liposomes are of considerable interest for drug delivery and for delivery of bioactive compounds in food products. Hydrophilic compounds can be stored in the core of the liposome and hydrophobic compounds can be accommodated in the membrane, i.e. the lipid bilayer. The procedure is called liposomal encapsulation and is further discussed below. When a bioactive compound is incorporated in the liposome, care must be taken not to use too high energy input in the preparation. Sonication is, for example, known to be harmful to sensitive compounds.

Ethosomes are a more recent type of delivery vehicle. They are multilamellar liposomes composed of water, ethanol, and phospholipids. They may also contain cholesterol. Ethosomes are claimed to be particularly suitable for transdermal delivery of bioactives. Due to their high deformability, they are capable of permeating intact through the skin [15].

3.3 Production

The generic word lecithin is used indifferently to refer to two classes of products that in reality present different characteristics: the plant-derived lecithins, which account for the majority of the volumes consumed, and the egg yolk lecithins, which are used in higher value-added applications, such as nutritional supplements, pharmacy, and personal care.

Plant-derived lecithins are obtained from the degumming of vegetable oils, the main sources being in the order soybean, sunflower, rapeseed, and canola oils (although often used interchangeably, canola is the edible crop of rapeseed oil (low erucic acid) developed by using breeding techniques). Egg yolk lecithin has a more balanced and unique composition and contains specific fatty acids that are not present in plant-derived lecithin [16].

It is difficult to estimate with precision the global market for lecithin but for the year 2023 figures of the order of 1.5 to 2 billion USD in value and about 1.7 million tons in volume with a growth rate of 3–5% up to the year 2030 have been put forward.

The production of plant-derived lecithin involves different steps that can be performed in batch or continuously, as illustrated below:

- *Vegetable oil degumming:* In the degumming process, phospholipids and gums are removed from the crude vegetable oil. This can be performed as a batch or a continuous process. In the batch process, the crude oil is first added to the reactor, the temperature is raised to 70–82 °C, then 2–3% water is added, and the pH is adjusted with an acid such as acetic acid. Stirring is continued for 30–60 minutes. The water hydrates the phospholipids, forming a sludge that is sent to the degumming centrifuge. Continuous degumming is carried out in pipelines, where the raw crude oil meets a fixed amount of water or steam, forming the hydrate, which is then purified.
- *Isolation:* After centrifugation, the sludge containing the phospholipid is washed with water containing a chelating agent to remove residual sugars and metal ions, centrifuged, and dehydrated, for example, in a thin film evaporator or by vacuum evaporation at 65–70 °C. The dehydration can be performed continuously. The dehydrated jelly must be quickly cooled to below 50 °C to prevent discoloration. Dilute hydrogen peroxide is often added to the wet gel to prevent bacterial spoilage.

- *Decolorization:* Decolorization uses hydrogen peroxide, particularly effective to reduce yellow pigments, or benzoyl peroxide, which is capable of reducing red pigments, or combinations of the two. The optimum process temperature is 70 °C. Activated carbon is added if polyaromatic hydrocarbons must be removed. Sodium hypochlorite can also be used as bleaching agent.
- *Drying:* Batch vacuum drying is common, but care must be taken to avoid foam formation. For this reason, long drying times (three to four hours) and vacuum are used. Thin film evaporator is particularly effective in preventing discoloration and in removing residual acetic acid formed in the degumming step.
- *Refining:* This stage is necessary to remove residual oil, fatty acids, and other impurities. Lecithin is mixed with acetone, in which it is insoluble. The acetone dissolves the impurities, and the acetone solution is subsequently removed.

To produce powder lecithin, after refining the residual acetone is removed with hot nitrogen gas and the lecithin is dried in a fluidized bed dryer.

The production process is inherently simple, but in order to maximize lecithin's utility and functionality, the processing conditions all the way back to the bean or seed must be carefully controlled. Bean storage and handling, crude oil storage, refining pretreatments, drying processes, bleaching, chemical modification, and storage can all affect the quality and performance of lecithin.

Other treatments are used to enhance lecithin's functionality. Fractionation with ethanol enriches the phosphatidylcholine content. Enzymatic modification produces higher Hydrophilic-Lipophilic Balance (HLB) lecithins. Phospholipase A2 causes hydrolysis at the 2-carbon position of glycerol, while phospholipase A1 hydrolyzes at the 1-carbon position, which is usually occupied by a saturated fatty acyl chain. Type A1 produces more polar lysophosphatides than type A2. Higher HLB lecithins can also be obtained chemically by hydroxylation, while acetylation results in lower HLB lecithins.

Production of egg yolk lecithin uses a different approach:

- *Solvent extraction:* This has been the first method used to extract egg yolk lecithin. It may use a single solvent (usually ethanol) or mixed solvents (such as ethanol/acetone), in which case it takes advantage of the different solubilities of egg yolk phospholipids in ethanol and acetone at low temperature to obtain a phospholipid-rich fraction of high purity.
- *Supercritical carbon dioxide extraction:* This technology can extract lecithin with higher purity (95–98%) from egg yolk powder. Addition of 5–10% ethanol is necessary to achieve complete extraction of the phospholipids [17]. Sometimes subcritical extraction is used, i.e. a fluid extraction technology that uses a thermodynamic state at the edge of a supercritical state, which is above the critical pressure and below the critical temperature. Compared to the supercritical fluid method, the subcritical fluid method uses milder conditions and has less stringent requirements for high-pressure equipment [16].
- *Column chromatography:* Column chromatography on silica gel or alumina gel produces high-purity egg yolk lecithin, but the processed volume is small, and the solvent (usually ethanol) consumption is large.

As for the plant-derived lecithins, enzymatic hydrolysis can be applied to the isolated egg yolk-derived phospholipids.

3.4 Producers and Prices

In the 150 years since its discovery, lecithin has become a widely used additive and formulation ingredient, which nowadays is supplied from a variety of sources in different physical forms by a multitude of producers and distributors, ranging from giant multinationals to regional or specialty manufacturers. In the majority of cases, lecithin is part of a complex and diversified palette of food, feed, and industrial additives.

Types of commercial lecithin include liquids, of both high and low viscosity, blends of lecithin and vegetable oils, e.g. soy and sunflower, powders and granules, de-oiled grades, which allow a higher concentration of phospholipids, fractionated and enzymatically and chemically modified products. There is a growing demand for non–genetically modified organisms (non-GMO) and identity preservation (IP) grades for food, personal care, and medical applications. The IP label may be used when soybean oil is produced without the aid of genetic engineering, with documented IP at each stage of processing, independent third-party certification of conformance to defined IP procedures, and a maximum of 0.9% adventitious residual genetically modified (GM) DNA. Demand for non-GMO lecithin is particularly strong in the European Union, where food manufacturers want to avoid labeling their products as genetically modified. Products containing GM soy-derived lecithin would require such labeling. Some food manufacturers are using alternatives to lecithin, e.g. citric acid fatty esters, due to GMO concerns and to avoid allergen labeling.

Cargill Inc. from the United States is a global company producing, processing, storing, and trading grains and oilseed for the agrochemical, animal nutrition, food and beverage, pharma, and industrial sectors. It is a giant multinational with more than 150 years of experience, over 160 000 co-workers, and operations in 70 countries.

Cargill manufactures lecithins at three premises: Decatur (IL, USA), Hamburg (Germany), and Zaandam (the Netherlands). The product pellet includes:

- *Fluid lecithin:* naturally occurring mixtures of phospholipids with minor amounts of other lipids. Trade names: Tiptichin™, Leciprime™, Lecisot™
- *Tailored fluid:* especially designed for bakery applications. Trade names Emulfluid™, Metarin™
- *Modified fluid:* chemically or enzymatically modified
- *Fractionated, de-oiled, and fluid lecithin:* soy lecithin fractionated into increased content of individual phospholipid components to increase functionality. Trade names: Epikuron™ and Emofluid™
- *De-oiled:* especially suitable for chocolate. Trade names: Emulpure™, Lecigran™, Epikuron, Lecimulthin™
- *Canola lecithin:* for use in a wide variety of food applications, including chocolate and confectionery
- *Fractionated and powder lecithin:* soy lecithin fractionated to increase the content of individual phospholipid components for higher functionality
- *Powdered lecithin*
- *Special fluid lecithin:* the lecithin base is combined with special oils and co-emulsifiers to provide functionalities for specific applications
- *Health-promoting lecithin:* provides nutritional benefits
- *Tailored de-oiled:* trade names: Emultop™ and Metarin™
- *Specialties:* Ovatin™

Table 3.1 Lecithin producers and distributors.

Company	Location	Lecithin origin	Comments
BT Agro LTDA	Brazil	Soy	Food, feed, technical grades
Natural Food	Poland	Sunflower	Food grade
Henry Franc S.A.S.	France	Soy	Non-GMO, IP, liquid, powder, granules
Buxtrade GmbH	Germany	Soy	Food grade
Lecico GmbH	Germany	Sunflower, rapeseed	De-oiled, enzymatically modified
Imperial Oel Import GmbH	Germany		Food and technical grades
SoleChem SRL	Italy	Soy	E 322 direct food additive
Lasenor Emul SL	Spain	Soy, sunflower, rapeseed	Operations in Spain, India, Russia, Brazil, USA; Verolec R from soy, GMO and non-GMO, Giralec R from sunflower, Semilec R from rapeseed
Bajrang Agro Industries Pvt Ltd.	India	Soy	
Krishna Oil Extractions Pvt Ltd.	India	Soy	Food and technical grades
Maharashtra Oil Extraction Pvt Ltd.	India	Soy	Food and technical grades
Sonic Biochem Extraction Pvt Ltd.	India	Soy, sunflower	Sunflower and soy lecithin, non-GMO soy lecithin
Mac World Industries Ltd.	Malaysia	Soy	Minimum 99% phospholipids
Clarkson Specialty Lecithins (formerly Clarkson Soy Products LLC)	USA		Specialty lecithins, mechanically extracted, Kosher, Halal
National Lecithin Inc.	USA		
UPI Chem Inc.	USA		Lecithin hydrolysate
Cape Crystal Production Inc.	USA		Soy and sunflower lecithins are part of a product pellet of gums, stabilizers, emulsifiers, gellifiers, acidifiers, kitchen tools
Shaanxi Guanje Technology Co., Ltd.	China	Soy, sunflower, egg yolk	Injectable grade; 20%, 50%, 70%, 80%, 90%, 95%, 98% phospholipid
Aurora Industry Co. Ltd. (AUCO)	China	Soy	Water-soluble hydrolyzed lecithin

(Continued)

Table 3.1 (Continued)

Company	Location	Lecithin origin	Comments
Henan Tianfu Chemical Co., Ltd.	China	Soy	Minimum 98% phospholipids
Jiaxing Reuze Import Export Co., Ltd.	China	Soy	Food grade
Henan Chiatai Biochemistry Trading Co., Ltd.	China	Soy	Food-grade liquid lecithin
Tian Jia Chemical Co., Ltd.	Japan	Soy	Food, feed, nutritional supplements, cosmetics, pharmaceuticals
Jedwards International Inc.	USA		**Distributor**
Keyser & Mackay	The Netherlands		**Distributor**. Serves Belgium, the Netherlands, France, Poland, Germany, Switzerland
Carl Roth GmbH	Germany	Soy, egg yolk	**Distributor**. Liquid minimum 97% phospholipids, for biochemistry

Besides this global supplier, there are scores of producers, some with international dimensions but many covering a regional market. Some companies are serving niche specialties.

Table 3.1 gives examples of lecithin suppliers, both producers and distributors. Lecithins are generally part of a larger product pellets that include other additives or even professional tools or equipment. The list is not complete; however, it illustrates the large variety of producers and products.

Lecithins are sold at different prices, depending upon application, physical form, specification, and volume, but as a broad approximation, there are three price bands:

- *Nutritional:* The lecithin is sold in capsules or as a liquid, a powder, or as granules in plastic bottles, typically from 100 g to 1 kg. Prices range from 130 to 210 euros per kg. Special grades for use in biochemical or medical research can attain 1600 to 2300 euros per kg. Egg yolk lecithins are 1.5 to 2 times the price of plant-derived lecithins.
- *Domestic and small producers:* Lecithins for domestic or small food and industrial use producers are sold in 1 to 5 kg plastic buckets or 10 to 20 L plastic drums or 25 kg bags. Prices range from 30 to 70 euros per kg.
- *Large volume producers:* The size per order ranges from 1000 kg up to 25 000 kg and prices range from 1.5 euros per kg for the low quality, industrial grades to 4 euros per kg for the specialty grades. Top quality grades sell for 6 to 8 euros per kg.

3.5 Existing Applications

3.5.1 Liposomes for Delivery of Bioactive Substances

Liposomes have an established use for oral delivery of drugs and for delivery of a variety of active substances in food and cosmetic products [18, 19]. The concept is versatile. As mentioned above, hydrophilic compounds can be housed in the aqueous interior of the liposome and hydrophobic substances can be accommodated in the nonpolar environment of the phospholipid bilayer. Liposomal delivery of drugs is illustrated in Figure 3.6.

Initially, liposomes were used for targeted drug delivery and this still represents an important application, but over the years, new uses for liposomal delivery have emerged. A problem that was soon discovered with liposomes for drug delivery applications was that they were rapidly destroyed by the body's immune system. That problem was to a large extent solved by pegylating the liposomes, i.e. attaching long poly(ethylene glycol) (PEG) chains to the liposomes. These so-called "stealth liposomes" were found to have a much-increased circulation time *in vivo* [21–23]. Figure 3.7 shows a pegylated liposome.

It is well recognized and scientifically proven that many traditional oral dietary and nutritional preparations have limited efficiency because of low absorption and insufficient

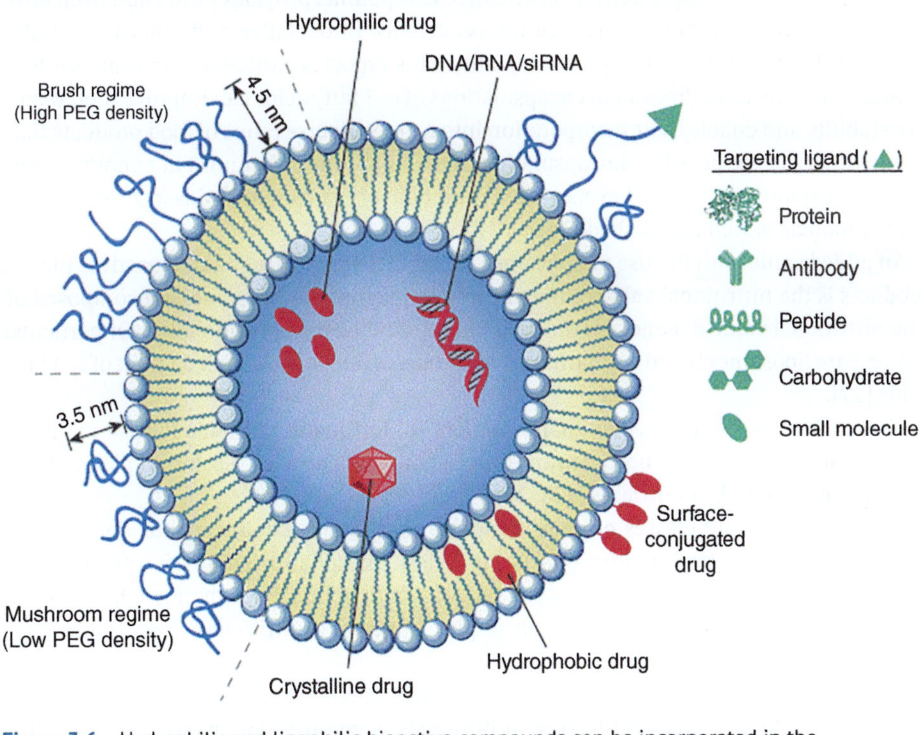

Figure 3.6 Hydrophilic and lipophilic bioactive compounds can be incorporated in the aqueous core and in the hydrophobic phospholipid bilayer, respectively, of liposomes. (From Kumar et al. [20] / International Journal of Indigenous Herbs and Drugs (IJIHD) / CC BY 4.0.)

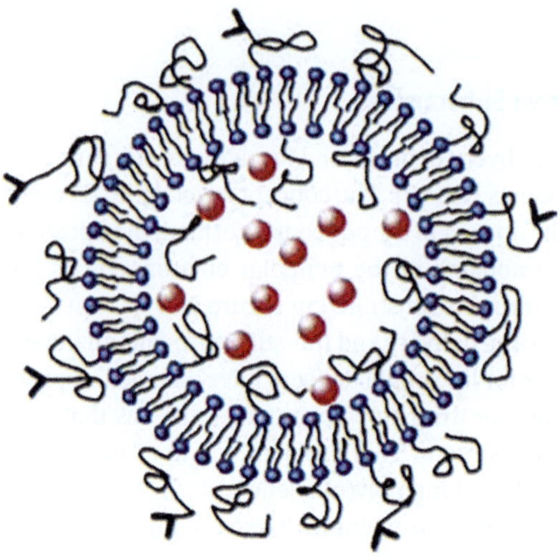

Figure 3.7 A pegylated liposome, also called a "stealth liposome," loaded with a hydrophilic drug. The PEG chains typically have a molecular weight of 1500–8000 Da. (From Mouritsen et al. [23] / Lipids, curvature, and nano-medicine. / John Wiley & Sons / CC BY 4.0.)

bioavailability. The encapsulation of lipophilic and hydrophilic nutrients within liposomes is an effective method of bypassing the destructive elements of the gastric systems and may result in a more efficient delivery of nutrients to cells and tissues [24].

The liposomal method of delivering sensitive compounds provides protection from environmental conditions and can reduce the occurrence of oxidative and other undesirable reactions that negatively affect product quality. This aspect is particularly relevant for food products. For instance, liposomal encapsulations of ω-3 fatty acids can improve their oxidative stability and enable their incorporation into a more diverse range of food products than would otherwise be possible. Encapsulation of the lipophilic vitamin D3 in nanoliposomes has been explored for use in beverages [25] and delivery of cannabinoids in food and beverage products is well established [26].

An added value with the use of liposomes for the delivery of drugs and bioactives in food products is the nutritional value of the phospholipids that the liposomes are composed of. The nutritional value depends on the type of fatty acid in the phospholipids. Of particular interest are liposomes based on marine animal sources due to their high content of ω-3 fatty acids [27].

Liposomes can also be used as "nano-carriers" of fertilizing nutrients to treat malnourished or sick plants. It has been demonstrated that such particles "soak into plants leaves more easily than naked nutrients" [28, 29].

Out of the different classes of phospholipids, phosphatidylcholine stands out as being particularly valuable from a nutritional point of view. Phosphatidylcholine is a precursor for choline, which is essential for the biosynthesis of acetylcholine. Phosphatidylethanolamine and phosphatidylserine are also known to play important nutritional roles, as has been summarized by Alhajj et al. [19].

The use of liposomal drug-delivery systems for cosmetics started in the late 1980s, and liposomes are today a frequent feature in hundreds of commercial preparations. The first such product on the market may have been CAPTURE®, an antiaging product from Dior, for which it was stated that the use of liposomes allowed modulation of skin penetration, as well as a controlled release of argan oil [30].

Liposomes form an occlusive layer adhering to the skin surface, which facilitates the delivery of active ingredients into the deeper skin layers while protecting it from external agents. This allows, for example, formulation of water-resistant sun protective products with water-soluble ultraviolet filter. Liposomes encapsulating CoQ10, a powerful antioxidant, are also beneficial to the skin. Moisturizers containing gel extracts of *aloe vera* leaf encapsulated in liposomes were found to have better antiwrinkle properties than the gel alone. The antioxidant CoQ10 also has a better effect on the skin when encapsulated. Larger vitamins such as vitamin E, water-soluble vitamins, fat-soluble amino acids, and even chemically modified active ingredients often improve their performance when encapsulated into liposomes [31].

The cosmetics and foodstuff sectors normally use refined lecithin as starting material for production of liposomes. The pharmaceutical sector, on the other hand, tends to use synthetic phospholipids as starting material because these make regulatory issues more straightforward.

3.5.2 Lecithin in Food, Feed, and Nutrition

Besides the use of lecithin in liposomes, lecithin has a variety of applications within the food and feed sectors. It is widely employed as an emulsifier for water-in-oil emulsions. For oil-in-water emulsions, it must be combined with a more hydrophilic surfactant, which may be a fatty alcohol ethoxylate, a sorbitan ester ethoxylate or a sugar-based surfactant. Lysolecithin is much less hydrophobic than lecithin and is suitable as emulsifier for oil-in-water emulsions, also without a cosurfactant.

Lecithin is used as a dispersant and as a wetting agent in food and feed formulations. It is also employed as a palletization improver in animal feed formulations.

In confectionery, lecithin reduces viscosity, replaces more expensive ingredients, controls sugar crystallization and the flow properties of chocolate, helps in the homogeneous mixing of ingredients, improves shelf life for some products, and can be used as a coating material. In emulsions and fat spreads, such as margarines with a high fat content, it stabilizes emulsions, reduces spattering (splashing and scattering of oil droplets) during frying, improves texture of spreads, and enhances flavor release [32]. In doughs and baking, it reduces fat and egg requirements, helps even out distribution of ingredients in the dough, stabilizes fermentation, increases volume, protects yeast cells in the dough when frozen, and acts as a release agent to prevent sticking, which simplifies cleaning. It improves wetting properties of hydrophilic powders (such as low-fat proteins) and lipophilic powders (such as cocoa powder), controls dust, and helps stabilize dispersions in water [33].

In addition to consuming lecithin through the diet, some people take lecithin supplements. Whether this procedure is really beneficial is debated. However, there are indications that lecithin supplements can be beneficial for controlling cholesterol levels. A study that included 30 people with high cholesterol levels showed that supplementing with 500 mg soy lecithin in the form of capsules daily for two months reduced the total cholesterol by 42% and the low-density lipoprotein (LDL) cholesterol by over 56% compared to placebo treatment [34]. LDL is often referred to as "bad" cholesterol because having too much LDL in the blood will increase the risk of heart disease. It is believed that lecithin lowers cholesterol by stimulating the secretion of bile acid, which is produced from the breakdown of cholesterol [35]. However, other studies have given somewhat contradictory results, indicating that lecithin does not have any effect on either total cholesterol or LDL cholesterol in people with high cholesterol levels but only in people with low levels of cholesterol [36]. More and larger studies are needed to determine with certainty how lecithin supplements impact blood cholesterol levels.

Choline is a nutrient needed for cellular health, immune function, metabolism, lipid transport, and more. Choline is especially important during pregnancy and breastfeeding, as it is necessary for fetal and infant brain development. The benefit from lecithin supplement, as a source of choline, has not been proven but a meta-analysis found that higher maternal choline intake was likely to be associated with better child neurocognition and neurodevelopment [37]. Infants who were born prematurely might have supplemental lecithin added to human milk to prevent fat loss. Women who are breastfeeding might experience clogged milk ducts and supplemental lecithin has been used to clear the blocked ducts.

In a study in 89 middle-aged Japanese women, it was found that supplementation with 1200 mg of soy lecithin per day for eight weeks significantly reduced diastolic blood pressure levels compared to a placebo [36]. Diastolic blood pressure measures the pressure in the arteries when the heart rests between beats. High blood pressure increases the risk of several conditions, including heart disease. The researchers who conducted the study noted that, although lecithin seemed to have heart disease-reducing potential, they were not exactly sure how the lecithin supplement affected the blood pressure and arterial stiffness.

People have used lecithin for treating Alzheimer's disease, Parkinson's disease, and other brain disorders, but there is no good scientific evidence to support these uses. On the contrary, a 2003 review of randomized trials found no benefit of lecithin in people with dementia [38].

3.5.3 Lecithin in Personal Care Formulations

Apart from the use of lecithin in liposomes for cosmetic applications, which was discussed above, lecithin is used as an oil-in-water emulsifier and as a wetting agent in a wide range of cosmetic formulations. Being an entirely natural product, it fits well with the "green" or "organic" trend within the personal care segment.

Lecithin is an efficient emollient, i.e. a substance that increases the moisture level. Emollients are good for dry skin and for skin that has been excessively exposed to soap and facewash products. Lecithin is therefore often used as an ingredient in skin care products.

Hydrogenated lecithin is lecithin that has been subject to hydrogenation to transform unsaturated acyl chains into saturated. Hydrogenated lecithin is sometimes claimed to be an even better emollient than standard lecithin. Based on available skin toxicity data, both lecithin and hydrogenated lecithin are considered safe for use in rinse-off and in leave-on cosmetic products at concentrations up to 15%, the highest concentration tested in clinical irritation and sensitization studies [39].

3.5.4 Lecithin in Other Industrial Areas

Lecithin, in particular soybean lecithin, has an established use in the paint and coatings industry as a wetting, grinding, and dispersion agent for pigments. It is used in both solvent-based and water-based coatings, including latex paints. The yellow iron oxide pigment is a typical example for which lecithin is an efficient dispersant [40], but it is used for dispersing other pigments such as iron blue and carbon black as well. Being an efficient dispersing agent, lecithin is also marketed as an antisettling agent, i.e. a compound that prevents formation of hard pigment deposits at the bottom of the paint can.

Lecithin is an antioxidant and used as such in the coatings industry in formulations of antirust paints.

Lecithin is also used in formulations for printing inks, where it acts both as a stabilizer of emulsions and as a pigment dispersant.

Mixtures of lecithin and Tween 80, which is an ethoxylated sorbitan ester, are effective dispersants for marine crude oil spills. The mechanism behind the effect is believed to be that spontaneous emulsification takes place with the help of the energy from the motion of the waves. Lecithin-rich mixtures result in water-in-oil emulsions and mixtures rich in Tween 80 give oil-in-water emulsions, in accordance with the hydrophilic–lipophilic balance of such mixtures [33].

3.6 Regulatory

Lecithin is classified as generally recognized as safe (GRAS) by the United States Food and Drug Administration (FDA). It is authorized as a food ingredient in accordance with the Code of Federal Regulations and in the Food Chemicals Codex.

In the European Union, the use of lecithin is regulated under EC Directive 96/77 EC and EC 2008/84. Lecithin is designated as food additive E322 ("Approved additives and E numbers," EU's Food Standards Agency) with the description given in Table 3.2.

Lecithin E322 is listed as antioxidant and emulsifier in the Codex Alimentarius INS 322 and according to it, it can be added to a variety of foods (most often without a maximum authorized level), in medical preparations, and in baby food complements (maximum 5000 mg/kg of body weight).

In Japan, yolk lecithin is reported with number 339 in the list of food additives authorized by the Ministry of Health, Labour and Welfare. Sunflower lecithin (Topcithin™ from Cargill) was approved by the same Ministry as food additive E322 on April 10, 2014. Previously, Japan was the only country where it had not been approved for use in food products.

In the year 2000, the European Commission passed the regulation (EC) 50/2000, which requires labeling of foodstuffs and food ingredients containing additives and flavorings that have been genetically modified or have been produced from GMOs. This regulation also includes lecithin. Because it is nearly impossible to detect the origin of products such as lecithin, the European regulations require those who wish to sell lecithin in Europe to use a meticulous, but essential system of IP [41].

The Chinese standard GB 2760-2014 specifies the application principles of food additives and lists additives that are allowed to be used in food in China. Besides, it also stipulates the scopes of application and maximum level or residue limits of food additives. Food-grade soybean lecithin is offered in China as a food additive by many Chinese and non-Chinese suppliers.

Table 3.2 Lecithin E322 description.

Definition	Lecithins are mixtures of fractions of phosphatides obtained by physical procedures from foods of animal or vegetable origin. It also includes hydrolyzed products obtained by the use of suitable harmless enzymes. The final product must not show any sign of residual enzyme activity. Lecithins can be easily bleached in aqueous medium with hydrogen peroxide. This oxidation must not chemically alter the lecithin phosphates
Composition	Lecithin: not less than 60% acetone-insoluble substances Hydrolyzed lecithin: not less than 56% acetone-insoluble substances
Description	Lecithin: liquid, viscous semi-liquid, or brown powder Hydrolyzed lecithin: light brown to brown viscous liquid or paste
Purity	
Loss on drying	Not more than 2% (105 °C, 1 h)
Toluene-insoluble substances	Not more than 0.3%
Acid value	Lecithin: not more than 35 mg KOH/g Hydrolyzed lecithin: not more than 45 mg KOH/g
Peroxide value	Less or equal to 10
Arsenic	Not more than 3 mg/kg
Lead	Not more than 2 mg/kg
Mercury	Not more than 1 mg/kg

References

1 van Hoogevest, P., Tiemessen, H., Metselaar, J.M. et al. (2021). The use of phospholipids to make pharmaceutical form line extensions. *European Journal of Lipid Science and Technology* 123: 2000297.

2 Singer, S.J. and Nicolson, G.L. (1972). The fluid mosaic model of the structure of cell membranes. *Science* 175: 720–731.

3 Nicolson, G.L. (2014). The fluid-mosaic model of membrane structure: still relevant to understanding the structure, function and dynamics of biological membranes after more than 40 years. *Biochimica et Biophysica Acta -Biomembranes* 1838 (6): 1451–1466.

4 Milad, N. and Morissette, M.C. (2021). Revisiting the role of pulmonary surfactant in chronic inflammatory lung diseases and environmental exposure. *European Respiratory Review* 30: 210077.

5 Olmeda, B., Villén, L., Cruz, A. et al. (2010). Pulmonary surfactant layers accelerate O_2 diffusion through the air-water interface. *Biochimica et Biophysica Acta -Biomembranes* 1798: 1281–1284.

6 Gobley, T. (1850). Recherches chemique sur les œufs de carpe. *Journal de Pharmacie et de Chemie*, 3rd series 17: 401–430. (in French).

7 Gobley, T. (1874). Sur la lécithine et la cérébrine. *Journal de Pharmacie et de Chimie*, 4th series 19: 346–353. (in French).

8 SIGMA. 1998. Product information. https://www.sigmaaldrich.com/deepweb/assets/sigmaaldrich/product/documents/321/842/l4129pis.pdf.

9 Marsh, D. (1990). *CRC Handbook of Lipid Bilayers*. Boca Raton, FL: CRC Press.

10 Gonçalves, R.A., Holmberg, K., and Lindman, B. (2023). Cationic surfactants: a review. *Journal of Molecular Liquids* 375: 121335.

11 da Cruz Francisco, J., Topgaard, D., Sivik, B. et al. (2004). Phase behavior of the system lecithin-water. The effects of addition of the hydrocarbon 1,8-cineole and supercritical carbon dioxide. *Journal of Supercritical Fluids* 31: 255–262.

12 Bergenståhl, B. and Fontell, K. (1983). Phase equilibria in the system soybean lecithin/water. *Progress in Colloid and Polymer Science* 68: 48–52.

13 Safran, S.A., Pincus, P., and Andelman, D. (1990). Theory of spontaneous vesicle formation in surfactant mixtures. *Science* 248: 354–356.

14 Safran, S.A., Pincus, P.A., Andelman, D. et al. (1991). Stability and phase behavior of mixed surfactant vesicles. *Physical Review A* 43: 1071–1078.

15 Touitou, E., Godin, B., Dayan, N. et al. (2001). Intracellular delivery mediated by an ethosomal carrier. *Biomaterials* 22: 3053–3059.

16 Zhao, F., Li, R., Liu, Y. et al. (2023). Perspectives of lecithin from egg yolk: extraction, physicochemical properties, modification, and applications. *Frontiers in Nutrition* 9: 1082671.

17 Ubeyitogullari, A. and Rizvi, S.S.H. (2020). Production of high-purity phospholipids concentrate from buttermilk powder using ethanol-modified supercritical carbon dioxide. *Journal of Dairy Science* 103: 8796–8807.

18 Hanley, L., Ghazani, S.M., and Marangoni, A.G. (2024). Giant multilamellar and large unilamellar lecithin vesicles for the encapsulation and oral delivery of cannabinoids. *Food Chemistry* 433: 137291.

19 Alhajj, M.J., Montero, N., Yarce, C.J. et al. (2020). Lecithins from vegetable, land, and marine animal sources and their potential applications for cosmetic, food, and pharmaceutical sectors. *Cosmetics* 7: 87.

20 Kumar, M., Singh, A., Pandey, S. et al. (2021). Liposomes: type, preparation, and evaluation. *International Journal of Indigenous Herbs and Drugs* 6: 17–22.

21 Blume, G. and Cevc, G. (1990). Liposomes for the sustained drug release in vivo. *Biochemica et Biophysica Acta –Biomembranes* 1029: 92–97.

22 Klibanov, A.L., Maruyama, K., Torchilin, V.P. et al. (1990). Amphipathic polyethyleneglycols effectively prolong the circulation time of liposomes. *FEBS Letters* 268: 235–237.

23 Mouritsen, O.G. (2011). Lipids, curvature, and nano-medicine. *European Journal of Lipid Science and Technology* 113: 1174–1187.

24 Bender, D.A. (2009). *Nutritional Biochemistry of the Vitamins*, 2nde. Cambridge, U.K.: Cambridge University Press.

25 Mohammadi, M., Ghanbarzadeh, B., and Hamishehkar, H. (2014). Formulation of nanoliposomal vitamin D3 for potential application in beverage fortification. *Advanced Pharmaceutical Bulletin* 4: 569–575.

26 Iftikhar, A., Zafar, U., Ahmed, W. et al. (2021). Applications of *Cannabis Sativa* L. in food and its therapeutic potential: from a prohibited drug to a nutritional supplement. *Molecules* 26: 7699.

27 Burri, L., Hoem, N., Banni, S., and Berge, K. (2012). Marine ω-3 phospholipids: metabolism and biological activities. *International Journal of Molecular Sciences* 13: 15401–15419.

28 Karny, A., Zinger, A., Kajal, A. et al. (2018). Therapeutic nanoparticles penetrate leaves and deliver nutrients to agricultural crops. *Scientific Reports* 8: 7589.

29 Temming, M. (2018). Nanoparticles could help rescue malnourished crops. *Science News* 2018-05-18.

30 van Hoogevest, P., Fahr, A., Cornier, J. et al. (2019). Phospholipids in cosmetic carriers. In: *Nanocosmetics: From Ideas to Products* (ed. J. Cornier, C.M. Keck, and M. Van de Voorde), 95–140. Cham, Switzerland: Springer.

31 Soni, V., Chandel, S., Jain, P., and Grumezescu, A.M. (2016). Role of liposomal drug-delivery system in cosmetics. In: *Nanobiomaterials in Galenic Formulations and Cosmetics: Applications of Nanobiomaterials* (ed. A.M. Grumezescu), 93–120. Amsterdam: Elsevier.

32 Gunstone, F.D., Harwood, J.L., and Dijkstra, A.J. (ed.) (2007). *The Lipid Handbook*, 3rde. Boca Raton, FL: CRC Press.

33 Riehm, D.A., Rokke, D.J., Paul, P.G. et al. (2017). Dispersion of oil into water using lecithin-tween 80 blends: the role of spontaneous emulsification. *Journal of Colloid and Interface Science* 487: 52–59.

34 Mourad, A.M., de Carvalho Pincinato, E., Mazzola, P.G. et al. (2010). Influence of soy lecithin administration on hypercholesterolemia. *Cholesterol* 2010: 824813.

35 Guitiérrez-Medéz, M., Chavez-Garay, D.R., and Leal-Ramos, M.Y. (2022). Lecithins: a comprehensive review of their properties and their use in formulating microemulsions. *Journal of Food Biochemistry* 46: e14157.

36 Hirose, A., Terauchi, M., Osaka, Y. et al. (2018). Effect of soy lecithin on fatigue and menopausal symptoms in middle-aged women: a randomized, double-blind, placebo-controlled study. *Nutrition Journal* 17: 4.

37 Obeid, R., Derbyshire, E., and Schön, C. (2022). Association between maternal choline, fetal brain development and child neurocognition: systematic review and meta-analysis of human studies. *Advances in Nutrition* 13: 2445–2457.

38 Higgins, J.P.T. and Flicker, L. (2003). Lecithin for dementia and cognitive impairment. *Cochrane Database of Systematic Reviews* 3: https://doi.org/10.1002/14651858.CD001015.

39 Fiume, Z. (2021). Final report on the safety assessment of lecithin and hydrogenated lecithin. *International Journal of Toxicology* 20 (Suppl. 1): 21–45.

40 Haramagatti, C.R., Dhande, P., Bhavsar, R. et al. (2018). Role of surfactants on stability of iron oxide yellow pigment dispersions. *Progress in Organic Coatings* 120: 260–265.

41 Davison, J. and Bertheau, Y. (2007). EU regulations on the traceability and detection of GMOs: difficulties in interpretation, implementation, and compliance. *CAB Reviews: Perspectives in Agriculture, Veterinary Science, Nutrition and Natural Resources* 2: https://doi.org/10.1079/PAVSNNR20072077.

4

Bile Acids

4.1 Introduction

Bile acids are steroid acids found predominantly in the bile (or gall) of mammals and other vertebrates. The bile acid synthesis occurs in liver cells, which synthesize primary bile acids (cholic acid and chenodeoxycholic acid in humans) via cytochrome P450-mediated oxidation of cholesterol in a multistep process. Cholic, chenodeoxycholic, and deoxycholic acid account for over 90% of human biliary bile acids [1, 2].

4.2 Chemical Structures and Properties

Bile acids have the condensed ring system characteristic of steroids and a side chain at C-17, which for most substances ends with a carboxyl group but which may also end with a sulfonate group. All bile acids have a hydroxyl group in 3-position but unlike the saponins, which may also be steroids, that hydroxyl group is not functionalized. Bile acids may also have hydroxyl groups at other positions of the steroid skeleton. Figure 4.1 shows the structure of cholic acid, the most abundant of the bile acids.

In the intestine, bile acids become conjugated with taurine or glycine residues, which are attached to the carboxyl group via an amide bond. The taurine and glycine conjugates of cholic acid are called taurocholic acid and glycocholic acid, respectively. In the bile of mammals, taurocholic acid is exclusively present in the deprotonated form and glycocholic acid, and being a weaker acid, is predominantly present in that form. The salts are called taurocholate and glycocholate. The amide bond in the conjugated bile salts makes these more soluble than normal bile acids, which is of importance for their biological function. The structures of the two important bile salt conjugates are given in Figure 4.2.

Bile acids, the conjugated as well as the unconjugated, are anionic amphiphiles. The carboxyl group (or sulfonate group for the taurocholic acid) is the polar headgroup and the steroid skeleton is the hydrophobic tail. Like synthetic surfactants, they form micelles at a certain concentration, the critical micelle concentration (CMC). As discussed in some detail in the chapter on saponins, the rigid skeleton makes micelle formation less favorable than for conventional surfactants with a more flexible hydrophobic tail. Consequently, the CMC is relatively high.

Natural Surfactants: Science, Technology, and Industrial Aspects, First Edition. Guido Bognolo and Krister Holmberg.

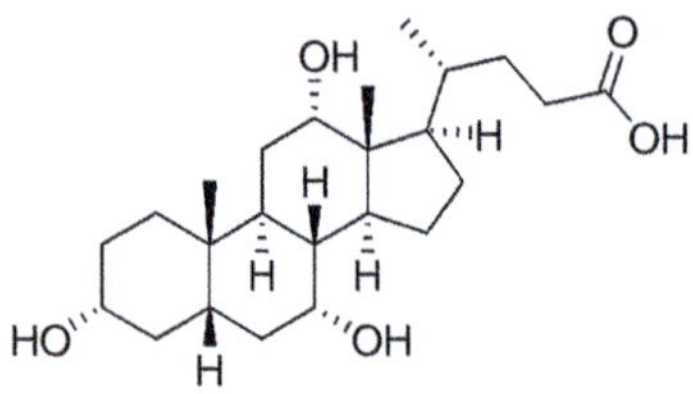

Figure 4.1 Cholic acid. (From Wikimedia Commons.)

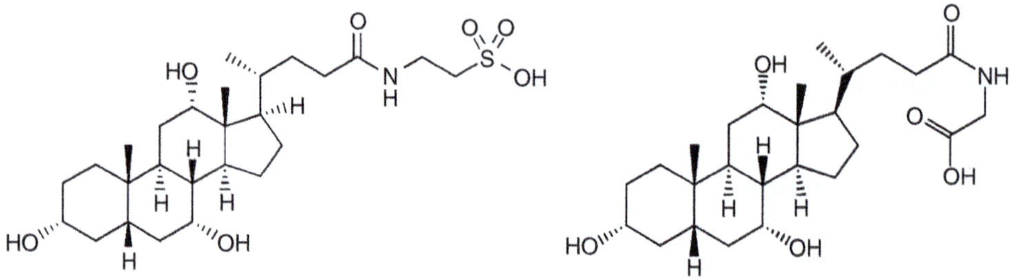

Figure 4.2 Left: taurocholic acid; right: glycocholic acid. (From Wikimedia Commons.)

4.3 Producers

Bile acids are available commercially in limited volume produced by synthesis starting from bovine-derived raw materials (e.g. chenodeoxycholic acid manufactured by **ICE Pharma**, Basaluzzo, Italy) or isolated from animal bile, for example, sodium taurocholic acid and its related products dehydrocholic acid, apocholic acid, nordeocholic acid, hyocholic acid, hyodeoxycholic acid, and glycodehydrocholic acid (from **BOC Sciences**, Shirley, NY, USA).

Sigma-Aldrich (St. Louis, MO, USA) offers a bile acid standard, and a kit for the bile acids test LS-MC, as well as components of the kit, notably the sodium salts of glycochenodeoxycholic acid, glycocholic acid, taurochenodeoxycholic acid, taurocholic acid, and taurodeoxycholic acid. **MilliporeSigma** (a branch of Sigma Aldrich) offers bile salts and bile salt mixtures.

4.4 Applications

The micelles of bile acids and bile acid conjugates solubilize triglycerides and other hydrophobic molecules and the solubilization aids lipase in the digestion process. The role of bile acids in fat digestion, as well as in the regulation of cholesterol (homeostasis), has been known for a long time [3, 4]. Over the last decades, however, it has become clear that bile acids are not simply digestive detergents and the primary route governing cholesterol catabolism. Bile acids are now recognized as signaling molecules with systemic endocrine functions, involved in the regulation of various metabolic processes [5, 6]. Through activation of various signaling pathways, bile acids regulate not only their own synthesis and enterohepatic circulation, but also triglyceride, cholesterol, glucose, and energy homeostasis. Bile acids and bile acid receptors have been shown to play an important role in the metabolic regulation [7]. Bile acids, as well as salts of conjugates such as taurocholate and glycocholate, are used for various pharmaceutical purposes. Outside the life science field, there seems not to be any use of this class of biosurfactants, however.

References

1 Hofmann, A.F. (1999). The continuing importance of bile acids in liver and intestinal disease. *Archives of Internal Medicine* 159: 2647–2658.

2 Hofmann, A.F., Hagey, L.R., and Krasowski, M.D. (2010). Bile salts of vertebrates: structural variation and possible evolutionary significance. *Journal of Lipid Research* 51: 226–246.

3 Hofmann, A.F. and Borgström, B. (1964). The intraluminal phase of fat digestion in man: the lipid content of the micellar and oil phases of intestinal content obtained during fat digestion and absorption. *Journal of Clinical Investigation* 43: 247–257.

4 Insull, W. Jr., (2006). Clinical utility of bile acid sequestrants in the treatment of dyslipidemia: a scientific review. *Southern Medical Journal* 99: 257–273.

5 Houten, S.M., Watanabe, M., and Auwerx, J. (2006). Endocrine function of bile acids. *EMBO Journal* 25: 1419–1425.

6 Chiang, J.Y.L. (2013). Bile acid metabolism and signalling. *Comprehensive Physiology* 3: 1191–1212.

7 Lefebvre, P., Cariou, B., Lien, F. et al. (2009). Role of bile acids and bile acid receptors in metabolic regulation. *Physiological Reviews* 89: 147–191.

5

Rhamnolipids

5.1 Introduction

Rhamnolipids, also named rhamnose lipids, are glycolipids that contain one or two L-rhamnose units linked to a hydrophobic chain via a glycosidic bond. Rhamnose is a so-called deoxy sugar or more specifically a 6-deoxy-hexose. The main classification of rhamnolipids is based on the number of rhamnose units (the glycone part), which is most commonly two, but which may also be one. In the di-rhamnolipids, the two rhamnose units are linked via an α-1,2-glycosidic bond.

The hydrophobic part, the aglycone, is built up of β-hydroxy alkanoic acids, which may be saturated or unsaturated and with typical chain lengths of ten carbon atoms; however, the acyl chains can be both shorter (down to C8) and longer (up to C24). In most rhamnolipids, the aglycone consists of two β-hydroxy alkanoic acid chains connected by an ester bond. Rhamnolipids with just one β-hydroxy alkanoic acid chain as aglycone linked to either a monomeric or a dimeric rhamnose also exist [1]. The complex nature of the aglycone in terms of chain lengths and degree of unsaturation leads to a rich variety of rhamnolipids in nature.

Rhamnolipids were first reported by Bergström et al. [2] and isolated by Jarvis and Johnson in 1949 from *Pseudomonas aeruginosa* cultures [3]. *Pseudomonas* species produce rhamnolipids under growth-limiting conditions on hydrophobic carbon sources, such as hydrocarbons and seed oils, and *Pseudomonas* remains to date the principal organism for rhamnolipid production, although later studies have identified several other bacterial species that produce the glycolipid [4].

5.2 Chemical Structures and Properties

Figure 5.1 shows the structures of typical mono- and di-rhamnolipids. The most common chain length of the β-hydroxy fatty acids that make up the aglycone is C10. The full names of two rhamnolipids shown in Figure 5.1 with C10 acyl chains are L-rhamnosyl-L-rhamnosyl-β-hydroxydecanoyl-β-hydroxy decanoate (the di-rhamnolipid) and L-rhamnosyl-β-hydroxydecanoyl-β-hydroxy decanoate (the mono-rhamnolipid). Using surfactant terminology, the rhamnolipid structures mean a highly branched

Natural Surfactants: Science, Technology, and Industrial Aspects, First Edition. Guido Bognolo and Krister Holmberg.

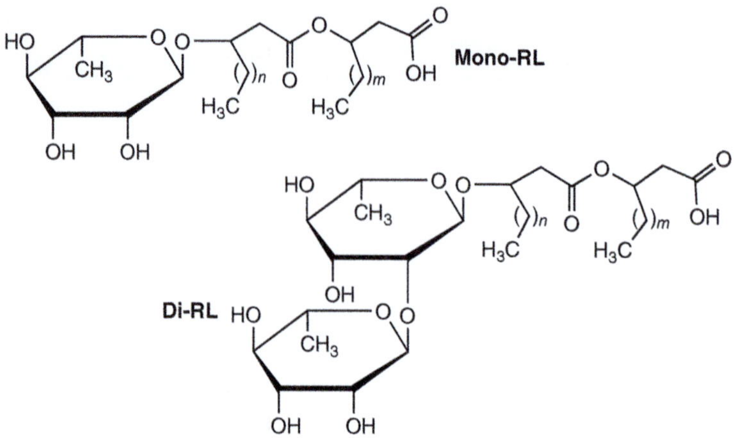

Figure 5.1 Typical structures of mono- and di-rhamnolipids (mono-RL and Di-RL, respectively). "m" and "n" may be the same or different and typical values are 6–8 but may be both smaller and larger.

C20 hydrophobic tail with an ester bond in the middle and with a carboxyl group at one end and a monomeric or a dimeric rhamnose unit as the nonionic polar headgroup.

The rhamnolipid structure, particularly the one with a dimeric rhamnose as glycone, is similar to that of the acidic form of sophorolipids. Like the acidic form of sophorolipids, as well as many other biosurfactants, the surface chemical properties are very pH dependent. At low pH, when the carboxyl group is protonated, the rhamnolipid can be regarded as a very hydrophobic nonionic surfactant with a branched C20 tail and a monomeric or a dimeric rhamnose as polar headgroup. An ester bond in a surfactant is relatively "neutral" when it comes to hydrophilic–lipophilic balance [5]. At higher pH, when the carboxyl group is deprotonated, the rhamnolipid will have two polar headgroups, one anionic and one nonionic, separated by a branched hydrophobic chain. The chain is flexible and such amphiphiles will adopt their conformation at an interface, for instance an oil–water interface, such that the two polar headgroups are directed into the aqueous phase and the connecting chain is in the oil phase. Alignment of alternating anionic and nonionic groups on the aqueous side of an oil–water interface and at an air–water interface is favorable and mixtures of anionic and nonionic surfactants are frequently used in formulations for cleaning, emulsion polymerization, and many other applications. Such surfactant combinations can reduce the surface tension to low values and be efficient stabilizers of oil–water interfaces.

Another characteristic feature that rhamnolipids share with sophorolipids is the susceptibility to chemical hydrolysis. The ester bond between the two hydroxy alkanoic acids is highly sensitive to alkali and moderately sensitive to acid and the acetal linkages (the glycosidic bonds) between the glycone and the aglycone and between the two rhamnose units in di-rhamnolipids are moderately sensitive to acid. Very low pH is not common in surfactant applications, but moderately high pH is not uncommon, for instance in cleaning. If the ester bond connecting the two hydroxy alkanoic acids is cleaved, then two new amphiphilic molecules will be generated, a rhamnolipid with a hydrophobic tail half the length of the original one and a hydroxy alkanoic acid salt. This mixture will obviously have physical and chemical characteristics different from those of the original rhamnolipid.

The physicochemical properties of a mono-rhamnolipid and a di-rhamnolipid with the same length of the aglycone will be very different as was shown by Chen et al. [6]. The authors investigated adsorption of a mono- and a di-rhamnolipid with C10 acyl chains at the air–water interface using surface tension measurements and neutron reflectivity as the main tools. Measurements were made at the natural pH in 0.5 M NaCl, in buffer pH 7, in buffer pH 9, and in ultrapure water. The neutron reflectivity measurements were made using deuterium-labeled rhamnolipids, obtained by growing *Pseudomonas aeruginosa* in D_2O and deuterated glycerol, followed by separation of the two species and purification.

Figure 5.2 shows surface tension plots for the mono- and the di-rhamnolipids. The kink on the curves indicates the critical micelle concentration (CMC). As can be seen, the values are lower for the mono-rhamnolipid than for the di-rhamnolipid, which is according to expectations considering that the extra rhamnose moiety in the latter should impart a certain hydrophilicity to the amphiphile. The surface tension at the CMC is also considerably lower for the mono-rhamnolipid than for the di-rhamnolipid.

Commercial rhamnolipids are often mixtures of the mono- and the di-form, and Chen et al. also included three such mixtures in their study: 30:70, 50:50, and 70:30. Table 5.1 shows values of CMC, surface tension at the CMC, and area per molecule at the air–water interface (assuming a Gibbs prefactor of 1). An interesting observation made from the neutron reflectivity data was that in all three mixed systems, at concentrations well above the CMC, the interface was dominated by the more surface-active mono-rhamnolipid. This is not according to standard thermodynamic treatment using regular solution theory, but the behavior is similar to that of a mixture of one more hydrophobic and one more hydrophilic nonionic surfactant: the more hydrophobic amphiphile, i.e. the one with the shorter oxyethylene chain, dominates at the air–water interface [7].

From the chemical structures (see Figure 5.1), rhamnolipids are expected to be anionic at high pH and progressively more nonionic as the pH is reduced because of protonation of the carboxylate group. In the work of Chen et al., two different methods were used to obtain the area per molecule at the air–water interface: surface tension measurements and neutron reflectivity. Neutron reflectivity gives a definite value, while the surface tension data need a prefactor in the Gibbs equation, which is 1 for nonionic and 2 for fully dissociated anionic surfactants. The values of area per molecule in Table 5.1 have been calculated using a value of 1 for the prefactor. The values obtained from surface tension data in pure water, as well as in 0.5 M NaCl, are the same as the neutron reflectivity values when a prefactor close to 1 is used. This is a strong indication that under natural conditions the rhamnolipids, both the mono- and the di-form, behave largely as nonionic surfactants, which means that the carboxylate group is either protonated or not dissociated. The fact that the CMC values in pure water and in 0.5 M NaCl are almost the same is another indication that the two rhamnolipids in water without added buffer should be regarded as nonionic amphiphiles. For normal anionic surfactants, a change from pure water to 0.5 M NaCl gives a dramatic reduction in CMC due to shielding of electrostatic repulsive forces.

The data for pH 9 buffer is somewhat different. A prefactor between 1 and 2 is required for the values of area per molecule from surface tension data to match the values obtained

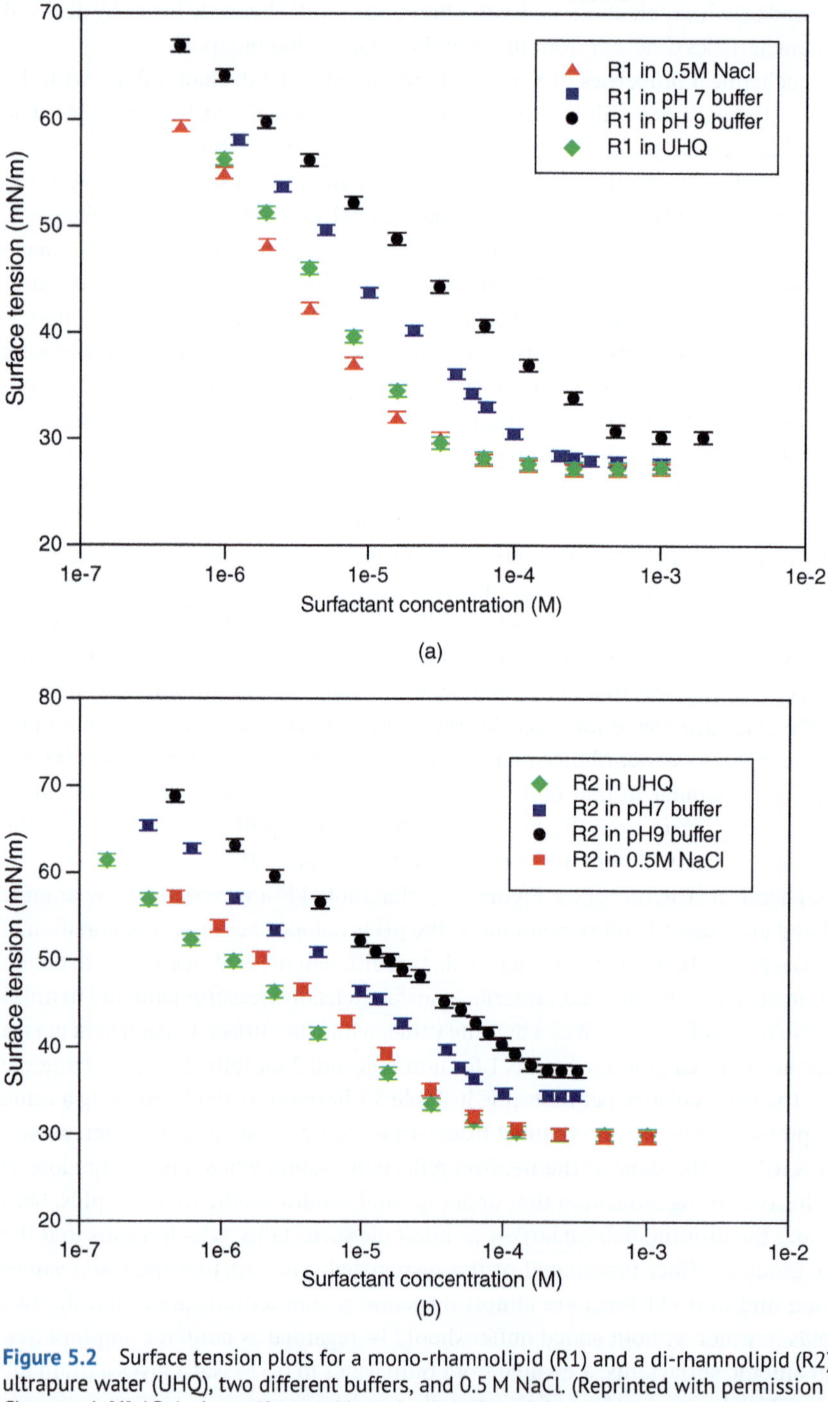

Figure 5.2 Surface tension plots for a mono-rhamnolipid (R1) and a di-rhamnolipid (R2) in ultrapure water (UHQ), two different buffers, and 0.5 M NaCl. (Reprinted with permission from Chen et al. [6] / Solution self-assembly and adsorption at the air-water interface of the monorhamnose and dirhamnose rhamnolipids and their mixtures / reproduced with permission American Chemical Society.)

Table 5.1 Surface tension parameters.

Compound	CMC (mM)	Surface tension (mN/m)	Area per molecule ($Å^2$)
Mono-RL in 0.5 M NaCl	0.03	27.8	52
Mono-RL in pH 7 buffer	0.18	28.7	66
Mono RL in pH 9 buffer	0.36	31.2	77
Mono-RL in ultrapure water	0.04	27.9	53
Di-RL in 0.5 M NaCl	0.08	30.4	79
Di-RL in pH 7 buffer	0.11	34.7	77
Di-RL in pH 9 buffer	0.18	37.4	80
Di-RL in ultrapure water	0.07	30.3	84
Mono-RL:di-RL 30:70 in pH 9 buffer	0.26	34.7	83
Mono-RL:di-RL 50:50 in pH 9 buffer	0.34	33.5	78
Mono-RL:di-RL 70:30 in pH 9 buffer	0.34	32.4	76

(RL stands for rhamnolipid; Adapted from Chen et al [6].)

from neutron reflectivity. This indicates that at pH 9, the rhamnolipids are more anionic and the carboxylate group is partly dissociated, which is according to expectations.

As can be seen from Table 5.1, the area per molecule is much smaller for the mono-rhamnolipid than for the di-rhamnolipid. This is not unexpected and can be explained by the larger di-rhamnose headgroup imposing greater packing constraints at the interface. This, in turn, leads to a higher adsorbed amount of the mono-rhamnolipid than of the di-rhamnolipid. This is similar to what has been observed for nonionic surfactants, where the area per molecule increases with increasing length of the oxyethylene chain [8].

pKa values in the range 5.5–6.0 have been reported for mono-rhamnolipids and di-rhamnolipids [9, 10]. Since the natural pH of rhamnolipids is 6–7, the biosurfactant in its natural environment should carry a negative charge [11]. However, the SANS data from Chen et al. indicated that the degree of ionization of the micelles is small, which is consistent with the data from the surface tension measurements discussed above. This indicates that the rhamnolipids, both the mono-form and the di-form, are only weakly ionic even at pH 9 due to low degree of dissociation. This is practically important information that should be considered in formulation work.

In dilute solutions (<20 mM), both mono-rhamnolipids and di-rhamnolipids form small globular micelles with aggregation numbers of 50 and 30, respectively. At higher concentrations, the mono-rhamnolipid has a predominantly planar structure, forming vesicles, whereas the di-rhamnolipid with its larger headgroup remains globular (with a modest increase in the aggregation number) [6]. The size and shape of the micelles differ somewhat depending on the length of the acyl chains.

The type of aggregate formed in aqueous solution will depend strongly on the pH. Wu et al. showed that for both mono- and di-rhamnolipids at concentrations just above the

CMC, a decrease in pH gave a micelle to vesicle transition. At very low pH, the mono- and the di-rhamnolipid formed vesicles with a diameter of 350 nm and 460 nm, respectively [12].

Somewhat different values of area per molecule have been reported, for instance by Eismin et al. [13] and by Hogan et al. [14]. The difference in values obtained by the different groups can at least partly be explained by different choices of prefactor in the Gibbs adsorption equation. The CMC values for mono- and di-rhamnolipids reported in the literature also differ, sometimes considerably. This may be due to differences in the length of the acyl chains, but it may also be caused by incomplete purification in the workup process [15].

At high concentrations, surfactants normally form liquid crystalline phases. This type of phase behavior has been thoroughly investigated for some biosurfactants but rhamnolipids seem not to have been studied in detail in this respect [16]. However, photographs under cross-polarized light indicate that both mono- and di-rhamnolipids form lamellar liquid crystals at high concentration in water [17].

5.3 Biosynthesis

In the bioprocess using *Pseudomonas aeruginosa* as microorganism and glucose as carbon source, the synthesis of the hydrophobic moiety consumes glucose to produce the acyl coenzyme A (acyl-CoA) [18], which catalyzes the formation of the fatty acid β-hydroxyethyl ACP. The unique rhamnosyltransferase RhlA in *Pseudomonas* transforms the acid into 3-(3-hydroxyalkanoyloxy) alkanoic acid (HAA) [19, 20].

Glucose is also the source of the hydrophilic moiety deoxythymidine diphosphate (dTDP)-L-rhamnose. The process takes place in a series of enzyme-catalyzed steps:

Glucose → Glucose-1-phosphate →dTDP-D-glucose→dTDP-6-deoxy-D-4-hexulose

dTDP-6-deoxy-D-4-hexulose → dTDP-6-deoxy-L-Lixo-4-hexulose →dTDP-L-rhamnose

The enzymes required for the catalysis are present in most bacteria [21]. The gene RhIB in *Pseudomonas aeruginosa* catalyzes the formation of the link between (dTDP)-L-rhamnose and HHA giving the mono-rhamnolipid and further intervention of the gene RhIC produces the di-rhamnolipid [22, 23].

The biosynthesis of rhamnolipids is a complicated process and is regulated by many pathways [24]. Supplying glucose in the medium does not necessarily guarantee rhamnolipid production. Fermentation control is still required as rhamnolipid production is an inducible process. Bacterial quorum sensing (QS) is one of the regulators that affects biosynthesis.

Bacterial communication relies on versatile chemical signaling molecules called autoinducers, which regulate bacterial gene expression. QS allows individual bacteria within colonies to coordinate and carry out colony-wide functions. The QS system is complicated and the regulators that affect rhamnolipid production have been reviewed in several publications [25–27].

Ochsner et al. investigated the QS influence on the rhamnolipid biosynthesis. A mutant strain (65E12) of *Pseudomonas aeruginosa*, unable to produce rhamnolipid biosurfactants and lacking rhamnosyltransferase activity, was genetically complemented by using a *Pseudomonas aeruginosa* PG201 wild-type gene library. A single complementing cosmid was isolated on the basis of surface tension measurements of subcultures of the transconjugants by using a sib selection strategy (a method of sequential fractionation of a heterogeneous sample that can be applied to isolate a sequence, a gene, or a gene family from a complete library). The subcloning of the complementing cosmid clone yielded a 2-kb fragment capable of restoring rhamnolipid biosynthesis, rhamnosyltransferase activity, and utilization of hexadecane as a carbon source in mutant 65E12 [26]. The same authors studied the capability of a newly identified gene (RhlI) to restore rhamnolipid production in a *Pseudomonas aeruginosa* strain carrying a disrupted RhlI gene that was unable to produce rhamnolipids and lacked rhamnosyltransferase activity. Rhamnolipid synthesis was restored by introducing a wild-type RhlI gene into such strains or, alternatively, by adding either the cell-free spent supernatant from a *Pseudomonas aeruginosa* wild-type strain or synthetic *N*-acyl homoserine lactones [27].

5.4 Biotechnological Production

Rhamnolipids are produced from resting cells in a two-step process. In the first step, the *Pseudomonas aeruginosa* cells are produced and harvested. In the second step, the biomass is used for producing the rhamnolipid under growth-limiting conditions [28, 29].

Despite having been discovered as early as 1946, there was no large-scale commercial production of rhamnolipids until 2016, the main reason being technological hurdles:

- Pathogenic nature of the producing microorganism
- Relatively low productivity
- Unreliable fermentation process, causing substrate and/or product inhibition
- Severe foam formation
- Difficult separation of the product from the biomass
- Impurities in the product
- Use of organic solvents in the downstream process

In general, natural bioprocesses are characterized by low yield, and improving the yield from fermentation processes was among the topics investigated since the early days of the biosurfactants. On the positive side, the surface-active properties identified in the early works unveiled the potential of rhamnolipids not only for the classical targets of biosurfactants, such as microbial enhanced oil recovery and soil remediation, but also for several other applications, notably skin care and hard surface cleaners.

The rhamnolipid yield is mainly dependent on the choice of the producing microorganism, the medium composition, and the fermentation conditions [30] and ranges from about 3–4 g/L to more than 100 g/L, depending on the cultivation strategies. Until the early 2020s, the highest yield reported was 112 g/L, achieved from *Pseudomonas aeruginosa* fermentation using soybean oil as a carbon source in an N-limited and Ca^{2+}-free medium at 30 °C and pH 6.3 [31–33].

5.4.1 Choosing the Bacteria

The commercial production of rhamnolipids starts with the selection of the fermentation strain, which is important because the yield of rhamnolipids varies considerably from species to species [34].

Pseudomonas aeruginosa has been found to produce rhamnolipids in good yields and has been used as a model strain to understand the genes that are critical for the biosynthesis of the biosurfactant. Production of rhamnolipids in *Pseudomonas* involves several steps. The precursors are the hydrophilic component (dTDP)-L-rhamnose and the hydrophobic moiety 3-(3-hydroxyalkanoyloxy)alkanoic acid (HAA). The sugar moiety can be synthesized from D-glucose, using enzymes present in many bacteria. The hydrophobic moiety can be synthesized through the fatty acid synthesis pathway, starting with two-carbon units [18].

There are three key enzymes involved in the synthesis of HAA and of the mono- and di-rhamnolipids: RhlA, RhlB, and RhlC; these are found almost exclusively in *Pseudomonas* species and *Burkholderia* species. The former is pathogenic while the latter is not, but so far *Burkholderia* has not been found to produce rhamnolipids in significant yields, despite the presence of the critical enzymes.

RhlA catalyzes the formation of HHA and RhlB is a rhamnosyltransferase catalyzing the reaction between HHA and dTPD-1-L-rhamnose to form mono-rhamnolipids. RhlA and RhlB are encoded by the *rhlAB* operon that is tightly controlled by the QS systems. RhlC, encoded by the *rhlC* operon, is a rhamnosyltransferase II that produces di-rhamnolipids by joining a mono-rhamnolipid with one unit of (dTPD)-L-rhamnose.

Several nonpathogenic strains are also able to produce rhamnolipids in large volumes when these three genes are introduced. As an example, *Pseudomonas putida* is the bacterium used by Evonik in its fermentation process for industrial production of rhamnolipids.

Bacteria screening is a strategy to select strains that give high yields. Random mutagenesis or metabolic engineering provides interesting opportunities. Random mutagenesis through chemical or radiation treatment may improve the yield of rhamnolipids. For example, a *Pseudomonas aeruginosa* EBN-8 mutant obtained by mutagenesis using gamma radiation had better growth on oil refinery wastes than the wild type [35]. Mutagenesis of *Pseudomonas aeruginosa* by using the chemical mutagen *N*-methyl-*N'*-nitro-*N*-nitrosoguanidine (MNNG) generated a mutant that produced rhamnolipids in yields in the range 70–120 g/L [33].

As discussed above, the biosynthesis of rhamnolipids in *Pseudomonas aeruginosa* requires the precursors HAA and (dTDP)-L-rhamnose, where HAA is synthesized from β-hydroxydecanoyl-ACP. As the biosynthesis pathways for both β-hydroxydecanoyl-ACP and (dTDP)-L-rhamnose are present in many bacteria, it is possible to construct nonpathogenic recombinant strains by introducing the genes important for rhamnolipid biosynthesis through metabolic engineering tools. This has been exploited to produce rhamnolipids in commercially relevant volumes from nonpathogenic strains. It is equally possible to alter the type and number of homologues and congeners, as well as their relative proportions. Many examples are given in a paper by Chong and Li [34].

5.4.2 Fermentation Conditions

Other than on the choice of bacterial strain, the rhamnolipid yield is to a large extent dependent on the medium composition and the fermentation parameters. Yields of rhamnolipids ranging from 3.9 to 79 g/L depending on the cultivation strategies are reported in the literature [34].

5.4.2.1 Medium Composition

Medium composition includes several variables: sources of carbon, nitrogen, and phosphorous, oligo elements, carbon to nitrogen ratio, carbon to phosphorus ratio, etc. Water-insoluble sources generally produce rhamnolipids in higher titers.

Carbon Sources Three types of carbon sources are commonly used in biosurfactant production: carbohydrates, oils and fats, and hydrocarbons. In order to decrease costs, various types of wastes have also been investigated, but apparently for the time being refined sources are irreplaceable to produce the yields needed for an economically viable industrial production.

Within the carbohydrate group, simple sugars, starch, and plant sugar-based carbohydrates are the major carbon sources used in biosurfactant production. Glucose is the typical example of a carbon source that can easily be metabolized by microorganisms through the glycolysis pathway for the generation of energy and is often reported to give high yields of product [36]. Glucose is currently the only carbon source used in the industrial production of rhamnolipids by the company Evonik. The yields obtained with carbon sources other than carbohydrates are generally considerably lower.

Macro Nutrients The macro nutrients are nitrogen and phosphorus. It has been reported that genes are activated to produce biosurfactants from various *Pseudomonas* species only under nitrogen-limiting conditions, and that nitrate has advantages over other nitrogen sources [37]. For example, with *Pseudomonas aeruginosa* strain 44T1, production of rhamnolipid started after 30 hours of growth, when the culture reached limiting concentrations of nitrogen, and continued production for up to 58 hours of fermentation [38]. An increase in the production of rhamnolipids by *Pseudomonas aeruginosa* LBI was observed following nitrogen depletion [39]. Some authors have observed that the synthesis of rhamnolipids by *Pseudomonas aeruginosa* occurs when the source of nitrogen has been depleted and its growth cycle is at its latency phase, and that the production by *Pseudomonas* sp. DSM 2874 is inhibited when nitrogen sources are added [31]. On the other hand, an optimum C/N ratio of between 6 and 13 has been proposed, which is a nonlimiting situation [40]. Nurfarahin et al. present a comprehensive review of the effects of different C/N ratios on the synthesis of rhamnolipids and biosurfactants in general [41].

Potassium dihydrogen phosphate, KH_2PO_4, and dipotassium hydrogen phosphate, K_2HPO_4, are usually added to the production medium to maintain the desired pH throughout the cultivation process. Mulligan et al. studied the changes in cell metabolism at the onset of rhamnolipid production, looking at the evolution of three enzymes: transhydrogenase (PATH), glucose-6-phosphate dehydrogenase (G6PD), and alkaline phosphatase (APase). The results showed that biosurfactant production was induced during the shift in metabolism, in particular, depletion of phosphate, induction of APase activity, and a decrease in PATH activity. It appears that a shift in phosphate metabolism coincided with rhamnolipid production in *Pseudomonas aeruginosa* [42].

Oligoelements Work by Guerra-Santos et al. reports that limitation of magnesium, calcium, and potassium might enhance rhamnolipid production by *Pseudomonas aeruginosa* [43]. Calcium ions work as common mediators in signal-delivering processes from the cell surface into the intracellular space of microorganisms. They are important for the maintenance of cell structure, motility, transport, and cell differentiation processes in bacteria [44]. Both potassium and calcium ions play a crucial role in balancing the osmotic pressure and controlling the cell's membrane potential, preventing lysis of the cell in the medium [45].

The addition of some micronutrient minerals or trace elements could have a positive effect on biosurfactant production. Zn, Cu, Mo, B, and Mn are required in the production of rhamnolipids from various *Pseudomonas* strains, for example *Psedomonas aeruginosa* RS29. *Pseudomonas aeruginosa* PTCC1637, on the other hand, uses only Zn, Cu, and Mn [41].

5.4.2.2 Fermentation Parameters

Important fermentation parameters are feeding profile, pH, temperature, and aeration rate/dissolved O_2.

Fed-batch culture is an operational technique in processes where one or more nutrients are fed to the bioreactor during cultivation and in which the product remains in the reactor until the end of the run. A fed-batch generally results in higher yield than batch. An advanced sequential fed-batch fermentation process with high cell densities was developed for the efficient fermentative production of rhamnolipids [46]. With this approach, the rhamnolipid production reached over 665 g/L by *Pseudomonas aeruginosa* PAO1 during a fermentation period of 18 days, with a productivity of 1.54 g/L h. This represents more than a fivefold improvement over the regular fed-batch fermentation. The advanced sequential fed-batch fermentation process was very efficient due to the high cell density and the presence of optimal concentrations of $NaNO_3$ and $FeSO_4 \cdot 7H_2O$, 18 g/L and 2 mg/L, respectively.

The pH should be neutral or slightly alkaline in the initial phase, and then slightly acidic. By this procedure, the cell growth in the middle and late stages of the fermentation is suppressed, which favors rhamnolipid production [47].

The effect of temperature is different from strain to strain, and the optimum temperature must be determined individually.

It has been shown that higher levels of dissolved O_2 give higher rhamnolipid yield [47, 48].

Coupling the bioreactor with a tandem defoaming system gives numerous advantages. In situ product removal assists in preventing product inhibition and often gives higher rhamnolipid titers together with other benefits, such as obtaining rhamnolipid of higher purity (which means lower purification cost) and preventing overflowing of microbial cultures during fermentation [49, 50]. The fermentation-defoaming tandem consists of two tanks, a regular fermenter and a foam collector, linked by pump, pipes, and valves. The overflow liquid and escaped foam from the regular fermenter are gathered in the foam collector, where the foam is made to collapse by adjusting the pressure and stirring. The recovered fluid is pumped back to the fermenter to continue the fermentation process [51].

5.4.2.3 Extraction and Purification

The downstream processing in many biotechnological processes, including production of biosurfactants, can contribute to as much as 60% of the total production costs [52–54]. The ease of product recovery will depend on factors such as water solubility (which may depend on pH and ionic strength) and location of the product (intracellular, extracellular, or bound to cell membrane). As rhamnolipids are recovered from the fermentation broth, they will be part of a complex mixture containing unfermented substrates, salts, amino acids, proteins, and other metabolic products. The downstream purification of rhamnolipids is therefore difficult, especially for low rhamnolipid titers [55]. In addition, there is the complication of the high surface activity of the rhamnolipid itself, which promotes unwanted foaming and emulsification, as well as solubilization of hydrophobic metabolites.

Despite such complexities, it is still possible to reach relatively high purity using various strategies such as foam fractionation [55]. The product purification costs can be significantly lowered, or even eliminated, if a cell-free fermentation broth can be used in place of purified rhamnolipids or if the rhamnolipid-producing microorganisms can be used as such. Microbial enhanced oil recovery, cleaning of oil-contaminated soil, and additives for animal feed are examples of such possible applications.

Precipitation with acid or ammonium sulfate is a common purification method, as it generally gives high-quality isolated material with relatively simple and inexpensive recovery methods. The crude precipitate may subsequently be centrifuged and/or solvent extracted [55]. Methanol/acetone/chloroform, methanol/chloroform, diethyl ether, and ethyl acetate are examples of solvents that have been used for the purpose [56].

Shah et al. compared the four different downstream recovery methods: acid precipitation, zinc sulfate precipitation, ammonium sulfate precipitation, and solvent extraction [57]. Solvent extraction gave the highest yield of rhamnolipid, 7.5 g/L, while acid precipitation gave the lowest, 3.5 g/L.

Other recovery methods for purification of rhamnolipids have been reported, which include foam fraction, ultrafiltration, adsorption chromatography, and ion exchange chromatography but apparently none has a definite advantage over the other [53]. The optimization of one parameter goes at the expense of another, such that the choice depends on the availability of equipment and the intended use (if any) of the rhamnolipid.

5.4.3 Reducing Raw Material Costs

Despite its many attractive features, rhamnolipids have difficulties in competing with the regular surfactants made by organic synthesis because of the high production cost resulting from low yield, low productivity, complexity of the isolation/purification, and small production scale. Rhamnolipids share this situation with most other biosurfactants. One strategy to increase the competitiveness of rhamnolipids is to reduce the raw material costs by using inexpensive carbon sources, such as wastes or cheap renewable substrates. The different renewable feedstocks that can be used as carbon and nitrogen sources for rhamnolipid fermentation with reduced substrate cost are discussed in detail in several reviews [58–62].

5.4.3.1 Waste Cooking Oils (WCO)

Modern society produces enormous volumes of different types of waste, which may still have a considerable residual value or which may be difficult to dispose of. Waste cooking oils (WCO) possess both features due to the fatty acid chains that give a high energetic value. The estimated worldwide amount of WCO is 40–45 million tons and a substantial fraction of this causes serious pollution problems, such as drinking water contamination, release of noxious gases, and pest growth [63–65].

Pseudomonas aeruginosa M4, a halotolerant strain with high soybean degradation ability, was selected and studied by Shi et al. for its capability to efficiently utilize WCO and produce rhamnolipids. An oil degradation of above 85% was obtained when optimal conditions

(5% inoculum at 35 °C and pH 8) were used. A rhamnolipid concentration of 1.12 g/L was obtained after four days when soybean oil was the sole carbon source [66].

Sharma et al. used another bacterial strain, *Pseudomonas aeruginosa* MTCC7815, and obtained a rhamnolipid concentration of 11 g/L after five days of incubation at 25 °C and pH 8 when 90% of the oil had been consumed. The authors concluded that *Pseudomonas aeruginosa* is suitable for biosurfactant production at a commercial scale along with waste remediation in an economic way [67].

In a work by Zhu et al., rhamnolipid production was investigated using waste frying oil as the sole carbon source. By culture in shaking flasks, a naturally isolated strain synthesized rhamnolipid at a concentration of 12 g/L, and its mutant after treatment by UV light increased the productivity to 25 g/L. Fermentation was also conducted in a 50-L bioreactor with a productivity of over 20 g/L. Besides the higher yield compared to the other works, it is interesting how the use of UV increased the productivity of the original strain [68].

A fed-batch feeding strategy was explored for the production of rhamnolipids from WCO. Two feeding strategies were explored: pulse and substrate uptake rate (SUR). In both, the feeding started when the dissolved oxygen profile showed a sharp increase, an indication that the WCO had reached the maximum concentration that could be consumed. In the pulse feeding, the WCO was added in discrete amounts over one hour with three to six hours interval between feedings. In the SUR feeding, the culture was fed continuously, according to the rate calculated from the pulse feeding strategy. The rhamnolipid yield was 8.4 g/L for the pulse feeding and 14.4 g/L for the SUR feeding. In comparison, batch feeding gave 2.5 g/L [69]. The higher yield achievable via a fed-batch over a batch process was later confirmed by Sharma et al. who obtained a rhamnolipid production of 16 g/L and a consumption of 94% of the WCO with the fed-batch process, while the batch process gave 11 g/L and a WCO consumption of 92% [70].

5.4.3.2 Other Sources

Pérez-Armendáriz et al. explored the potential of waste canola oil as a low-cost and environmentally friendly substrate to produce rhamnolipids by *Pseudomonas aeruginosa*. The highest observed yield was 3.6 g/L when *Pseudomonas aeruginosa* was cultured for 14 days with 3% waste canola oil and 4 g/L of $NaNO_3$. Although the yield was low, it was similar to or even higher than that obtained using fresh canola oil. A 30-fold increase in production yield was observed when the nitrogen source was changed from ammonium sulphate to sodium nitrate, confirming that optimization of the feed conditions can produce major benefits [71].

The production of rhamnolipids by growing *Pseudomonas aeruginosa* SSL-4 on waste engine oil was investigated over a range of parameters: pH 5–11, incubation temperature 25–40 °C, carbon source concentration 1–7%, carbon/nitrogen (C/N) ratio 10–40, and carbon/phosphorous (C/P) ratio 10–40. At the optimum conditions (pH 7, temperature 30 °C, carbon source concentration 1%, C/N ratio 10, and C/P ratio 10), the yield of rhamnolipid produced was 21.4 g/g of nutrient and 72% of the oil was consumed from samples of oil-contaminated soil [72]. This is a good illustration of how biosurfactant production can go hand in hand with a soil remediation process. A related example is rhamnolipid production from *Pseudomonas aeruginosa* growing on soapstock. Soapstock from alkali refining of soybean oil is a source of fatty acids, but it also presents a handling, storage, and disposal problem. Benincasa et al. showed that rhamnolipids could be produced in good yield by batch cultivation in a mineral salt medium with soapstock as the sole carbon source. The biosurfactant production increased after nitrogen depletion [39].

Orange peels, carrot peels, lime peelings, coconut oil cake, and banana waste were used to feed *Pseudomons aeruginosa* MTCC 2297. The best results were found with orange peels, which produced 9.18 g/L rhamnolipid [73].

Rhamnolipid production with *Pseudomonas aeruginosa* from soybean oil waste in a recycling bioreactor as opposed to a shake flask or nonrecycling flow was investigated by Lotfabad et al. The biosurfactant concentration in the bioreactor including the recycling flow, which was switched on and off during a six-hour period, reached 25.5 g/L, which was 42% and 22% higher than what was obtained with a shake flask and a non-recycling-flow bioreactor, respectively [74].

Rashedi et al. studied the production of rhamnolipids from molasses, a waste product from the sugar industry, and found that the biosurfactant production increased with an increase in the concentration of molasses, reaching a maximum production when 7% (v/v) of molasses was used. Further increase in the concentration of molasses did not affect the yield significantly; however, the yield of biomass increased under these conditions [75].

In a review from 2021, Varjani et al. discuss the structure and nature of rhamnolipid production using different feedstock and waste materials like bakery waste, shrimp shell, rice grains, orange peels, cassava wastewater, lignocellulose residues, whey, and dairy wastes, reporting yields ranging from 0.5 to 15 g/L [76].

5.4.4 Overcoming Pathogenicity

The major hurdle confronting rhamnolipids is the pathogenicity of the most efficient production strain, *Pseudomonas aeruginosa*. This makes it necessary to employ a complex purification procedure to completely eliminate the microorganism, and even if that is successful, the possibility that small amounts of a pathogenic bacterium may be accidentally present in the biosurfactant will severely affect its attractiveness for application in sensitive areas such as pharma, food, and biomedicine.

Indeed, this was one of the major problems confronting Evonik Industries (Essen, Germany), reportedly the first company to produce rhamnolipids on an industrial scale in June 2016 [77]. In response, Evonik adopted the strategy of transplanting the genes of the pathogenic *Pseeudomonas aeruginosa* responsible to produce the rhamnolipid onto *Pseudomonas putida*. Unlike *Pseudomonas aeruginosa*, *Psedomonas putida* is not a pathogen, which means it is less risky – and costly – to handle. *Pseudomonas putida* is also a common organism in the biotech world; it is well characterized and not difficult to manipulate. In 2014, Evonik filed a patent that describes a method of preparing a rhamnolipid by (a) contacting a recombinant cell with a medium containing a carbon source, and (b) culturing the cell under suitable conditions for preparation of the rhamnolipid from the carbon source by the cell. The recombinant cell has been genetically modified such that, compared to the wild-type, the cell has an increased activity of at least one of the three enzymes α/β hydrolase, rhamnosyltransferase I, and rhamnosyltransferase II. The carbon source, according to the patent, is a C4 molecule (butane, 1-butanol, 2-butanol, *tert*-butanol, butanone, or butyric acid) [78].

According to Evonik, their process leaves no organisms and no detectable DNA and uses no genetically modified feedstock crops. A major advantage of the process is that it uses sugars as the carbon source [79].

In 2016, there was an expression of interest related to rhamnolipid production for *Burkholderia thailandensis* E264, a Gram-negative bacterium first isolated from the soils and stagnant waters of central and northeastern Thailand [80, 81]. The interest was prompted by a number of features. First, *Burkholderia thailandensis* E264 is considered an essentially nonpathogenic, biosafety level 1, organism [82]. Second, it produces rhamnolipids similar to those produced by *Pseudomonas aeruginosa,* although more hydrophobic. Feeding with glycerol and canola oil produced mainly di-rhamnolipid. The most abundant species, C14-C14 di-rhamnolipid constituted 70–77% of the total rhamnolipid yield. The C14 species dominated also for the mono-rhamnolipid category. Third, the genome of *Burkholderia thailandensis* has two identical gene clusters containing copies of the *rhlA*, *rhlB*, and *rhlC* genes [83]. This is unlike the corresponding gene orthologs in *Pseudomonas aeruginosa* where there is a single operon containing the *rhlA* and *rhlB* genes (as well as two regulator genes, *rhlI* and *rhlR*). The *rhlC* gene is located at a different site in the *Pseudomonas aeruginosa* genome, although it is still coexpressed with the other *rhl* genes [84]. This leaves open the possibility of an overproduction of rhamnolipid if the expression of the two *rhl* gene operons in *Burkholderia thailandensis* is controlled by two independent promoter systems.

Burkholderia thailandensis E264 produced rhamnolipid from winery residues (nonfermented grape marcs) and olive oil pomaces, but in inadequate quantities (1.07 g/L and 0.3 g/L, respectively) [85]. Another study investigated the possibility of producing rhamnolipid with *Burkholderia thailandensis* E264 from the agrochemical byproducts corn steep liquor and olive mill wastewater. Rhamnolipids were indeed produced, but at a yield of only 253–269 mg/L [86]. Thus, it seems that the potential identified in the original study with *Burkholderia thailandensis* E264 does not materialize, even if the growth conditions that have been used in the above-mentioned works have not been optimal for the bacterium.

A study published in 2018 reports the successful production of mono-rhamnolipids from sucrose by recombinant *Saccharomyces cerevisiae* strains. *Saccharomices cerevisiae* is a widely used microorganism with generally recognized as safe (GRAS) status. Six enzymes from *Pseudomonas aeruginosa* involved in mono-rhamnolipid biosynthesis were functionally expressed in the yeast. The study indicated the possibility of producing rhamnolipids by a GRAS eukaryotic microorganism from an inexpensive substrate, sucrose [87].

5.5 Rhamnolipids Through Organic Synthesis

Organic synthesis is an alternative to the biotechnological route to produce rhamnolipids and different processes have been developed and published. Westerduin et al. published a total synthesis already in 1988 [88] and this work was later followed by another group using a somewhat different synthesis strategy [89]. Both approaches are versatile and can be used to produce saturated mono- or di-rhamnolipids. However, the procedures are not readily amenable to production at a larger scale as they are labor intensive (>18 steps per diastereomer), use expensive reagents, and give a low final yield.

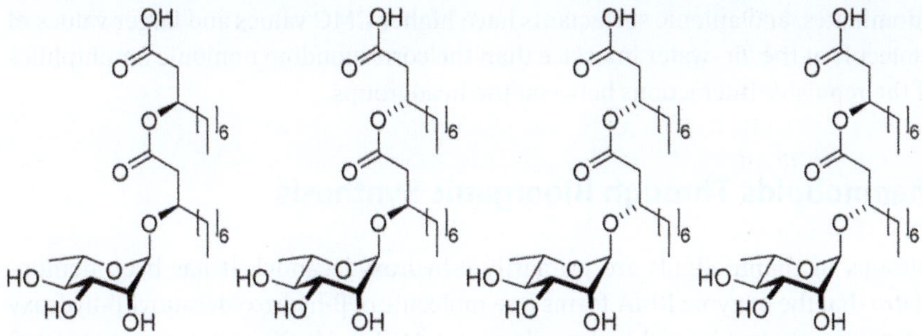

Figure 5.3 Four diastereomers of the mono-rhamnolipid α-rhamnopyranosyl-β-hydroxydecanoyl-β-hydroxy decanoate (Rha-C10-C10). (Reprinted with permission from Pacheco et al. [90] / Synthesis and characterization of four diastereomers of monorhamnolipids / reproduced with permission of American Chemical Society.)

A much more straightforward procedure to prepare mono-rhamnolipids was published in 2017 by Pacheco et al. [90]. Starting from peracetylated rhamnose, four diastereomers of the most common mono-rhamnolipid, α-rhamnopyranosyl-β-hydroxydecanoyl-β-hydroxy decanoate, see Figure 5.3, were prepared by a flexible and scalable synthesis methodology using considerably fewer steps than the previously published procedures.

The chromatographic separation of the diastereomeric mixture was made with peracetylated rhamnosides, i.e. with the three remaining hydroxyl groups on the sugar ring acetylated. A silica column was used, and the mobile phase was a 1:1 mixture of ethyl acetate and hexane +1% acetic acid. The individual diastereomers were subsequently deacetylated.

The same group later refined the synthesis for single-tailed mono-rhamnolipids [91]. The new procedure involved the use of the Reformatsky condensation for the synthesis of the 3-hydroxyalkanoate esters, which are key building blocks. The resulting synthesis is environmentally friendly and uses inexpensive starting materials.

Organic synthesis should probably be seen as a viable alternative to biotechnological production of rhamnolipids, at least of mono-rhamnolipids, in particular when there is a demand for high purity. Organic synthesis provides the ability to tailor surfactant performance through choice of alkyl chain length; it provides considerably simpler workup and purification than the biotechnological route, and it eliminates the risk of contamination with a pathogenic microorganism.

The surface activity of the four diastereomers discussed above was evaluated by surface tension plots conducted at pH 4 and pH 8. The values for CMC and for surface tension at the CMC were almost the same for the four substances. The value for area per molecule at the air–water interface was also almost identical, indicating similar orientation at the surface.

The CMC values for the diastereomers were considerably lower at pH 4 than at pH 8 and so were the values for area per molecule. This is the expected behavior. At pH 4, the rhamnolipid can be regarded as a nonionic amphiphile (although there is a contribution of anionic character because of a small fraction of deprotonated carboxyl group). At pH 8, the anionic character dominates, and anionic surfactants have higher CMC values and larger values of area per molecule at the air–water interface than the corresponding nonionic amphiphiles because of the repulsive interactions between the headgroups.

5.6 Rhamnolipids Through Bioorganic Synthesis

The acyl groups in rhamnolipids are primarily β-hydroxydecanoyl. It has been demonstrated *in vitro* that the enzyme RhlA forms one molecule of β-hydroxydecanoyl-β-hydroxy decanoate from two molecules of β-hydroxydecanoyl-ACP and is the only enzyme required to generate the lipid component of rhamnolipids. It has also been shown that RhlA has a specific affinity for 10-carbon substrates, which, as mentioned above, are the most common in both mono- and di-rhamnolipids [92].

Mono-rhamnolipids can then be obtained by reaction between the lipid component and the rhamnose derivative rhamnose-dTDP in a reaction catalyzed by the enzyme RhlB and di-rhamnolipids are obtained in a subsequent step, catalyzed by RhlC [93]. The reaction sequence is shown in Figure 5.4.

RhlA and RhlB can readily be expressed in *Escherichia coli* while expression of RhlC seems to be more problematic [93]. Bioorganic synthesis *in vitro* should be seen as another

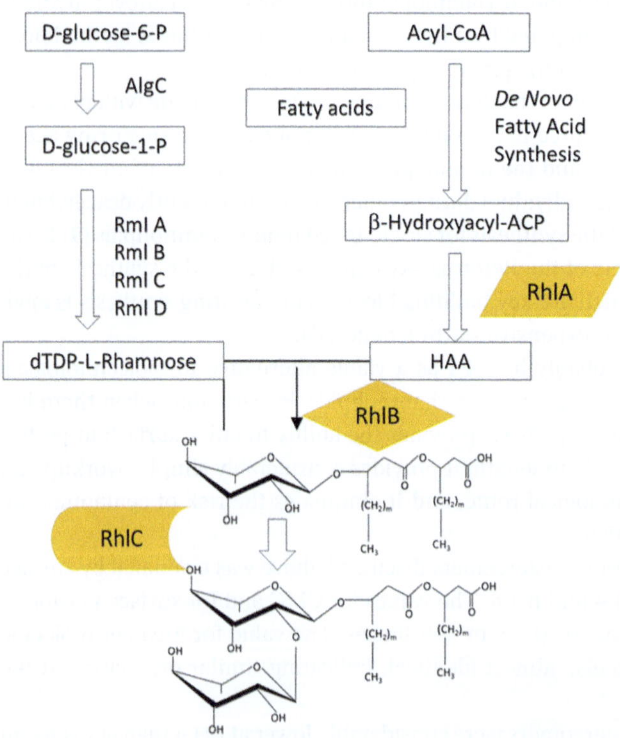

Figure 5.4 Bioorganic synthesis of mono- and di-rhamnolipids catalyzed by the enzymes RhlA, RhlB, and RhlC. dTDP stands for thymidine diphosphate. (From Chong et al. [34] / Springer Nature / CC BY 4.0.)

alternative to the biotechnological route. Like the organic synthesis route discussed above, it is a way to avoid the risk of contamination by a pathogenic microorganism.

5.7 Biodegradability, Toxicity, and Antimicrobial Effect

Mohan et al. evaluated the biodegradation of a rhamnolipid under aerobic, nitrate reducing, sulphate reducing, and anaerobic conditions using a respirometer. The results indicated that the rhamnolipid was biodegradable under all conditions while a typical nonionic surfactant, Triton X-100, was partially biodegradable under aerobic conditions and non-biodegradable under nitrate reducing, sulphate reducing, and anaerobic conditions [94]. In a study by Renkin, a commercial rhamnolipid (JBR 425 from Jeneil Biosurfactant Co. LLC [Milwaukee, WI, USA]) was compared to anionic and nonionic synthetic surfactants for aquatic toxicity in *Daphnia magna*, *Pseudokirchneriella subcapitata*, and *Tetrahymena thermophila*. The biodegradability was examined according to OECD 301F and the manometric respirometry test and the methodology proposed by Gerike were used to estimate the space for stable metabolites. On *Daphnia magna* the rhamnolipid was an order of magnitude less toxic than the anionic and nonionic synthetic surfactants, on *Tetrahymena thermophila* it was comparable to the anionic but better than the nonionic surfactant and on *Pseudokirkneriella. subcapitata* it was better than the nonionic but inferior to the anionic. The rhamnolipid was readily biodegradable and gave no bioaccumulation [95].

In a study from 2019, the biodegradability (CO_2 respirometry), acute toxicity (microtox assay), embryo toxicity (zebrafish assay), and cytotoxicity (xCELLigence and MTS assays) of the four mono-rhamnolipid diastereomers shown in Figure 5.3 and prepared by organic synthesis (Rha-C10-C10) were compared with those of a biosynthesized mono-rhamnolipid mixture (bio-mRL). All the Rha-C10-C10 diastereomers and the bio-mRL were inherently biodegradable ranging from 34% to 92% mineralized. The microtox assay showed that all the Rha-C10-C10 diastereomers and the bio-mRL were slightly toxic according to the US EPA ecotoxicity categories with five-minute EC_{50} values ranging from 40 to 88 μM. The zebrafish assay showed that out of 22 developmental endpoints tested, mortality was only observed at 120 hours post fertilization; all the Rha-C10-C10 diastereomers and the bio-mRL caused significant mortality at 640 μM, except Rha-C10-C10 (*R,R*), which showed no developmental effects. xCELLigence and MTS showed IC_{50} values ranging from 103 to 191 μM for the human lung cell line H1299 after 72 hours of exposure [14].

Johann et al. investigated mono-rhamnolipids produced by a recombinant *Pseudomonas putida* strain using the Green Toxicology concept for acute and mechanism-specific toxicity in an ecotoxicological test battery [96]. Acute toxicity tests with the invertebrate *Daphnia magna* and with zebrafish embryos (*Danio rerio*) were performed and the mutagenicity was tested by means of the Ames fluctuation assay. The results indicate that the acute toxicity to daphnids and zebrafish embryos caused by mono-rhamnolipids is comparable to or lower than that of regular synthetic surfactants. No frameshift mutation or base substitutions were observed in the Ames fluctuation assay. Rhamnolipids showed very low toxicity to the germination of *Aspergillus niger* spores and to the growth of *Candida albicans*. Simulations indicated that rhamnolipids have no estrogenic potential. The authors' conclusion was that mono-rhamnolipids are an environmentally friendly alternative to synthetic surfactants, as the ecotoxicological potential is low.

The antimicrobial effect of rhamnolipids was investigated by Haba et al. [97]. It was found that the rhamnolipid produced by *Pseudomonas aeruginosa* 47 T2 NCBIM 40044 with waste frying oil as carbon source was strongly microbicidal. Low minimum inhibitory concentration (MIC) values were determined for the bacteria *Serratia marcescens* (4 μg/mL), *Enterobacter aerogenes* (8 μg/mL), *Klebsiella pneumoniae* (0.5 μg/mL), *Staphylococcus aureus* and *Staphylococcus epidermidis* (32 μg/mL), and *Bacillus subtilis* (16 μg/mL), as well as for the phytopathogenic fungal species *Chaetomium globosum* (64 μg/mL), *Penicillium funiculosum* (16 μg/mL), *Gliocladium virens* (32 μg/mL), and *Fusarium solani* (75 μg/mL). In general, rhamnolipids have a more detrimental effect on the growth of Gram-positive than Gram-negative bacteria [98].

5.8 Suppliers, Formulators, Products, and Prices

Evonik Industries (Essen, Germany) produces commercial quantities of rhamnolipids at the Slovenská Ľupča site in Slovakia, where it has a dedicated plant to produce the biosurfactant. Since 2019, Evonik has established a partnership with Unilever.

In 2019, **Unilever** (London, UK) launched the liquid hand dishwashing brand "Quix" in Chile, the first appearance in the world of a brand containing rhamnolipid. Following the success of Quix and of a second brand, "Sunlight," in Vietnam, Unilever has expressed the intention to roll out rhamnolipid across its global hand dishwash portfolio. The company later launched the Clean Future initiative, which has set the goal to replace virgin fossil carbon in all its cleaning product formulations with renewable or recycled carbon by 2030.

Biotensidon is a Swiss company founded in 2011 with research and technical facilities in Karlsruhe (Germany). It provides services in technology and equipment development to produce industrial biological raw materials. It was the first company to develop a cost-effective process to produce rhamnolipid on an industrial scale (5000 tons per year).

In January 2023, **Holiferm** Ltd. (Manchester, UK) and **Sasol Chemicals** (Hamburg, Germany), a business unit of Sasol Ltd., announced a collaboration to produce and market rhamnolipids and mannosylerythritol lipids. This collaboration expands the partnership announced in March 2022 for the production of sophorolipids.

TeeGene Biotech (Wilton, UK) has the capability to produce rhamnolipids and other biosurfactants in facilities adapted to scale-up. The company is a spinout venture from Teesside University with the vision to become the leader in the production of biosurfactants.

The surfactant company **Stepan** (Northfield, IL, USA) purchased the rhamnolipid technology NatSurFact from **Logos Technologies** Inc. (Fairfax, VA, USA) and is supplying samples to NatSurFact's existing customers. NatSurFact's rhamnolipid products use vegetable oils as carbon source. Stepan has plans for expanded production. Jointly with **Jeneil Biosurfactant Co.** LLC (Milwaukee, WI, USA), it markets a rhamnolipid-based fungicide for the agricultural market.

AGAE Technologies (Corvallis, OR, USA) produces commercial volumes of rhamnolipids. In September 2022, the company announced the discovery of selected strains and media and optimized fermentation conditions that gave record-high yields of rhamnolipids at low production cost. It was stated that the novel fermentation method increases the yield by nearly 200% compared to the conventionally optimized fermentation conditions and at the same time reduces production costs considerably (AGAE Technologies press release, September 12, 2022).

Biosurfactants LLC (Tampa, FL, USA) produces rhamnolipids from *Pseudomonas aeruginosa*, which it promotes for the food and crop protection (fungicides) industries. It claims to be using artificial intelligence to study the effect of biosurfactants for different applications.

Jeneil Biotech Inc. (Saukville, WI, USA) sells two rhamnolipids, JBR 320 and JBR 425, and a rhamnolipid-based fungicide, Zonix ™, which it markets together with Stepan.

MarkNature (Fullerton, CA, USA) sells Purified Rhamnolipid Liquid, a technical grade rhamnolipid for 70 euros and 260 euros for 200 g and 1 kg, respectively. A 30% powder technical grade is priced at 34 euros and 140 euros for 200 g and 1 kg, respectively.

Besides the companies mentioned above, there are a number of smaller players in the United States that have been working on glycolipid processes for years. In most cases, they do not file patents protecting their methods, relying instead on trade secrets.

In China, there is a considerable industrial activity in rhamnolipids. Examples of producers are given in Table 5.2.

In the United States, rhamnolipids are found in a growing number of personal care products such as creams, serums, and shampoos. Examples of brands using rhamnolipids are Joico, René Furterer, Briogeo, and MayLū. The MayLū brand covers personal care products formulated by **AGAE Technologies** (Source: https://mayluclean.com/pages/about-us) and rhamnolipids are typical ingredients in the formulations. In general, it seems that products containing rhamnolipids are doing well on the market and their market share is growing. (Source: Gravel AI, Chemical Demand Intelligence Platform,

Table 5.2 Rhamnolipid producers in China.

Supplier	Product description	Price (USD)
Shaanxi Pioneer Biotechnology Co., Ltd. (Xi'an, Shaanxi)	Rhamnolipid, 90%	10/g
Shaanxi Rixin Environmental Protection Technology (Xi'an, Shaanxi)	Rhamnolipid as EOR additive	3.60–3.90/kg
Zishang Chemical Co. Ltd. (Jinan, Shandong)	Rhamnolipid, 28–34%	5.60–12.33/kg[a)] 9.80–96/kg[b)]
Healthife LLC (Xi'an, Shaanxi)	Rhamnolipid biosurfactant	25–50/kg
Xi'an Sanwei Technological Dev. Co., Ltd. (Xi'an, Shaanxi)	Rhamnolipids as flocculants and as EOR additive[c)]	1.60–3.60/kg

a) For cleaning/general industrial.
b) For cosmetic/pharmaceutical.
c) Minimum order 1000 kg.
(*Source:* Alibaba / reproduced from Alibaba / https://www.alibaba.com / Last accessed 03.12.2023.)

London, UK, consulted on November 21, 2023) Below are examples of other personal care products that contain rhamnolipids:

- Perfecting Day Cream SPF 30 from Murad
- Honest Beauty Makeup Remover Facial Wipes from Ninth Avenue
- The Ordinary Glycolipid Cream Cleanser from Lyko
- Signature Skin Mini Kit from FaceGym
- The INKEY List Makeup Prep 101 Kit from Sephora
- Phyto Nature Oxygen Cream from Dermalogica
- Calm Cleansing Capsules from Booni Doon

5.9 Existing Applications

The toxicological and environmental profiles combined with the high surface activity of rhamnolipids are major assets for cleaning applications. As mentioned above, Evonik's rhamnolipid is used by Unilever in the hand dishwashing brands Quix in Chile and Sunlight in Vietnam (Unilever, personal communication). One of the important criteria when consumers assess the value of a hand dishwashing product is its foaming and foaming persistency during the washing task. A thick, persistent foam conveys the impression of a strong cleaning power. Another important criterion is the mildness to skin, given the long exposure of the hands to the washing liquor and the inconvenience of using protective gloves. Rhamnolipids score excellently in both respects. There are further developments to be expected, given the collaboration between the two companies and the Unilever Clean Future initiative.

The cosmetic industry is already exploiting and could exploit much more the unique properties of rhamnolipids, like foaming and mildness to skin for a variety of formulations: facial cleansers, hair and body shampoos, oral care, shower gels, intimate care, and baby wipes. Examples of cosmetic products that contain biosurfactants, including rhamnolipids, have been given in the section "Producers and formulators." In oral care products, rhamnolipids present the advantage of a pleasant taste, thus eliminating the need for taste-masking ingredients.

INCI names are systematic names developed by the International Nomenclature of Cosmetics Ingredients (INCI) to identify cosmetic ingredients. INCI names often differ from chemical names. "Rhamnolipids" is the INCI registered name for glycolipids produced by *Pseudomonas aeruginosa* and consisting of rhamnose linked to a β-hydroxy alkanoic acid moiety. Having a registered INCI name is an asset. It contributes to the awareness of their existence across the personal care industry.

As mentioned in the section "Chemical structure and properties," at concentrations above the CMC, rhamnolipids self-assemble into a variety of structures, such as micelles with various shapes, vesicles, and lamellar liquid crystals. The type of structure depends on the rhamnolipid concentration, pH, temperature, and the presence of electrolytes and organic matter. This is not unique to rhamnolipids; all biosurfactants form these structures above the CMC, and the topic has been reviewed by Kitamoto et al. [99]. These supramolecular structures can be exploited in cosmetic and pharmaceutical formulations, and it is likely that this is already happening; however, published evidence is missing.

Soil remediation and microbial enhanced oil recovery have been among the first applications suggested for glycolipids in general and for rhamnolipids in particular and there is an abundant academic literature on the subject. Among others, Gudiña et al. reported the use of a rhamnolipid mixture obtained from growing *Pseudomonas aeruginosa* on an inexpensive culture medium based on corn steep liquor and molasses for the removal of crude oil from sand [100]. The oil removal efficacy of the rhamnolipid mixture was equal to or better than what was obtained with two conventional surfactants used for the purpose. The same research group had previously obtained similar results with another type of biosurfactant, a lipopeptide [101].

Rhamnolipids have been explored to enhance the rate of biodegradation of oil sludge from crude oil tank bottoms. It was found that addition of rhamnolipids together with nutrients such as nitrogen, phosphorus, and potassium to a bacterial culture enhanced the degradation of hydrocarbons considerably. After 56 days of treatment, n-alkanes up to C11 were completely degraded; higher alkanes were partially degraded. Thus, rhamnolipids can be useful additives in the bioremediation process [102]. They are also of interest for the treatment of marine pollutions, i.e. pollution of the seas and coasts by crude oils resulting from oil tanker discharges and accidents [31]. It is quite possible that rhamnolipids are today used in a commercial scale by the oil industry in environmentally sensitive situations, for example as oil-spill dispersants, for equipment cleaning, or for soil remediation, albeit not in the mega volumes originally (and wishfully) expected when they were eyed for enhanced oil recovery. The application is too fragmented to allow clear identification of existing users.

In agriculture, rhamnolipids are already commercialized for their fungicidal activity, and it can be expected that their use will expand since their biodegradability and low toxicity reduce the risks for the users and the environment compared to many of the alternatives.

There is no doubt that the secure supply of larger volumes of rhamnolipids, especially at prices reduced by economies of scale and by advances in producing strains, fermentation technology, and raw materials usage will foster their use for the applications already discovered. A further impulse will come from a better understanding of the products, their properties, and the formulation technology. These combined effects will broaden the cost/effectiveness reach for rhamnolipids, possibly opening opportunities also for their use in areas discussed in the following section.

5.10 Potential Applications

If approved for direct food contact, rhamnolipids could be used as stabilizers and emulsifiers, for example, for bakery and confectionery products. Furthermore, they are a source of rhamnose for the synthesis of flavors [28].

With respect to pharmaceutical and therapeutic applications, rhamnolipids are nontoxic and have antimicrobial effects against pathogenic bacteria such as *Staphylococcus aureus* and *Listeria monocytogenes* and they also have antifungal properties [97, 103–105]. Rhamnolipids also exhibit a pronounced microbicidal effect against enveloped viruses, such as those belonging to the *Herpesviridae* and *Coronaviridae* families, including SARS-CoV-2 [106].

Due to their antimicrobial activity, the documented acceleration of wound healing, and the effectiveness against various dermatologic diseases, rhamnolipids can be regarded as an active pharmaceutical ingredient [107]. Based on their pronounced surface activity, rhamnolipids may also be used (and may already be to a limited extent) as an adjuvant in pharmaceutical products, e.g. as an emulsifier or as an amphiphile in microemulsion-based formulations.

Several studies have shown that rhamnolipids can affect the secretion of immune-related cytokines and stimulate the release of a large number of immune factors. These results have implications for cystic fibrosis, which is characterized by repeated *Pseudomonas* infections in the lungs, which means that the inflamed lungs of patients with cystic fibrosis have high levels of rhamnolipids [108–110].

Rhamnolipids are being considered as replacement for antibiotics in poultry. Zhang et al. studied the effect of a rhamnolipid supplement in the feed of broilers and observed that rhamnolipid-fed broilers had higher bacterial richness and higher relative abundances of several bacteria in the gut: *Megasphaera hypermegale*, *Lachnospiraceae bacterium*, *Clostridium spiroform*, and *Alistipes obesi*. The results indicate that rhamnolipid supplementation improves growth performance, benefits the intestinal villus morphology, regulates host immune function, and raises intestinal volatile fatty acid content, and the relative abundance of the gut microbiota in broiler chickens. The effect of rhamnolipids on the growth performance of the broilers is dose-related, however. Whereas a dose of 1 g/kg significantly increased the growth after 56 days compared with antibiotics, a dose of 0.5 g/kg gave approximately the same growth as obtained with the polypeptide antibiotic bacitracin [111].

The synergistic effects of surface active agents and actives are well-known in the crop protection industry and the effect may range from better foliar adhesion, spreading, and penetration of the actives to enhanced root uptake. Knowing the mechanism of action of the active enables the choice of the surfactant with the functionalities to maximize the effect. The low surface tension and the low CMC values of many rhamnolipids should significantly improve the spreading/wetting of the aqueous formulations of the actives. The excellent biodegradability of rhamnolipids will prevent accumulation in the soil, which is a well-known problem with the conventional surfactants used as adjuvants in crop protection formulations and the relatively high area per molecule should reduce the risks of phytotoxicity. There appear not to be any studies in the open literature exploring the adjuvancy potential of rhamnolipids (as there are for sophorolipids, see for example [112]) but this does not mean that the subject has not been addressed. In any case, the potential should not be ignored.

Macrophomina phaseolina is one of the most harmful plant pathogens, infecting many species. The fungus has been found in peanuts, cabbage, peppers, peas, soybeans, sunflowers, sweet potatoes, alfalfa, sesame, potatoes, sorghum, wheat, and maize and causes progressive dejection, early death, loss of power, and loss of function. Reddy et al. investigated mango kernel oil as a substrate for *Pseudomonasa aeruginosa* DR1 biosurfactant production and assessed the antifungal efficacy of the biosurfactant against various phytopathogens, including *Macrophomina phaseolina*. They reported a mycelial growth inhibition of over 60% [113].

Being able to claim both surfactant and disinfectant action would make rhamnolipids even more valuable for both household and industrial cleaning applications where disinfection is an issue. However, in many countries, disinfectant claims are strictly regulated (in the United States by the Environmental Protection Agency and in the

European Union by the European Chemicals Agency) and require a considerable volume of supporting data. Companies do not venture with such claims until they gather the data the agency requires.

Pairing different biosurfactants opens interesting options for household cleaning formulations. For example, it has been reported that a water-based all-purpose cleaner containing 1% rhamnolipid and 2% sophorolipid was an excellent product for oily surfaces. The two biosurfactants were claimed to complement each other, the rhamnolipid being the better cleanser and the sophorolipid the better emulsifier [79]. Nguyen et al. used mixtures of three biosurfactants, a rhamnolipid, a sophorolipid, and lecithin, to produce microemulsions of various oils [114]. Winsor I, II, and III type microemulsions were obtained with all the oils and ultralow interfacial tension was obtained at optimal salinity. The detergency, measured as hexadecane removal, was excellent. An interesting observation was that the phase behavior did not change much with variations in temperature and electrolyte concentration.

One limiting factor for a broader use of biosurfactants in general and of rhamnolipids in particular may come from the regulatory. Food, feed, agriculture, and pharmacy are heavily regulated and, in most instances, even formulation additives, so-called adjuvants, must be registered before their use is allowed. The amount of information that must be provided depends upon the application, but even in the less demanding protocols, the costs and times involved are considerable. This means that a high degree of certitude of market success is required before the investment is made.

5.11 Regulatory

Rhamnolipids are registered with the European Chemicals Agency (ECHA) as "Fermentation products of glucose with *Pseudomonas putida*, Pseudomonadaceae." The EC number is 943-175-7.

The US Environmental Protection Agency (EPA) Safer Chemical Classes classifies rhamnolipids in the "green circle" category: the chemical has been verified to be of low concern based on experimental and modeled data.

The US Federal Register reports rhamnolipid biosurfactants under Title 40, 180–412 "Tolerances and exemptions for pesticide chemical residues in food" following the EPA, final rule document 9963, pages 16796–16800 that reads: "This regulation establishes an exemption from the requirement of a tolerance for residues of the biochemical, rhamnolipid biosurfactant, on all food commodities when applied/used as a fungicide (*omissis*). This regulation eliminates the need to establish a maximum permissible level for residues of rhamnolipid biosurfactant." It is effective as of March 31, 2004.

References

1 Develter, D.W.G. and Fleurackers, S. (2010). Sophorolipids and rhamnolipids. In: *Surfactants from Renewable Resources* (ed. M. Kjellin and I. Johansson), 213–238. Chichester, UK: Wiley.

2 Bergström, S., Theorell, H., and Davide, H. (1946). Pyolipic acid, a metabolic product of *Pseudomonas pyocyanea*, active against *Mycobacterium tuberculosis*. *Archives of Biochemistry* 10: 165–166.

3 Jarvis, F.G. and Johnson, M.J. (1949). A glyco-lipide produced by *Pseudomonas aeruginosa. Journal of the American Chemical Society* 71: 4124–4126.

4 Abdel-Mawgoud, A.M., Lépine, F., and Déziel, E. (2010). Rhamnolipids: diversity of structures, microbial origin and roles. *Applied Microbiology and Biotechnology* 86: 1323–1336.

5 Stjerndahl, M. and Holmberg, K. (2005). Hydrolysable nonionic surfactants: stability and physicochemical properties of surfactants containing carbonate, ester and amide bonds. *Journal of Colloid and Interface Science* 291: 570–576.

6 Chen, M.L., Penfold, J., Thomas, R.K. et al. (2010). Solution self-assembly and adsorption at the air–water interface of the monorhamnose and dirhamnose rhamnolipids and their mixtures. *Langmuir* 26: 18281–18292.

7 Penfold, J. and Thomas, R.K. (2022). Neutron reflection and the thermodynamics of the air-water interface. *Physical Chemistry Chemical Physics* 24: 8553–8577.

8 Lu, J.R., Li, Z.X., Su, T.J. et al. (1993). Structure of adsorbed layers of ethylene glycol mono dodecyl ether surfactants with one, two, and four ethylene oxide groups, as determined by neutron reflection. *Langmuir* 9: 2408–2416.

9 Lebrón-Paler, A., Pemberton, J.E., Becker, B.A. et al. (2006). Determination of the acid dissociation constant of the biosurfactant monorhamnolipid in aqueous solution by potentiometric and spectroscopic methods. *Analytical Chemistry* 78: 7649–7658.

10 Abbasi, H., Noghabi, K.A., Hamedi, M.M. et al. (2013). Physicochemical characterization of a monorhamnolipid secreted by *Pseudomonas aeruginosa* MA01 in aqueous media. An experimental and molecular dynamics study. *Colloids and Surfaces B* 101: 256–265.

11 Zhang, Y., Placek, T.L., Jahan, R. et al. (2022). Rhamnolipid micellization and adsorption properties. *International Journal of Molecular Science* 23: 11090.

12 Wu, L.-M., Lai, Q., Mei, P. et al. (2019). Comparative studies of the surface/interface properties and aggregation behavior of mono-rhamnolipid and di-rhamnolipid. *Colloids and Surfaces B* 181: 593–601.

13 Eismin, R.J., Munusamy, E., Kegel, L.L. et al. (2017). Evolution of aggregate structure in solutions of anionic monorhamnolipids: experimental and computational results. *Langmuir* 33: 7412–7424.

14 Hogan, D.E., Tian, F., Malm, S.W. et al. (2019). Biodegradability and toxicity of monorhamnolipid biosurfactant diastereomers. *Journal of Hazardous Materials* 364: 600–607.

15 Guzmán, E., Ortega, F., and Rubio, R.G. (2024). Exploring the world of rhamnolipids: a critical review of their production, interfacial properties, and potential application. *Current Opinion in Colloid and Interface Science* 69: 101780.

16 Davidson, P. (2024). Liquid crystalline behavior of concentrated aqueous solutions of biosurfactants. *Current Opinion in Colloid and Interface Science* 69: 101777.

17 Özdemir, G., Peker, S., and Helvaci, S.S. (2004). Effect of pH on the surface and interfacial behavior of rhamnolipids R1 and R2. *Colloids and surfaces A* 234: 135–143.

18 Abdel-Mawgoud, A.M., Lépine, F., and Déziel, E. (2014). A stereospecific pathway diverts b-oxidation intermediates to the biosynthesis of rhamnolipid biosurfactants. *Chemistry & Biology* 21: 156–164.

19 Cabrera-Valladares, N., Richardson, A.P., Olvera, C. et al. (2006). Monorhamnolipids and 3-(3-hydroxyalkanoyloxy)alkanoic acids (HAAs) production using *Escherichia coli* as a heterologous host. *Applied Microbiology and Biotechnology* 73: 187–194.

20 Déziel, E., Lépine, F., Milot, S., and Villemur, R. (2003). rhlA is required for the production of a novel biosurfactant promoting swarming motility in *Pseudomonas aeruginosa*: 3-(3-hydroxyalkanoyloxy)alkanoic acids (HAAs), the precursors of rhamnolipids. *Microbiology* 149: 2005–2013.

21 Olvera, C., Goldberg, J.B., Sánchez, R. et al. (1999). The *Pseudomonas aeruginosa* algC gene product participates in rhamnolipid biosynthesis. *FEMS Microbiology Letters* 179: 85–90.

22 Ochsner, U.A., Fiechter, A., and Reiser, J. (1994). Isolation, characterization, and expression in *Escherichia coli* of the *Pseudomonas aeruginosa* rhlAB genes encoding a rhamnosyltransferase involved in rhamnolipid biosurfactant synthesis. *Journal of Biological Chemistry* 269: 19787–19795.

23 Rahim, R., Ochsner, U.A., Olvera, C. et al. (2001). Cloning and functional characterization of the *Pseudomonas aeruginosa* rhlC gene that encodes rhamnosyltransferase 2, an enzyme responsible for di-rhamnolipid biosynthesis. *Molecular Microbiology* 40: 708–718.

24 Dusane, D.H., Zinjarde, S.S., Venugopalan, V.P. et al. (2010). Quorum sensing: implications on rhamnolipid biosurfactant production. *Biotechnology and Genetic Engineering Reviews* 27: 159–184.

25 Ochsner, U.A., Hembach, T., and Fiechter, A. (1996). Production of rhamnolipid biosurfactants. *Advances in Biochemical Engineering/ Biotechnology* 53: 89–118.

26 Ochsner, U.A., Koch, A.K., Fiechter, A., and Reiser, J. (1994). Isolation and characterization of a regulatory gene affecting rhamnolipid biosurfactant synthesis in *Pseudomonas aeruginosa*. *Journal of Bacteriology* 176: 2044–2054.

27 Ochsner, U.A. and Reiser, J. (1995). Autoinducer-mediated regulation of rhamnolipid biosurfactant synthesis in *Pseudomonas aeruginosa*. *Proceedings of the National Academy of Sciences* 92: 6424–6428.

28 Trummler, K., Effenberger, F., and Syldatk, C. (2003). An integrated microbial/enzymatic process for production of rhamnolipids and L-(+)-rhamnose from rapeseed oils with *Pseudomonas* sp. DSM 2874. *European Journal of Lipid Science and Technology* 105: 563–571.

29 Eslami, P., Hajfarajollah, H., and Bazsefidpar, S. (2020). Recent advancements in the production of rhamnolipid biosurfactants by *Pseudomonas aeruginosa*. *RSC Advances* 10: 34014–34032.

30 Kaskatepe, B., Yildiz, S., Gumustas, M. et al. (2015). Biosurfactant production by *Pseudomonas aeruginosa* in kefir and fish meal. *Brazilian Journal of Microbiology* 46: 855–859.

31 Lang, S. and Wullbrandt, D. (1999). Rhamnose lipids—biosynthesis, microbial production and application potential. *Applied Microbiology and Biotechnology* 51: 22–32.

32 Ma, K.-Y., Sun, M.Y., Dong, W. et al. (2016). Effects of nutrition optimization strategy on rhamnolipid production in a *Pseudomonas aeruginosa* strain DN1 for bioremediation of crude oil. *Biocatalysis and Agricultural Biotechnology* 6: 144–151.

33 Giani, C., Wullbrandt, D., Rothert, R., et al. (1996). *Pseudomonas aeruginosa* and its use in a process for the biotechnological preparation of L-rhamnose. US Patent 5 501 966 A, filed 19 January 1995 and issued 26 March 1996.

34 Chong, H. and Li, Q. (2017). Microbial production of rhamnolipids: opportunities, challenges and strategies. *Microbial Cell Factories* 16: 137.

35 Raza, Z.A., Rehman, A., Khan, M.S., and Khalid, Z.M. (2007). Improved production of biosurfactant by a *Pseudomonas aeruginosa* mutant using vegetable oil refinery wastes. *Biodegradation* 18: 115–121.

36 Archana, S., Tomar, G.S., and Srinikethan, G. (2016). Studies on production of biosurfactant from *Pseudomonas aeruginosa* (MTCC7815) & its application in microbial enhanced oil recovery. *Research Journal of Chemical and Environmental Sciences* 4: 84–91.

37 Wu, T., Jiang, J., He, N. et al. (2019). High-performance production of biosurfactant rhamnolipid with nitrogen feeding. *Journal of Surfactants and Detergents* 22: 395–402.

38 Xia, W.J., Luo, Z.B., Dong, H.P., and Yu, L. (2013). Studies of biosurfactant for microbial enhanced oil recovery by using bacteria isolated from the formation water of a petroleum reservoir. *Petroleum Science and Technology* 31: 2311–2317.

39 Benincasa, M., Contiero, J., Manresa, M.A., and Moraes, I.O. (2002). Rhamnolipid productions by *Pseudomonas aeruginosa* LBI growing on soapstock as the sole carbon source. *Journal of Food Engineering* 54: 283–288.

40 Saikia, R.R., Deka, S., Deka, M., and Banat, I.M. (2012). Isolation of biosurfactant-producing *Pseudomonas aeruginosa* RS29 from oil-contaminated soil and evaluation of different nitrogen sources in biosurfactant production. *Annals of Microbiology* 62: 753–763.

41 Nurfarahin, A.H., Mohamed, M.S., and Phang, L.Y. (2018). Culture medium development for microbial-derived surfactants production - an overview. *Molecules* 23: 1049.

42 Mulligan, C.N., Mahmourides, G., and Gibbs, B.F. (1989). The influence of phosphate metabolism on biosurfactant production by *Pseudomonas aeruginosa*. *Journal of Biotechnology* 12: 199–209.

43 Guerra-Santos, L.H., Käppeli, O., and Fiechter, A. (1986). Dependence of *Pseudomonas aeruginosa* continuous culture biosurfactant production on nutritional and environmental factors. *Applied Microbiology and Biotechnology* 24: 443–448.

44 Dominguez, D.C. (2004). Calcium signaling in bacteria. *Molecular Microbiology* 54: 291–297.

45 Cinátl, J. (1969). Inorganic-organic multimolecular complexes of salt solutions, culture media and biological fluids and their possible significance for the origin of life. *Journal of Theoretical Biology* 23: 1–10.

46 Jiang, J., Zhang, D., Niu, J. et al. (2021). Extremely high-performance production of rhamnolipids by advanced sequential fed-batch fermentation with high cell density. *Journal of Cleaner Production* 326: 129382.

47 Zhu, L., Yang, X., Xue, C. et al. (2012). Enhanced rhamnolipids production by *Pseudomonas aeruginosa* based on a pH stage-controlled fed-batch fermentation process. *Bioresource Technology* 117: 208–213.

48 Kahraman, H. and Erenler, S.O. (2012). Rhamnolipid production by *Pseudomonas aeruginosa* engineered with the *Vitreoscilla* hemoglobin gene. *Applied Biochemistry and Microbiology* 48: 188–193.

49 Gong, Z., Peng, Y., and Wang, Q. (2015). Rhamnolipid production, characterization and fermentation scale-up by *Pseudomonas aeruginosa* with plant oils. *Biotechnology Letters* 37: 2033–2038.

50 Medina-Moreno, S.A., Jiménez-Islas, D., Gracida-Rodríguez, J.N. et al. (2011). Modeling rhamnolipids production by *Pseudomonas aeruginosa* from immiscible carbon source in a batch system. *International Journal of Environmental Science and Technology* 8: 471–482.

51 Gong, Z., Yang, G., Che, C. et al. (2021). Foaming of rhamnolipids fermentation: impact factors and fermentation strategies. *Microbial Cell Factories* 20: 77.

52 Desai, J.D. and Banat, I.M. (1997). Microbial production of surfactants and their commercial potential. *Microbiology and Molecular Biology Reviews* 61: 47–64.

53 Mukherjee, S., Das, P., and Sen, R. (2006). Toward commercial production of microbial surfactants. *Trends in Biotechnology* 24: 509–515.

54 Mnif, I. and Ghribi, D. (2016). Glycolipid biosurfactants: Main properties and potential applications in agriculture and food industry. *Journal of the Science of Food and Agriculture* 96: 4310–4320.

55 Beuker, J., Steier, A., Wittgens, A. et al. (2016). Integrated foam fractionation for heterologous rhamnolipid production with recombinant *Pseudomonas putida* in a bioreactor. *AMB Express* 6: 11.

56 Çakmak, H., Güngörmedi, G., Dikmen, G. et al. (2017). The true methodology for rhamnolipid: various solvents affect rhamnolipid characteristics. *European Journal of Lipid Science and Technology* 119: 1700002.

57 Shah, M.U.H., Sivapragasam, M., Moniruzzaman, M., and Yusup, S.B. (2016). A comparison of recovery methods of rhamnolipids produced by Pseudomonas aeruginosa. *Procedia Engineering* 148: 494–500.

58 Thakur, V., Baghmare, P., Verma, A. et al. (2024). Recent progress in microbial biosurfactants production strategies: applications, technological bottlenecks, and future outlook. *Bioresource Technology* 408: 131211.

59 Gudina, E.J., Rodrigues, A.I., de Freitas, V. et al. (2016). Valorization of agro-industrial wastes towards the production of rhamnolipids. *Bioresource Technology* 212: 144–150.

60 Li, Q. (2017). Rhamnolipid synthesis and production with diverse resources. *Frontiers of Chemical Science and Engineering* 11: 27–36.

61 Henkel, M., Müller, M.M., Kügler, J.H. et al. (2012). Rhamnolipids as biosurfactants from renewable resources: concepts for next-generation rhamnolipid production. *Process Biochemistry* 47: 1207–1219.

62 Marchant, R. and Banat, I.M. (2012). Microbial biosurfactants: challenges and opportunities for future exploitation. *Trends in Biotechnology* 30: 558–565.

63 Hosseinzadeh-Bandbafha, H., Li, C., Chen, X. et al. (2022). Managing the hazardous waste cooking oil by conversion into bioenergy through the application of waste-derived green catalysts: a review. *Journal of Hazardous Materials* 424C: 127636.

64 Yu, N., Xing, D., Li, W. et al. (2017). Electricity and methane production from soybean edible oil refinery wastewater using microbial electrochemical systems. *International Journal of Hydrogen Energy* 42: 96–102.

65 Radelyuk, I., Tussupova, K., Klemeš, J.J., and Persson, K.M. (2021). Oil refinery and water pollution in the context of sustainable development: developing and developed countries. *Journal of Cleaner Production* 302: 126987.

66 Shi, J., Chen, Y., Liu, X., and Li, D. (2021). Rhamnolipid production from waste cooking oil using newly isolated halotolerant *Pseudomonas aeruginosa* M4. *Journal of Cleaner Production* 278: 123879.

67 Sharma, S., Datta, P., Kumar, B. et al. (2019). Production of novel rhamnolipids via biodegradation of waste cooking oil using *Pseudomonas aeruginosa* MTCC7815. *Biodegradation* 30: 301–312.

68 Zhu, Y., Gan, J.-J., Zhang, G.-L. et al. (2007). Reuse of waste frying oil for production of rhamnolipids using *Pseudomonas aeruginosa* zju.u1M. *Journal of Zhejiang University-SCIENCE A* 8: 1514–1520.

69 Samsu, A., Yusof, S.Z.M., Awang, M.S. et al. (2014). Production of valuable biosurfactant, rhamnolipid, from waste cooking oil by indigenously isolate *Pseudomonas aeruginosa* USM-AR2 1,2. *Advances in Environmental Biology* 8: 33–38.

70 Sharma, S., Verma, R., Dhull, S. et al. (2021). Biodegradation of waste cooking oil and simultaneous production of rhamnolipid biosurfactant by *Pseudomonas aeruginosa* P7815 in batch and fed-batch bioreactor. *Bioprocess and Biosystems Engineering* 45: 309–319.

71 Pérez-Armendáriz, B., Cal-y-Mayor-Luna, C., El-Kassis, E.G. et al. (2019). Use of waste canola oil as a low-cost substrate for rhamnolipid production using *Pseudomonas aeruginosa*. *AMB Express* 9: 61.

72 Gaur, S., Gupta, S., Jha, P.N., and Jain, A. (2023). Rhamnolipid production by *Pseudomonas aeruginosa* SSL-4 on waste engine oil (WEO): Taguchi optimization, soil remediation, and phytotoxicity investigation. *Environmental Technology* 45: 4536–4549.

73 George, S. and Jayachandran, K. (2009). Analysis of rhamnolipid biosurfactants produced through submerged fermentation using orange fruit peelings as sole carbon source. *Applied Biochemistry and Biotechnology* 158: 694–705.

74 Lotfabad, T.B., Ebadipour, N., and Roostaazad, R. (2016). Evaluation of a recycling bioreactor for biosurfactant production by Pseudomonas aeruginosa MR01 using soybean oil waste. *Journal of Chemical Technology and Biotechnology* 91: 1368–1377.

75 Rashedi, H., Assadi, M.M., Bonakdarpour, B. et al. (2005). Environmental importance of rhamnolipid production from molasses as a carbon source. *International Journal of Environmental Science and Technology* 2: 59–62.

76 Varjani, S., Rakholiya, P., Ng, H.Y. et al. (2021). Bio-based rhamnolipids production and recovery from waste streams: status and perspectives. *Bioresource Technology* 319: 124213.

77 Krauter, J (2016). Evonik Commercializes Biosurfactants. Evonik Press Release. http://Corporate.evonik.com; " Evonik commercializes biosurfactants" at http://corporate.evonik.com/en/media/press_releases/pages/news-details.aspx?newsid=60276 (2016).

78 Thum, O., Engel, P., Gehring, C., et al. (2014). Methods of producing rhamnolipids. European Patent 2,949,214 A1, filed 26 May 2014 and issued 2 December 2015 and European Patent 3,148,335, filed 27 April 2015 and issued 1 January 2020.

79 Bettenhausen, C. (2020). Rhamnolipids rise as green surfactants. *Chemical & Engineering News* 98 (23): 23–24.

80 Brett, P.J., deShazer, D., and Woods, D. (1998). *Burkholderia thailandensis* sp. nov., a *Burkholderia pseudomallei*-like species. *International Journal of Systematic Bacteriology* 48: 317–320.

81 Funston, S.J., Tsaousi, K., Rudden, M. et al. (2016). Characterising rhamnolipid production in *Burkholderia thailandensis* E264, a non-pathogenic producer. *Applied Microbiology and Biotechnology* 100: 7945–7956.

82 Koh, S.F., Tay, S.T., Sermswan, R. et al. (2012). Development of a multiplex PCR assay for rapid identification of *Burkholderia pseudomallei*, *Burkholderia thailandensis*, *Burkholderia mallei* and *Burkholderia cepacia* complex. *Journal of Microbiology Methods* 90: 305–308.

83 Dubeau, D., Déziel, E., Woods, D.E., and Lépine, F. (2009). *Burkholderia thailandensis* harbors two identical *rhl* gene clusters responsible for the biosynthesis of rhamnolipids. *BMC Microbiology* 9: 263.

84 Perfumo, A., Rudden, M., Smyth, T.J.P. et al. (2013). Rhamnolipids are conserved biosurfactants molecules: implications for their biotechnological potential. *Applied Microbiology and Biotechnology* 97: 7297–7306.

85 Chebbi, A., Tazzari, M., Rizzi, C. et al. (2021). *Burkholderia thailandensis* E264 as a promising safe rhamnolipids' producer towards a sustainable valorization of grape marcs and olive mill pomace. *Applied Microbiology and Biotechnology* 105: 3825–3842.

86 Correia, J., Gudiña, E.J., Lazar, Z. et al. (2022). Cost-effective rhamnolipid production by *Burkholderia thailandensis* E264 using agro industrial residues. *Applied Microbiology and Biotechnology* 106: 7477–7489.

87 Bahia, F.M., de Almeida, G.C., de Andrade, L.P. et al. (2018). Rhamnolipids production from sucrose by engineered *Saccharomyces cerevisiae*. *Scientific Reports* 8: 2905.

88 Westerduin, P., de Haan, P.E., Dees, M.J. et al. (1988). Synthesis of methyl 3-[3-(2-*O*-α-L-rhamnopyranosyl-α-L-rhamnopyranosyloxy)decanoyloxy]decanoate, a rhamnolipid from *Pseudomonas aeruginosa*. *Carbohydrate Research* 180: 195–205.

89 Bauer, J., Brandenburg, K., Zähringer, U., and Rademann, J. (2006). Chemical synthesis of a glycolipid library by a solid-phase strategy allows elucidation of the structural specificity of immunostimulation by rhamnolipids. *Chemistry: A European Journal* 12: 7116–7124.

90 Pacheco, R.P., Eismin, R.J., Coss, C.S. et al. (2017). Synthesis and characterization of four diastereomers of monorhamnolipids. *Journal of the American Chemical Society* 139: 5125–5132.

91 Compton, A.A., Deodhar, B.S., Fathi, A., and Pemberton, J.E. (2020). Optimization of a chemical synthesis for single-chain rhamnolipids. *ACS Sustainable Chemistry & Engineering* 8: 8918–8927.

92 Zhu, K. and Rock, C.O. (2008). RhlA converts b-hydroxyacyl-acyl carrier protein intermediates in fatty acid synthesis to the b-hydroxydecanoyl-beta-hydroxydecanoate component of rhamnolipids in *Pseudomonas aeruginosa*. *Journal of Bacteriology* 190: 3147–3154.

93 Kiss, K., Ng, W.T., and Li, Q. (2017). Production of rhamnolipids-producing enzymes of *Pseudomonas* in *E. coli* and structural characterization. *Frontiers of Chemical Science and Engineering* 11: 133–138.

94 Mohan, P.K., Nakhla, G., and Yanful, E.K. (2006). Biokinetics of biodegradation of surfactants under aerobic, anoxic and anaerobic conditions. *Water Research* 40: 533–540.

95 Renkin, M. (2003). Environmental profile of sophorolipid and rhamnolipid biosurfactants. *Rivista Italiana Delle Sostanze Grasse* 80: 249–252.

96 Johann, S., Seiler, T.-B., Tiso, T. et al. (2016). Mechanism-specific and whole-organism ecotoxicity of mono-rhamnolipids. *Science of the Total Environment* 548-549: 155–163.

97 Haba, E., Pinazo, A., Jauregui, O. et al. (2003). Physicochemical characterization and antimicrobial properties of rhamnolipids produced by Pseudomonas aeruginosa 47T2 NCBIM 40044. *Biotechnology and Bioengineering* 81: 316–322.

98 Sotirova, A.V., Spasova, D.I., Galabova, D.N. et al. (2008). Rhamnolipid-biosurfactant permeabilizing effects on gram-positive and gram-negative bacterial strains. *Current Microbiology* 56: 639–644.

99 Kitamoto, D., Morita, T., Fukuoka, T. et al. (2009). Self-assembling properties of glycolipid biosurfactants and their potential applications. *Current Opinion in Colloid and Interface Science* 14: 315–328.

100 Gudiña, E.J., Rodrigues, A.I., Alves, E. et al. (2015). Bioconversion of agro-industrial by-products in rhamnolipids toward applications in enhanced oil recovery and bioremediation. *Bioresource Technology* 177: 87–93.

101 Pereira, J.F.B., Gudiña, E.J., Costa, R. et al. (2013). Optimization and characterization of biosurfactant production by Bacillus subtilis isolates towards microbial enhanced oil recovery applications. *Fuel* 111: 259–268.

102 Rahman, K.S.M., Rahman, T.J., Kourkoutas, Y. et al. (2003). Enhanced bioremediation of n-alkane in petroleum sludge using bacterial consortium amended with rhamnolipid and micronutrients. *Bioresource Technology* 90: 159–168.

103 Chen, J., Wu, Q., Hua, Y. et al. (2017). Potential applications of biosurfactant rhamnolipids in agriculture and biomedicine. *Applied Microbiology and Biotechnology* 101: 8309–8319.

104 Benincasa, M., Abalos, A., Oliveira, I., and Manresa, A. (2004). Chemical structure, surface properties and biological activities of the biosurfactant produced by *Pseudomonas aeruginosa* LBI from soap-stock. *Antonie Van Leeuwenhoek* 85: 1–8.

105 Thakur, P., Saini, N.K., Thakur, V.K. et al. (2021). Rhamnolipid the glycolipid biosurfactant: emerging trends and promising strategies in the field of biotechnology and biomedicine. *Microbial Cell Factories* 20: 1.

106 Giugliano, R., Buonocore, C., Zannella, C. et al. (2021). Antiviral activity of the rhamnolipids mixture from the Antarctic bacterium *Pseudomonas gessardii* M15 against herpes simplex viruses and coronaviruses. *Pharmaceutics* 13: 2121.

107 Stipcevic, T., Piljac, T., and Isseroff, R.R. (2005). Di-rhamnolipid from *Pseudomonas aeruginosa* displays differential effects on human keratinocyte and fibroblast cultures. *Journal of Dermatological Science* 40: 141–143.

108 Shryock, T.R., Silver, S.A., Banschbach, M.W., and Kramer, J.C. (1984). Effect of *Pseudomonas aeruginosa* rhamnolipid on human neutrophil migration. *Current Microbiology* 10: 323–328.

109 König, B., Bergmann, U., and König, W. (1992). Induction of inflammatory mediator release (serotonin and 12-hydroxyeicosatetraenoic acid) from human platelets by *Pseudomonas aeruginosa* glycolipid. *Infection and Immunity* 60: 3150–3155.

110 Bédard, M., McClure, C.D., Schiller, N.L. et al. (1993). Release of interleukin-8, interleukin-6, and colony-stimulating factors by upper airway epithelial cells: implications for cystic fibrosis. *American Journal of Respiratory Cell and Molecular Biology* 9: 455–462.

111 Zhang, B., Chen, G., Zhang, H. et al. (2021). Effects of rhamnolipids on growth performance and intestinal health parameters in Linnan yellow broilers. *Poultry Science* 100: 810–819.

112 Celligoi, M.A.P.C., Silveira, V.A.I., Hipólito, A. et al. (2020). Sophorolipids: a review on production and perspectives of application in agriculture. *Spanish Journal of Agricultural Research* 18: e03R01.

113 Reddy, K.S., Khan, M.Y., Archana, K. et al. (2016). Utilization of mango kernel oil for the rhamnolipid production by *Pseudomonas aeruginosa* DR1 towards its application as biocontrol agent. *Bioresource Technology* 221: 291–299.

114 Nguyen, T.T.L., Edelen, A., Neighbours, B. et al. (2010). Biocompatible lecithin-based microemulsions with rhamnolipid and sophorolipid biosurfactants: formulation and potential applications. *Journal of Colloid and Interface Science* 348: 498–504.

6

Sophorolipids

6.1 Introduction

Sophorolipids, also named sophorose lipids or by the acronym SLs, were first described by Gorin et al. in 1961 as products produced by the ascomycetous yeast species *Torulopsis magnoliae* (current name *Candida apicola*) [1]. Several other strains, such as *Starmerella bombicola*, *Candida albicans*, and *Cryptococcus curvatus*, have later been found to produce sophorolipids in high yields [2]. Because of its wide applicability, its favorable properties, and its high production yield, i.e. in the bioconversion from saccharides and/or vegetable oils, it is probably the biosurfactant that has received most attention from academia and industry [3].

6.2 Chemical Structures and Properties

Sophorolipids consist of the disaccharide sophorose to which a long-chain hydroxy fatty acid is linked via a β-glycosidic bond. The hydroxyl group in the acyl chain is either at the terminal (ω) or subterminal (ω−1) position. The carboxyl group is either free, in which case the molecule is anionic at neutral and alkaline pH, or esterified with the hydroxyl group in the 4″ position of the sophorose moiety. Then a cyclic ester, a lactone, is formed. Lactonization may also occur via the 6′- or 6″-position but that is rare. The lactonic form is dominant when wild strains are used for production. Commercial sophorolipids are usually mixtures of the lactonic form and the acidic form (also called open form) with a ratio of the lactonic to the acidic form typically being around 7:3 [4, 5]. The lactonic form is the more surface active and reduces the surface tension of water to low values. The acidic form is more water soluble and gives more foam.

The hydroxyl groups in 6′- and 6″-positions of the sophorose moiety may, or may not, be acetylated. The fully acetylated form is strongly hydrophobic, which means that it is difficult to handle in aqueous solutions. On the other hand, the acetylated form is of high biological interest because of pronounced antiviral and cytokine-stimulating effects [6]. Figure 6.1 shows the structures of the lactonic form and the acidic form of a deacetylated sophorolipid. There is an equilibrium between the two forms. Removal of water drives esterification, i.e. lactone formation. This means that the ratio between the two forms will depend not only

Natural Surfactants: Science, Technology, and Industrial Aspects, First Edition. Guido Bognolo and Krister Holmberg.

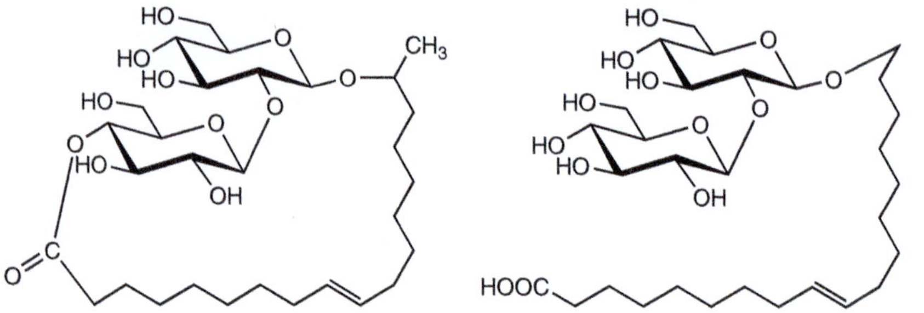

Figure 6.1 A deacetylated sophorolipid in (left) lactonic form; (right) acidic (open) form.

on the type of strain and carbon source used but also on the conditions during the work-up procedure.

The fatty acid typically contains 16 carbon atoms in a saturated chain or 18 carbon atoms in a chain that may be saturated, monounsaturated (oleic acid), or diunsaturated (linoleic acid). Small amounts of triunsaturated C18 acid (linolenic acid) may also be present. As mentioned above, the hydroxyl group may be at the terminal (ω) or subterminal (ω−1) position. The subterminal position is the most common.

Commercial sophorolipids may also contain a product in which two sophorose units are linked by the hydroxy fatty acid, one via an ester bond and the other via a glycosidic linkage. Figure 6.2 shows such a molecule based on 17-hydroxyoleic acid as fatty acid. This is an example of a so-called bolaform surfactant, an amphiphile with two polar headgroups

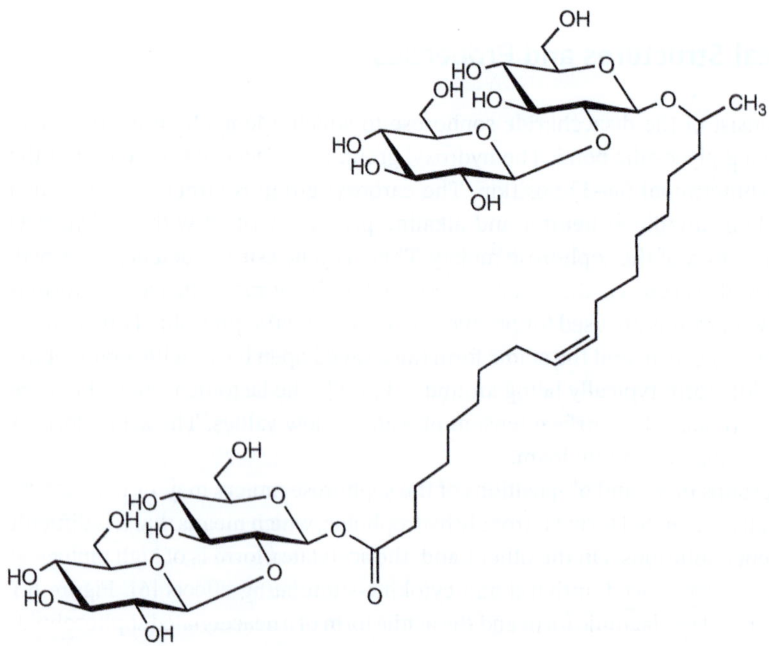

Figure 6.2 A bolaform sophorolipid based on 17-hydroxyoleic acid.

connected by a long hydrophobic tail. The amount of the bolaform sophorolipid is usually low but it has been found that an engineered form of *Starmerella bombicola* yeast gives high yields of this product [7]. A CMC value of 228 mg/L and a surface tension at the CMC of 36 mN/m have been reported. A CMC of 228 mg/L is at the higher end of the broad range of CMC values reported for most sophorolipids, see below, which means that the bolaform sophorolipids are not very surface active. This is according to expectation considering the structure.

The structure of the sophorolipid may vary with the choice of yeast used in the production and other structures than those shown in Figures 6.1 and 6.2 have been reported. For instance, *Rhodotorula bogoriensis* (formerly referred to as *Candida bogoriensis*) gives a sophorolipid where the fatty acid is hydroxylated in the 13-position. The glycosidic bond from the 2′-position of the sugar unit then connects to the 13-position of the fatty acid. This sophorolipid only exists in the acidic form. It is likely that steric hindrance makes the formation of the lactone difficult [8].

The fact that a commercial sophorolipid contains a sizable fraction with a free carboxyl group and that it usually contains one or two acetyl groups makes the pH dependence complex. At low pH, pH 5 or below, the carboxyl group is largely protonated, which is manifested in poor water solubility and poor surfactancy. Above pH 5, the carboxyl group becomes increasingly deprotonated, which improves the water solubility. At pH 8 and above, the ester bonds in 6′- and 6″-positions start to hydrolyze, which means that the hydrophobic acetyl groups will be replaced by hydrophilic hydroxyl groups, imparting increased water solubility. At high pH, the internal ester bond will also hydrolyze, transforming the lactonic form into the acidic form, which, in turn, means much improved water solubility. Thus, the water solubility and, consequently, the physicochemical properties of sophorolipids are unusually pH dependent [9].

The surface activity, and consequently the CMC values, reported for sophorolipids vary widely. Important parameters in this respect are the ratio of lactonic to acidic form and the type of carbon source used. Ashby et al. reported CMC values between 35 and 250 mg/L [10]. The surface tension at the CMC was around 35 mN/m, a value that has been reported by other groups as well [11, 12].

The acetic acid ester bonds, the acetal linkages, and the lactone bond are also susceptible to acid although the stability in acid is higher than in alkali. However, the glycosidic bonds between the glucose units in sophorose and between the sophorose and the hydroxy fatty acid will degrade at low pH. Thus, exposing a sophorolipid to low pH for an extended period of time will generate the hydroxy fatty acid in protonated form together with glucose and acetic acid.

The two main components of commercial sophorolipids, the acidic form and the lactonic form, both have a geometry that is unusual compared to conventional, synthetic surfactants. The vast majority of the surfactants that are used commercially today contain one polar headgroup, which prefers a polar environment, typically water, and one or two apolar chains, which are directed away from water, i.e. into oil, into air, or onto a hydrophobic solid surface. The acidic form of sophorolipids contains two polar headgroups, the carboxylate group and the sophorose unit. The first is an anionic group (at neutral or alkaline pH) and the second is a nonionic moiety. The two polar groups will prefer to be in water and the long

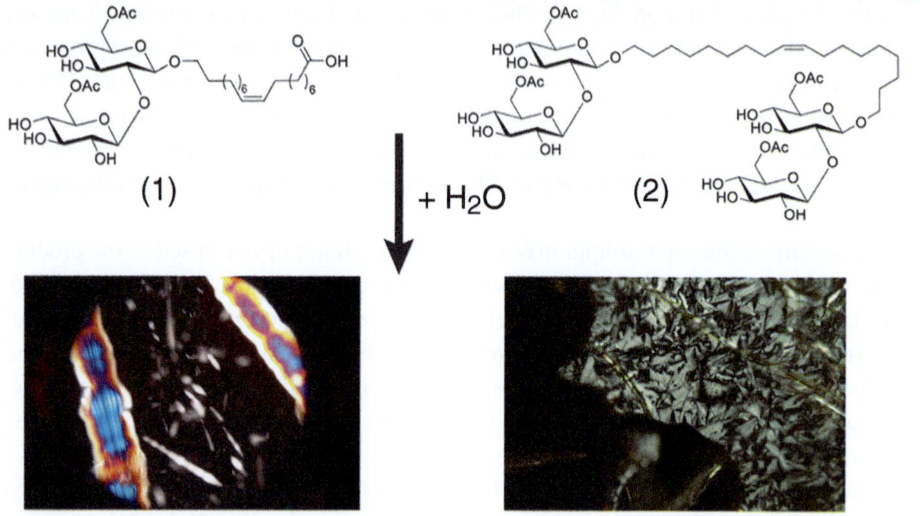

Figure 6.3 A sophorolipid in acidic (open) form (1) and a bolaform sophorolipid (2), generating lyotropic liquid crystals in water, as is shown by polarized-light microscopy. Sample (1) is clearly biphasic, with a small amount of solvent coexisting with the rod-like birefringent liquid crystalline phase, which is assumed to be lamellar. The optical texture of sample (2) suggests the presence of a liquid crystalline phase that is also believed to be lamellar. (Reproduced with permission from Saci et al. [13] / American Chemical Society.)

hydrocarbon chain that connects them will go into an apolar environment, which may be oil, air, or a hydrophobic solid surface.

Saci et al. have investigated the self-assembly properties of several sophorolipids in water at high concentrations (20–80 wt%) using polarized-light microscopy and X-ray scattering. Some of the sophorolipids formed lyotropic liquid crystalline phases with either hexagonal or lamellar geometry while others displayed thermotropic liquid-crystalline behavior [13, 14]. This is valuable knowledge for the understanding of their biological function, and it is also potentially important for the use of this class of biosurfactants for applications within the cosmetic sector, where the formulations often involve lyotropic liquid crystalline phases. Figure 6.3 shows polarized-light microscopy images of two different sophorolipids, both at a concentration of 70 wt%, which was the lowest concentration at which the biosurfactants formed liquid crystalline phases.

Thus, a sophorolipid can be regarded as a nonionic biosurfactant but with a certain fraction of anionic component included. The fact that it is not entirely a nonionic, or an anionic amphiphile need not be a practical disadvantage. Most formulations for applications within personal care, cleaning, crop protection, and metal working contain both anionic and nonionic amphiphiles. However, it is important that the ratio between the nonionic and the anionic components in a commercial sophorolipid is consistent from batch to batch. The degree of acetylation of the two primary hydroxyl groups (6′- and 6″-positions) of the sophorose units is also important for the physicochemical properties. The water solubility of the biosurfactant is considerably enhanced if the acetyl groups are converted into hydroxyl groups. The hydrophobic forms of sophorolipids are very soluble in ethanol and other polar solvents, however [15].

Thus, the physicochemical properties of sophorolipids depend to a very large extent on the ratio of the acidic form to the lactonic form. However, also the character of the long hydroxyacyl chain is important. As mentioned above, it may be saturated, it may have one double bond, usually a *cis* double bond in 9-position, or it may have two double bonds, usually *cis* bonds in 9- and 12-position. It may also have a third double bond, a *cis* bond in the 15-position, although that is not very common. Each *cis* double bond means a kink on the chain, which means that the volume of the chain will vary depending on the degree of unsaturation, which, in turn, will influence the self-assembly behavior. As an example, Dhasaiyan et al. investigated the self-assembly structures obtained with three sophorolipids, which were identical except for the number of *cis* double bonds in the chain. One chain was 17-hydroxyoleic acid (having a double bond in 9-position), the second was 17-hydroxylinoleic acid (with double bonds in 9- and 12-positions), and the third was 17-hydroxylinolenic acid (with double bonds in 9-, 12-, and 15-positions). All were in the acidic form, and all were fully deacetylated, i.e. had hydroxyl groups in the 6′- and 6″-positions. The oleic acid derivative formed ribbon-like structures, the linoleic acid derivative did not form any discernible structures, and the linolenic acid derivative gave multilamellar giant vesicles [16]. Low-angle X-ray diffraction combined with confocal laser scanning microscopy indicated that the sophorolipid molecules in the vesicle were arranged in tail-head–head-tail fashion, as illustrated in Figure 6.4.

From the above discussion, it is obvious that sophorolipids are extremely complex mixtures of species with varying degrees of surface activity and with different self-assembly characteristics. Values of critical micelle concentration (CMC), surface tension at the CMC, interfacial tension, wetting, and similar characteristics have been reported in the literature [17] but should be treated with some caution considering the complexity of this biosurfactant. Many of the sophorolipids studied are not pure, well-defined molecules and small amounts of a hydrophobic component, e.g. a diacetylated sophorolipid in the lactonic form, would drastically reduce both the CMC and the surface tension at the CMC.

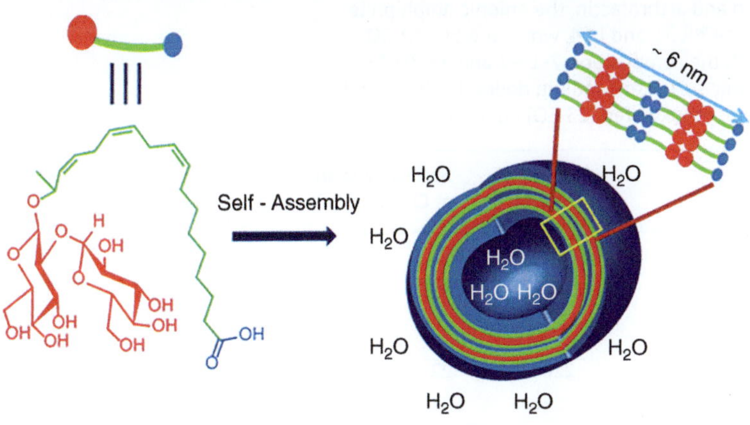

Figure 6.4 A fully deacetylated linolenic acid-based sophorolipid in acidic form self-assembles into multilamellar giant vesicles. (From Dhasaiyan et al. [16] / Vesicle structures from bolaamphiphilic biosurfactants: Experimental and molecular dynamics simulation studies on the effect of unsaturation on sophorolipid self-assemblies / reproduced with permission of John Wiley & Sons.)

However, Hirata et al. have made a study with well-defined sophorolipids with varying ratio between the lactonic and the acidic form [18]. The CMC of the acidic form was much higher than that of the lactonic form, 680 and 73 mg/L, respectively. This is not surprising since anionic surfactants are known to have much higher CMC values than nonionics. The surface tension at the CMC was about the same for the two forms, around 42 mN/m. An interesting observation was that the CMC value reached a minimum, 62 mg/L, at a ratio of the lactonic to the acidic form of 72:28. This ratio also gave the lowest surface tension at the CMC, 39.5 mN/m (still not very low). A minimum in CMC at a certain mixing ratio of surfactants is often referred to as synergism, implying that at this composition the mixture is more surface active than the individual components. It is interesting that the lactonic and the acidic form of sophorolipids exhibit synergism at a ratio that is typical of what is formed naturally by the most productive yeast strains, such as *Candida bombicola*.

Hirata et al. also investigated the foaming behavior, detergency, and rate of biodegradation of a sophorolipid and compared the values with those of two amphiphilic cyclic lipopeptides, surfactin and arthrofactin, as well as with sodium laurate (soap) and four common synthetic surfactants: sodium dodecyl sulfate, ethoxylated (23 EO) dodecanol and two EO-PO-EO block copolymers (2-16-2 and 13-30-13, with trade names Pluronic® L32 and L64, respectively). They also measured and compared the CMC and the surface tension at the CMC for these surfactants [19]. The sophorolipid was produced from *Candida bombicola* ATCC22214 using a jar-fermenter with a medium of 10% glucose, 10% soybean oil, 1% KH_2PO_4, 0.5% $MgSO_4$, 0.1% NaCl, 0.1% urea, and 0.25% yeast extract. The ratio of lactonic to acidic form was 7:3, which is a typical value for sophorolipids.

As can be seen from Table 6.1, the CMC and the surface tension at the CMC for the sophorolipid were rather similar to the values for the synthetic nonionic surfactant ethoxylated (23EO) dodecanol, also known as C12E23.

Table 6.1 Values of critical micelle concentration (CMC) and surface tension at the CMC of a sophorolipid, the cyclic lipopeptides surfactin and arthrofactin, the anionic amphiphile sodium laurate, Pluronic® L31 and L64, which are EO-PO-EO block copolymers with the composition 2-16-2 and 13-30-13, respectively, the anionic surfactant sodium dodecyl sulfate, and the nonionic surfactant ethoxylated (23 EO) dodecanol (C12E23).

Surfactant	CMC (mg/L)	Surface tension at CMC (mN/m)
Sophorolipid	43	37
Surfactin	36	28
Arthrofactin	9.0	27
Sodium laurate	2200	22
Pluronic® L31	—	45
Pluronic® L64	—	41
Sodium dodecyl sulfate	270	34
Ethoxylated (23 EO) dodecanol	42	42

(*Source:* Hirata et al. [19] / Novel characteristics of sophorolipids, yeast glycolipid biosurfactants, as biodegradable low-foaming surfactants / with permission of Elsevier.)

Table 6.2 Foam height (in mm) for the surfactants described in the head of Table 6.1. Water hardness is calculated as ppm $CaCO_3$ and a buffer (pH 8.94) containing 5 mM Na_2CO_3 and 90 mM $NaHCO_3$.

Surfactant	Distilled water, initial	Distilled water, after 5 min	100 ppm water hardness, initial	100 ppm water hardness, after 5 min
Sophorolipid	24	20	23	9
Surfactin	145	135	117	95
Arthrofactin	134	118	94	86
Sodium laurate	237	201	10	8
Pluronic® L31	18	1	10	0
Pluronic L64	27	8	15	0
Sodium dodecyl sulfate	202	195	227	202
Ethoxylated (23EO) dodecanol	69	60	65	56

(*Source:* Hirata et al. [19] / Novel characteristics of sophorolipids, yeast glycolipid biosurfactants, as biodegradable low-foaming surfactants / with permission of Elsevier.)

An interesting observation was that the sophorolipid was very low foaming both in distilled water and at high electrolyte concentration. Tests of foaming ability and of foam stability gave low values compared to the other surfactants. Detergency tests were also made in distilled water and in hard water (100 ppm hard water, containing 27 mg/L calcium salt and 8 mg/L magnesium salt). The difference in cleaning ability between the surfactants was relatively small in distilled water but some surfactants, in particular sodium laurate and the two block copolymers, lost efficacy at the higher electrolyte concentration, which the sophorolipid did not. The sophorolipid also proved to be good at wetting very hydrophobic surfaces. Low foaming surfactants with good wetting and cleaning ability in both soft and hard water are of interest for applications such as hard surface cleaning and machine dishwashing rinse aids. Table 6.2 shows the results from the foaming tests.

6.3 Biosynthesis

The general pathway for the biosynthesis of sophorolipids was described by Fleurakers and Van Bogaert and is summarized by Develter and Fleurakers [9].

Sophorolipids can be produced from feeds that are incorporated directly into the molecule as the hydrophobic moiety, i.e. the C16 or C18 fatty acid. This differs from production of most other amphiphilic glycolipids, e.g. rhamnolipids, where the biosynthesis occurs with a de novo synthesis of the hydrophobic part. The direct incorporation of the fatty acid is advantageous and gives a high productivity. Only these chain lengths are directly incorporated. Chains shorter than C16 are completely metabolized through β-oxidation, whereas chains longer than C18 have to be shortened by the same mechanism before they are incorporated. In fact, only six naturally occurring fatty acids are directly converted into sophorolipids: palmitic (C16), palmitoleic (C16:1), stearic (C18), oleic (C18:1), linoleic (C18:2), and α-linolenic (C18:3) [9].

When *n*-alkanes are used as carbon source, they are taken up by the yeast and oxidized to the corresponding alcohol by cytochrome P450 monooxygenase. The alcohol is further oxidized to the fatty acid by alcohol- and aldehyde-dehydrogenase. Starting from a fatty acid ester requires the intervention of an external lipase that converts the ester into the fatty acid. In a subsequent step, molecular oxygen and NADPH (nicotinamide adenine dinucleotide phosphate, reduced form) hydroxylate the intracellular fatty acid at the end of the

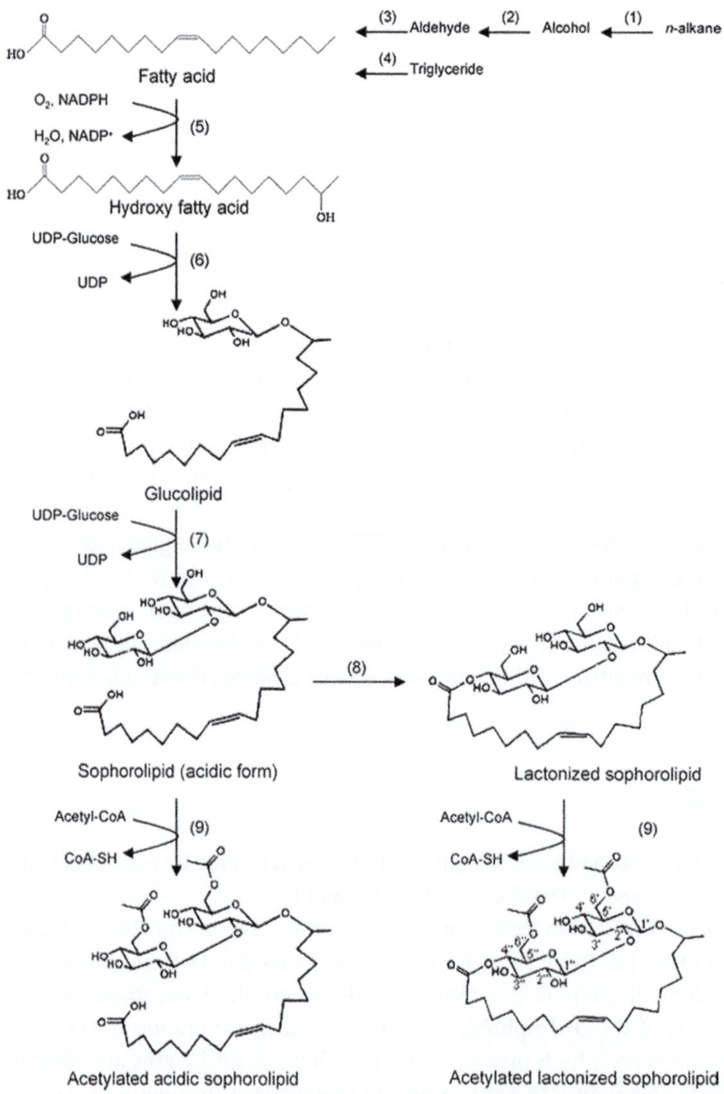

Figure 6.5 Sophorolipid synthesis pathway. (From Van Bogaert et al. [21] / Microbial production and application of sophorolipids / reproduced with permission of Springer Nature.)

tail (ω position) for C16 chains or at one carbon atom away from the end (ω−1 position) for C18 chains through the action of cytochrome P450 monooxygenase. To the resulting hydroxy fatty acid, a first glucose unit is added by the help of glucosyltransferase creating a glycosidic bond between the 1′-carbon of the glucose and the hydroxylated carbon of the hydroxy fatty acid. A second glucosyltransferase adds another molecule of glucose through the formation of a second glycosidic bond between the positions 2′ and 1″. The generated sophorolipid can be excreted as such or can undergo further modification, i.e. lactonization by linking the carboxyl group to the 4″-hydroxyl group, and/or acetylation of the 6′ and/or 6″ hydroxyl groups. The predominant component of secreted sophorolipids is the lactonic, doubly acetylated form [20]. The biosynthesis is shown in Figure 6.5.

The intracellular biosynthesis of sophorolipids produces a variety of structures, lactonic, acidic, non-, mono-, and di-acetylated, that are transported extracellularly by transmembrane transport proteins. Such transport is of primary importance as it allows reaching high production titres. The mechanism is well understood in the case of fungal biosurfactants, such as sophorolipids [22]. A final lactonization occurs extracellularly when the responsible enzyme *sblc* is secreted [17].

6.4 Production

Sophorolipids are produced by many nonpathogenic yeasts, principally from *Starmerella* genus (which was previously classified as *Candida*) and *Torulopsis* [2]. The fact that they are produced by nonpathogenic microorganisms is a considerable advantage with respect to assessment of hazards and classification. They share this advantage with the mannosylerythritol lipids. Many other biosurfactants are produced by pathogenic microorganisms. For instance, rhamnolipids, a related glycolipid, are produced by the opportunistic pathogen *Pseudomonas aeruginosa*.

Sophorolipids can be produced in high yields of over 400 g/L with a yield on total carbon sources of 0.73 g/g and a throughput efficiency of 2.4 g/L/h [23]. The high yield is the consequence of the relatively long fed-batch fermentation of resting cells with high content of oleic acid substrates. The fed-batch fermentation is carried out aerobically at a pH of 3.5 and 30 °C for 480 hours. *Candida bombicola* ATCC 22214 is mainly used as producing strain. The closely related *Candida apicola* IMET 42747 can also be used [21].

The yeasts used in the process are sub-cultured from the stock culture on freshly prepared agar slants, for example potatoes dextrose (PDA, composition (g/L): potato extract, 4.0; dextrose, 20.0; agar, 15.0) or glucose yeast peptone (GYPA, composition (g/L): glucose, 100; yeast extract, 30; peptone, 50; agar, 20), and incubated at 25 °C for 24–48 hours [24]. The microorganisms are then inoculated in the culture medium. A typical medium for the production of sophorolipids consists of over 10% glucose as the main source of energy and carbon. Nitrogen is supplied as yeast extract, sometimes supplemented by mineral nitrogen, potassium buffer, and sources of calcium, magnesium, and sodium. An example is given in Table 6.3.

The production of sophorolipids is considerably more efficient when two types of carbon sources, a hydrophilic (glycidic) and a hydrophobic (lipidic) one, are provided. Glucose is the preferred glycidic carbon source as it gives the highest yield. Many substrates can

Table 6.3 Example of medium composition for production of sophorolipids.

Ingredient	Concentration (g/L)
Glucose	120
Yeast extract	5
Sodium citrate x $3H_2O$	5
Ammonium chloride	1.5
Potassium dihydrogen phosphate	1
Dipotassium hydrogen phosphate	1.16
Magnesium sulfate x $7H_2O$	0.7
Sodium chloride	0.5
Calcium chloride	0.27

(*Source:* Lang et al. [25] / Production of native and modified sophorose lipids / reproduced from EurekaMag.)

act as lipidic carbon sources: oils, fatty acids, and their corresponding esters, alkanes, etc. Hexadecane and octadecane give the best production yields as they appear to be directly converted into the hydroxy fatty acids that can subsequently be incorporated into the sophorolipid molecule. The hydroxy fatty acids are also obtained from fatty acids or fatty acid methyl esters of appropriate chain length, with oleic acid (C18:1) giving the highest yields [21].

After the growth of the microorganism, the lipophilic substrate is added, either wholly, stepwise, or continuously. Additional glucose is added if its concentration drops below 1%.

Yeast extract is essential for both cell growth and sophorolipid formation. Substitution with ureum or peptone negatively influenced the yield of biomass and glycolipids [21]. The possibility of recovering yeast extract for use in a subsequent cycle resulted in a much reduced yield; thus, it is preferable to grow a new culture for each production.

Sophorolipid synthesis starts when the yeast cells enter the stationary phase (about 24–48 hours after inoculation) and is probably triggered by nitrogen starvation [26]. The optimal growth temperature of *Candida bombicola* ATCC 2214 is 28.8 °C (information from the National Collection of Yeast Cultures, UK). The optimal temperature for sophorolipid formation is 21 °C but for easier handling temperatures between 25 and 30 °C are used in most industrial fermentation processes. The quantity of sophorolipid obtained is nearly identical but at lower temperature the biomass growth is lower, and the glucose consumption rate is higher [27]. During the exponential growth phase, the pH drops to very low levels and must be maintained at pH 3.5 with NaOH for optimal sophorolipid production [28]. The low pH and the antimicrobial effect of sophorolipids protect the fermentation broth against contamination, even when fed-batch processes of more than 200 hours are run.

Oxygen supply is very important throughout the whole fermentation process. The optimal aeration for high sophorolipid yield, expressed in terms of oxygen transfer rate, is between 50 and 80 mM O_2/L/h [29].

In general, cells can be kept viable and producing in the stationary phase for a long time. Fermentation runs of over 450 hours have been reported [30]. However, an optimal production process takes about 10 days; upon longer incubation, high sophorolipid concentrations will hinder efficient aeration and stirring.

While in the sophorolipids the glycidic moiety is always sophorose, irrespective of the hydrophilic carbon source or the culture conditions, the fatty acid moiety is to a large extent influenced by the supplemented hydrophobic carbon source. In fact, the sophorose unit can be mono-or diacetylated and esterified with the carboxyl group of the fatty acid. It is not entirely clear what determines the distribution into the acidic or the lactonic form. It has been observed that the use of alkanes results in 85% or more of the diacetylated lactones. Such a high level of lactonization cannot be achieved with triglyceride oils or fatty acids without modifying the fermentation conditions [31]. The level of yeast extract in the medium also affects the balance between the acidic and the lactonic form. A concentration of 1 g/L gives mainly the lactonic form, whereas 20 g/L favors the acidic form of the sophorolipids [21].

Under laboratory conditions, the recovery of sophorolipids is normally done by extraction with ethyl acetate, which may however, co-extract residual hydrocarbons, leading to difficulties in further downstream operations. However, subsequent extraction with pentane, hexane, or *t*-butyl methyl ether completely removes these residues.

For both the hydrophobic and the hydrophilic carbon source, cheaper alternatives have been evaluated, for example waste streams, such as biodiesel co-product streams, waste frying oil, industrial fatty acid residues, restaurant waste oil, and soybean dark oil [32–36]. Unfortunately, many studies did not perform parallel cultivation with a conventional substrate, which makes it hard to evaluate the competitiveness of the alternative sources. One partial conclusion is that soybean dark oil seems to perform almost equally well as soybean oil. In experiments with restaurant waste oil, a yield of 34 g/L was obtained compared to 42 g/L for rapeseed oil.

In an attempt to reduce the substrate cost of the hydrophilic moiety, glucose was replaced by deproteinized whey concentrate. Although lactose was not consumed and no β-galactosidase activity could be detected, high levels of sophorolipids were obtained (280 g/L) [37]. The same researchers were able to increase the productivity to 422 g/L by applying a two-stage cultivation process. First, the oleaginous yeast *Cryptococcus curvatus* was grown on the whey, then these single-cell oil organisms were harvested, and the disrupted cells served as lipidic substrate for the *Candida bombicola* cells [38].

While the possibility to use cheap substrates has been amply demonstrated, this generally results in lower yields and the unit price of the sophorolipid would not be reduced. There are also other cost contributors that dilute the benefit of reduced raw material costs, e.g. energy, labor, and product isolation, that should be considered. However, the work of Daniel and coworkers seems interesting, and so does the use of molasses, especially when it can replace or decrease the amount of yeast extract that is also an expensive raw material.

In order to overcome the problem of product inhibition and to accelerate the rate of substrate utilization, an integrated system consisting of semicontinuous fermentation in a novel bioreactor with dual ventilation pipes and dual sieve plates was developed. After recovery of the sophorolipid, the fermentation broth containing residual glucose and oleic acid was recycled back into the bioreactor [39].

There is only little information in the literature on strain improvement by genetic engineering of the producing organisms. Roelants et al. showed that with overexpression or depletion, respectively, of lactone esterase, solely the lactonic form or solely the acidic form could be obtained. This would considerably simplify the downstream processing to obtain the specific pure forms, if these are in demand [40].

6.5 Production Costs

Ashby et al. developed a process model for the estimation of the economics of sophorolipids production with 90 000 tons/year of production capacity [41]. The equipment used in the sophorolipids biosynthesis model included a fermenter, storage tanks, decanter, centrifuge, vacuum dryer, spray dryer, and molecular sieve units. The costs of land acquisition, building construction, and net working capital were ignored in the model. The total capital cost was assessed at USD 51.4 million, comprising USD 17.1 million for equipment purchase and USD 34.3 million for installation cost.

In a 2021 article by Elias et al., two processes for the production of sophorolipids were compared. The processes were integrated into an existing plant to produce first-generation ethanol from sugarcane and differed in the preparation of the feedstock and the downstream treatment while the bioreactor stage was the same in both cases. The first process was based on aerobic cultivation of *Candida mucoides*, using xylene from the stream obtained after acid pretreatment of the sophorolipids fraction and subsequent recovery based on acid precipitation and liquid–liquid extraction. In the second process, hot water was used for pretreatment of the bagasse and two steps of ultrafiltration were employed for the recovery of the sophorolipids. The xylose-rich feedstock was fermented by the strain *Candida mucoides* UFMG-CM-Y6148 at 30 °C for 72 hours to produce the sophorolipids in a yield of 12.49 g/L. A life cycle analysis showed that the second approach gave the best economic and environmental performance. The bioreactor stage was responsible for most of the capital investment in both cases (62–66%), followed by the downstream treatment units (16–23%). The net present value was between USD 1.9 and 3.5 million while the internal rate of return was between 11.6 and 12.6% p.a., higher in both cases than the critical threshold of 11.0% p.a. Operating costs ranged from USD 5.6 and USD 11.3 million/year with direct production costs accounting for USD 4.7 and USD 9.8 million/year, respectively [42].

In a related study, Wang et al. carried out a techno-economic analysis of a process for the production of sophorolipids from food waste. They showed that 100 tons of food waste can produce 93 tons of sophorolipid crystals or 120 tons of sophorolipid syrup. Compared to the production of other bioproducts, e.g. lactic acid or succinic acid, the sophorolipid production was more profitable, because of the higher commodity price of sophorolipids. Four scenarios were compared:

- equipment made in the USA + sophorolipids crystals
- equipment made in the USA + sophorolipids syrup
- equipment made in China + sophorolipids crystals
- equipment made in China + sophorolipids syrup

Production of sophorolipid crystals made with equipment from China was found to be the most profitable option. The authors reported an energy consumption of respectively 11 and

17 MJ/kg for the biosurfactant syrup and crystals, which are two orders of magnitude lower than in the above-mentioned study by Elias et al. This was attributed to the preparation of the substrate, which is enzymatic in Wang's study versus a chemical process in the study by Elias. Enzymes carry lower energy consumption, but higher operating expense because of the cost of the enzymes. The two studies report similar minimum selling prices [23].

Martinez and coworkers studied sophorolipids production by solid-state fermentation using winterized oil cake and molasses as feedstock. Three support materials were compared: wheat straw, rice husk, and coconut fiber. It was demonstrated that sophorolipids recovery from the solid matrix was the major contributor to the operating costs, while the fermentation equipment shaped the capital costs. It was found that the physical properties of the substrates and supports were decisive of the process costs and more important than the feedstock price and the process yields [43].

6.6 Biodegradability and Toxicity

The biodegradation of sophorolipids is extremely fast. In the standard test for biodegradation according to OECD Guidelines for Testing of Chemicals (301C, Modified MITI Test), the sophorolipid reached 61% biodegradation after eight days. The requirement for a surfactant to be classified as "readily biodegradable" is to reach 60% within 28 days. Sophorolipids are totally degraded to one-carbon atom metabolites that do not accumulate [19].

The values for aquatic toxicity are also good. Acute toxicity (48 hours, EC_{50}) in the crustacean *Daphnia magna*, the ciliate *Tetrahymena thermophila*, and the microalga *Pseudokirchneriella subcapitata* is 10-fold lower than that of conventional surfactants [44]. Furthermore, the sophorolipid did not affect the reproduction of *Daphnia*.

The evaluation of chronic toxicity on *Daphnia magna* showed a no-observed-effect concentration (NOEC) of 11.3 mg/L, an order of magnitude higher than that of traditional surfactants. Furthermore, sophorolipids did not affect *Daphnia magna* reproduction capacity [11].

Crude and acidic sophorolipids are not irritating to the skin, do not trigger allergic reactions, and have an oral safety level greater or equal to 5 mL/kg body weight [45]. The cytotoxicity was evaluated by the 3-(4,5-dimethylthiazol-yl)-2,5-diphenyltetrazolium bromide (MTT) method with human epidermal keratinocytes. The cytotoxicity of the sophorolipid was found to be considerably lower than that of conventional anionic and nonionic surfactants and also lower than that of two lipopeptides that were included in the test [19].

6.7 Suppliers, Formulators, Products, and Prices

In Europe, **Evonik** (Essen, Germany) has extensive research and production facilities in the Slovak Republic and has announced a large expansion investment there (the nominal capacity is reported at 20 ktons per year). It has both sophorolipids and rhamnolipids in its product range, although its commercial interest seems to be directed more to the rhamnolipids. Unilever is a customer of Evonik. SOPHANCE® LA-A is a mild, acid-rich sophorolipid with high cleaning performance for personal hygiene. REWOFERM® SL ONE is a 60/40 acidic/lactonic sophorolipid recommended for household cleaning, laundry and dishwashing detergents, hard surface cleaners, etc.

In March 2022, **Sasol** (Hamburg, Germany) announced a partnership with **Holiferm** Ltd. (Manchester, UK), a UK-based start-up company, to develop and commercialize sophorolipids. On January 25, 2023, Holiferm Ltd. and Sasol Chemicals, a business unit of Sasol Ltd., announced a collaboration to produce and market rhamnolipids and mannosylerythritol lipids. The agreement includes a commitment from Sasol to purchase the majority of the production from a new Holiferm plant. In October 2023, Sasol launched two brands, CARINEX SL for personal care and LIVINEX SL for home care, technical, institutional, and industrial cleaning that are based on sophorolipids manufactured using Holiferm's proprietary technology.

At the 13th ICIS World Surfactants Conference on May 4–5, 2023 in Jersey City (USA), Sasol declared an expected production of around 1100 tons of sophorolipids per year (active concentration not specified), in a process that is described as "continuous rather than batch" but sees a market potential of 400 000 tons/year. Sasol plans to increase production to up to 3000 tons in 2024 and scale this to up to 20 000 tons by 2025. "Oh So Mild Shower Gel" is an ultra-mild formulation containing sophorolipid HF. (HF stands for high foaming.) The formulation is given in Table 6.4.

Preparation procedure:

1) Homogenize phase A
2) Add phase B to phase A and homogenize
3) Adjust pH to 6.0
4) Add phase C and homogenize

Table 6.4 Composition of Sasol Chemicals product "Oh So Mild Shower Gel."

Phase	Ingredient	Content (% w/w)	Function
A	Sodium laureth sulfate	14.2	Foaming agent, cleanser
	Aqua	30.0	
B	Glycolipid (sophorolipid)	8.3	Ultra-mild surfactant
	Dicapryl ether	0.5	Emollient
	Cocoamidopropylbetaine	13.1	Mild cleanser and foaming agent
	Glycerol	0.5	Moisturizer
	Panthenol	0.5	Active care agent
	Lauryl glucoside	1.0	Mild co-surfactant and foaming agent
	Sodium cocoamphoacetate	1.0	Mild co-surfactant and foaming agent
	Perfume/fragrance	0.3	
	Preservative	q.s.	
	Aqua	balance to 100	
C	Sodium chloride	2.5	Thickening agent

In 2021, **BASF** (Ludwigshafen, Germany) formed a strategic alliance with **Allied Carbon Solutions** (ACS) (Nazuma City, Japan), which includes an equity stake and makes BASF the single largest shareholder. It further entails an exclusive technology cooperation, commercial agreement, and product development for sophorolipids. ACS owns a small-scale reactor.

Also in 2021, BASF signed a strategic technology agreement with the above-mentioned Holiferm Technologies to jointly develop fermentatively accessible glycolipids for home care, personal care, and industrial applications.

Ecover (Malle, Belgium) manufactures ecologically sound cleaning products. For many years, it has been involved in research, production, and formulation of sophorolipids for home care products. Since 2017, it has been owned by **S.C. Johnson & Son**, Inc. (Racine, WI, USA). Ecover Belgium NV was the first company in Europe to launch a household cleaner based on biosurfactants. It sells a 45% aqueous solution of a high lactonic sophorolipid under the trade name SL 18.

In July 2022, **Solvay** (Brussels, Belgium), a major chemical multinational, introduced on the international market MIRASOFT® SL L60 and MIRASOFT® SL A60. Both are 100% biobased and biodegradable surfactants manufactured through a cost-efficient fermentation process, targeting a broad range of applications in hair and skin care. Mirasoft® SL L60, a sophorolipid with high lactonic content and Mirasoft® SL A60, an acid-rich sophorolipid, have both an active content of 55–65%. MIRASOFT® SL A60 is claimed to be suitable for sensitive skin and scalps, while MIRASOFT® SL L60 is also suitable for acne-prone skin, oily skin, and scalps, and acts against dandruff. The products are distributed by **Summit Cosmetic Europe** (Paris, France), an international distributor specializing in raw materials for the cosmetic industry.

In 2014, **Givaudan Active Beauty** (Vernier, Switzerland) acquired **Soliance** (Pleumer-Bodou, France), a company specializing in biotechnology and biosurfactants. On July 24, 2012, Givaudan REACH registered sophorolipids for a market volume of between 1 and 10 tons/year (Record ID 100.223.691). Givaudan Active Beauty promotes Sopholiance® S, which is sophorolipid-based and claimed to have antiacne and antimicrobial activity, odor-neutralizing properties, and to perform as sebum regulator. Sophogreen™ from the same company is rich in sophorolipids and acts as a solubilizer. (It is claimed to perform better than PEG-40 hydrogenated castor oil.)

Wheatoleo (Pomacle, France) is a company that develops specialty surfactants for the detergent, industrial, and plant protection markets. It sells Sophoclean and Sophorolipid ECO S®, which are described as low foaming, fast wetting, and good degreasing biosurfactants.

In the United Kingdom, the company **Holiferm** Ltd. (Manchester, UK) promotes a range of 60% active sophorolipids:

- HoliSurf HF® (high foaming) and HoliSurf LF® (low foaming) for personal care
- HoneySurf HF® (high foaming), HoneySurf L® (low foaming) for household, industrial cleaning, and crop protection
- HoneySurf AG® for household, industrial cleaning, and crop protection

Another UK-based company, **Alchemy Ingredients** (Bracknell, UK), offers Sophocoll®, a mild formulation for facial wash, shampoo, body wash, etc. that contains sophorolipids together with a collagen-derived surfactant.

TeeGene Biotech Ltd. is a venture company from Teesside University in the United Kingdom. It is specialized in the development and production of biosurfactants. It collaborates with **Biobase Europe Pilot Plant** (Ghent, Belgium) for the scale-up and commercial production of biosurfactants. It does not have a brand of its own, it simply offers services.

The company **Locus Ingredients** (Solon, OH, USA) sells a line of lactonic sophorolipids, CAS Number 1573124–58-9, under the trade name Amphi®. They are TSCA-approved and REACH-registered and intended for various industrial applications, where they act as wetting agents, dispersants, emulsifiers, etc. Amphi® M and Amphi® CL are hydrophobic amphiphiles that give no or low foam, while Amphi® CH is a relatively hydrophilic surfactant with medium foaming properties.

In 2023, Locus Ingredients signed an agreement to supply **Dow** (Midland, MI, USA) with sophorolipids and the products are sold under the trade name EcoSense™:

- EcoSense™ GL-60 HA is an acid-rich sophorolipid. It is a hydrophilic amphiphile with good emulsifying properties, intended for personal care applications.
- EcoSense™ GL-60 HL is a lactone-rich sophorolipid. It is a hydrophobic low-foaming amphiphile, also intended for personal care applications.
- EcoSense™ HC-2100 is an acid-rich sophorolipid, hydrophilic in character, intended for home care.
- EcoSense™ HC-2200 is a lactone-rich sophorolipid, hydrophobic in character, also intended for home care.

It is reported that the collaboration between Dow and Locus includes the building of a production plant at Dow's Terneuzen (The Netherlands) site.

In the United States, there are several small companies producing sophorolipids. **Actera Ingredients** (Newton, PA, USA) sells TeraBiome™, a sophorolipid recommended for personal care and household cleaning applications. **Biosynth International Inc.** (Gardner, MA, USA) offers an acidic sophorolipid, mix acetylated, at between USD 85 and 151 per gram, an acidic sophorolipid nonacetylated at USD 410 to 560 per gram and a lactonic sophorolipid nonacetylated at USD 675 to 1460 per gram, depending on quantity. **Cayman Chemical Company** (Michigan, USA) sells sophorolipids primarily to universities and research organizations.

MarkNature (Fullerton, CA, USA) is an online vendor for nature-related products offering a 50% liquid, acidic, and lactonic, acetylated sophorolipid at Euro 34, 110, and 430 for respectively 200 g, 1 kg, and 5 kg packages. They also sell a sophorolipid in extremely high purity grade at between Euro 460 and 520 per gram, depending on volume.

SyntheZyme™ (Rensselaer, NY, USA) is a green chemistry company developing bio-based chemical building blocks, bioplastics, and biosurfactants from renewable sources using proprietary engineered yeasts. SyntheZyme has developed intellectual property for the modification of sophorolipids, which is claimed to have significantly enhanced their value in use of a wide range of applications.

The company **Dispersa** (Laval, Québec, Canada) is a start-up aiming at producing biosurfactants from waste. In 2021, it launched PuraSurf™ HF and LF, sophorolipids produced using proprietary enzymatic technology.

Godrej Industries Ltd. (Mumbai, India) sells Biogod, which is a sophorolipid claimed to be an excellent emulsifier in a broad range of pH, salinity, and temperature conditions. The product Ginoguard PCB contains sophorolipids and glycerol esters of C8–C10 fatty acids. It is recommended as a preservative for personal care formulations.

Sihauli Chemicals Private Ltd. (Palghar, India) sells a sophorolipid industrial grade at 7–8 USD/kg.

The company **Altinbio Scientific Private** Ltd. (Navi Mumbai, India) sells a 50% active lactonic diacetylated sophorolipid for general biosurfactant applications.

In Japan, **Allied Carbon Solutions** Co. Ltd. (ACS) is specialized in developing and producing sophorolipids. It has established a mass production technology that increases yield and lowers production costs and has commercialized sophorolipids for detergents, skin care, livestock feed, etc. It has a capital and business alliance with BASF and is partner with the National Institute of Advanced Industrial Science and Technology (AIST) of Japan for joint research and development. ACS promotes:

- ACS-Sophor® for livestock and fisheries
- BioToLife® for personal care
- Sophomore®for home care

Also in Japan **Saraya** Co. Ltd. (Osaka, Japan) sells sophorolipids under the trade name Soforo. Soforo is the key ingredient in the household detergent Happy Elephant. The company **Kaneka** Co. (Tokyo, Japan) has diacetylated over 80% lactonic sophorolipids for cosmetics and toiletries.

Toyobo Co., Ltd. (Osaka, Japan) and **Kanebo Cosmetics** Inc. (Tokyo, Japan) market personal care products containing sophorolipids, as well as mannosylerythritol lipids.

The Korean company **MG Intobio** Co., Ltd. (Incheon, Korea) promotes the Sopholine range of sophorolipids for the personal care industry.

Several companies commercialize sophorolipids in China; a snapshot is given in Table 6.5. Prices for the cosmetic-grade lactonic form typically range from 5 to 12 USD/kg but may go as high as 70 USD/kg or more, depending on purity and quantity.

Table 6.5 Sophorolipid producers in China.

Company	Location	Product	Recommended use
Hubei Marvel Biotech Co., Ltd.	Wuhan, Hubei	Lactonic sophorolipids, 99% purity	
Baoding Zhongbao Chemical Industry Trading Co., Ltd.	Baoding, Hebei	Sophorolipids	Cosmetics, hair care, oral care, detergents
Shandong Qilu Biotechnology Group Co., Ltd.	Liaocheng, Shandong	Sophorolipids	
Natural Micron Chem Tech Co., Ltd.	Jinan, Shandong	Lactonic sophorolipids, 98–99% purity	General biosurfactant applications
Qingdao Chemo Technology Develop Co., Ltd.	Qingdao, Shandong	Lactonic sophorolipid, 98–99% purity	General biosurfactant applications
Jinan Huijinchuan Chemical Co., Ltd.	Jinan, Shandong	Lactonic sophorolipids, 99% purity	General biosurfactant applications
Weifang Ocean Trading Co., Ltd.	Weifang City, Shandong	Lactonic sophorolipids, 50% purity	

Shandong Longze Chemical Co., Ltd. (Dongying, Shandong, China) sells petroleum-grade sophorolipids for EOR at between 5000 and 8000 USD/ton.

In China, personal care grade sophorolipids sell at between 15 and 65 USD/kg. As can be seen, there is a broad range of selling prices, depending on product quality, purity, quantities, and suppliers, but even for the lowest quality grades, the selling prices remain high, which puts a strain on cost/effectiveness.

6.8 Existing Applications

Sophorolipids were the first biosurfactants available in commercial volumes and over the years the knowledge and understanding of their industrial value have increased and are probably superior to any other class of product synthesized by microorganisms.

6.8.1 Cosmetics

Sophorolipids have been shown to be multifunctional biosurfactants and are used as emulsifiers and solubilizers for skin care formulations and for hair and body shampoos, make-up removers, lip rouge, lip cream, and eye shadow, as well as in compressed powder cosmetics, given their skin compatibility and excellent moisturizing properties. Because of their antibacterial activity, they are used in products for treatment of acne, dandruff, and body odors. They also exhibit antioxidant properties and this, along with the antibacterial effect, is highly valued for the preservation of the formulation, both during storage and use.

Sophorolipids have a documented antimicrobial activity [8] and the first commercial sophorolipid exploited this property more than the other features of surface activity, mildness, and biodegradability that are now valorized. With time, sophorolipids have proven to be prominent ingredients with antimicrobial action in topical cosmetic creams and for oral hygiene [46, 47].

With all segments across the personal care industry highlighting sustainable advances, biosurfactants and not least sophorolipids are gaining attention for delivering functionality while meeting brands and consumer demand. Data from Innova Market Insight indicates that surfactants with a biodegradability association featured an 11% annual growth from 2019 to 2023, more than half of those launches being in the bath shower category.

6.8.2 Household Cleaning

Sophorolipids have been employed in domestic cleaning products, initially in Japan and Korea and later in Europe. They were introduced in 2009 in the Belgian market by the company Ecover NV. Ecover has since continued to promote ecological household products in which biosurfactants and especially sophorolipids play a pivotal role.

The household detergent market industry these days values convenience in use, safety (intended as hygiene/antimicrobial), nontoxicity, and cleaning performance as the most valuable attributes for a cleaning product. It is interesting that cleaning performance is just one of the four valuable key attributes. Sophorolipids, as formulation ingredients, offer safety and nontoxicity and are increasingly included in household detergents.

6.8.3 Source of Chemicals

Sophorolipids are used as a source of the rare ω- and (ω-1)-hydroxy fatty acids and the disaccharide sophorose, which is used as foodstuff in the production of the enzyme cellulase [8].

6.9 Potential Applications

Below are reported sophorolipid applications that can be inferred from laboratory experiments presented in the scientific literature. It is quite possible that some of the suggestions have already found commercial applications.

The leaching and chelating action of acidic sophorolipids are pH dependent and can promote their application in hydrometallurgical extraction of metals and minerals through foam flotation [48].

Enzymatic saccharification of vegetable biomass is substantially increased by the addition of sophorolipids, which would be of value for the production of bioethanol [49, 50].

Sophorolipids are promising ingredients in the food industry as they are excellent emulsifiers for edible oils and are stable at a wide range of pH, temperature, and salinity [12]. Sophorolipids from *Candida albicans* SC5314 and *Candida glabrata* CBS138 have been reported to have emulsifying properties for olive oil, soy oil, almond oil, and mustard [51]. The inhibitory and antimicrobial properties of sophorolipids have been amply documented and could improve both shelf-life of food and food safety [52]. In addition, it has been demonstrated that sophorolipids have taste-sensory properties, which opens numerous possibilities for practical applications to ameliorate bitter tastes in foods and drugs [53].

Reverse micelle extraction is a liquid–liquid extraction method that utilizes reverse micelles for separation. The technique can be used for recovery of antibiotics such as erythromycin and amoxicillin. Sophorolipids have been found suitable for the purpose and their use implies increased sustainability and environmental friendliness compared to the use of traditional synthetic surfactants [54].

In agriculture, petrochemical surface active agents are used as chemical adjuvants to increase the effectiveness of the active substance for crop protection. This leads to an increased intake of the active promoted by the wetting effect exerted by the surfactant, although other mechanisms are also possible. Nonselective herbicides such as Paraquat™ or Roundup™ consume large volumes of surfactants, as added-in or tank mix adjuvants. It can be expected that biosurfactants (especially those having low CMC and giving low surface tension) would produce similar effects, with the advantage of a better environmental profile.

Vaughn et al. studied the effect of sophorolipids as adjuvant for postemergence herbicides, i.e. herbicides used to control weeds in lawns after they have already emerged from the soil. Sophorolipids were tested together with the water-soluble postemergence herbicide phosphinothricin (ammonium glufosinate) and as emulsifying agent for the water-insoluble contact herbicide lemongrass oil [55]. The biosurfactants used were a lactone-rich sophorolipid from *Starmerella (Candida) bombicola* (Sb) and a mainly open-form sophorolipid from *Candida kuoi* (Ck). The reference synthetic surfactant was ethoxylated (15 EO) tallow amine, which is a highly effective adjuvant, present as a built-in adjuvant in many commercial formulations or used as tank-mix. The test plants were the broadleaf sicklepod (*Senna obstrusifolia*) and the needlelike leaf corn (*Zea mais* L) and the dosage was 1% added to a 1 mg/mL solution of phosphinothricin. The loss in fresh weight, dry weight, and the herbicide damage rate (HDR) were assessed. After three days, both Sb and Ck showed a significant increase in adjuvancy over the reference surfactant for the broadleaf sicklepod, with a higher decrease in fresh and dry weight and a significant increase in HDR. For needlelike corn, only Ck showed a small, yet interesting, advantage over the control. Ck, but not Sb, was superior to the reference surfactant as emulsifier for the lipophilic organic herbicide lemongrass oil.

The above study appears not to have been followed up by the same or other researchers, which is surprising in view of the results presented. However, there are other publications that demonstrate the beneficial effect of sophorolipids to stimulate plant growth. A US patent from 2017 describes the use of sophorolipids to increase the yield in the production of plants of agricultural interest. It suggests employing the sophorolipid at 100–500 g/ha to get a better formation of total biomass or improving harvestable parts in plants [56].

Plants of *Medicago sativa* and *Bidens pilosa* showed an increase in the growth of their shoots and roots when grown in soils augmented with sophorolipids. The authors speculate that there is a greater microbial activity in the soil after the addition of the biosurfactant [57].

As is discussed in the section "A glimpse into the future" in the chapter "General views on natural surfactants," microbial enhanced oil recovery (M-EOR) is a technique of considerable current interest. A surfactant-producing microorganism is injected into the reservoir to generate in situ production of biosurfactants, such as sophorolipids. The biosurfactant should reduce the interfacial tension between the crude and the injection water to very low levels, preferably 10^{-3} mN/m or lower, which is key for successful mobilization of the oil. A mixture of sophorolipids obtained from *Candida bombicola* ATCC 22214 was investigated in the laboratory including core flooding experiments under reservoir conditions. An additional 27% of the residual oil was recovered, which is a promising result [24]. Large field tests using different microorganisms for M-EOR are currently being conducted in China.

Related to M-EOR is the in situ production of biosurfactants for treatment of oil spills and cleaning of oil pipelines and oil storage tanks, as well as remediation of oily soil. Sophorolipids are attractive candidates for such purposes [58, 59].

6.10 Regulatory

Sophorolipids are ECHA registered as "Fermentation products of glucose and fatty acids, C18 (unsaturated), glycerol esters with yeast *Candida bombicola*, partially hydrolysed," EC list No.: 941-810-2. The substance causes serious eye irritation and should be labeled accordingly.

Sophorolipids are registered under the REACH regulations and may be manufactured or imported to the European Economic Area in quantities up to 1000 tons/year.

References

1 Gorin, P., Spencer, J., and Tulloch, A. (1961). Hydroxy fatty acids glycosides of sophorose from Torulopsis magnolia. *Canadian Journal of Chemistry* 39: 846–855.

2 Wang, H., Roelants, S.L.K.W., To, M.H. et al. (2019). *Starmerella bombicola*: recent advances on sophorolipid production and prospects of waste stream utilization. *Journal of Chemical Technology and Biotechnology* 94: 999–1007.

3 Kim, Y.-B., Yun, H.S., and Kim, E.-K. (2009). Enhanced sophorolipid production by feeding-rate-controlled fed-batch culture. *Bioresource Technology* 100: 6028–6032.

4 Madankar, C.S. and Borde, P.K. (2023). Review on sophorolipids – a promising microbial biosurfactant. *Tenside Surfactant Detergents* 60: 95–105.

5 Shu, Q., Lou, H., Wei, T. et al. (2021). Contributions of glycolipid biosurfactants and glycolipid-modified materials to antimicrobial strategy: a review. *Pharmaceutics* 13: 227.

6 Shah, V., Doncel, G.F., Seyoum, T. et al. (2005). Sophorolipids, microbial glycolipids with anti-human immunodeficiency virus and sperm-immobilizing activities. *Antimicrobial Agents and Chemotherapy* 49: 4093–4100.

7 Van Bogaert, I.N.A., Buyst, D., Martins, J.C. et al. (2016). Synthesis of bolaform biosurfactants by engineered *Starmerella bombicola* yeast. *Biotechnology and Bioengineering* 113: 2644–2651.

8 Van Bogaert, I.N.A., Zhang, J., and Soetaert, W. (2011). Microbial synthesis of sophorolipids. *Process Biochemistry* 46: 821–833.

9 Develter, D.W.G. and Fleurackers, S.J.J. (2010). Sophorolipids and rhamnolipids. In: *Surfactants from Renewable Resources* (ed. M. Kjellin and I. Johansson), 213–240. Chichester, UK: Wiley.

10 Ashby, R.D., Solaiman, D.K.Y., and Foglia, T.A. (2008). Property control of sophorolipids: influence of fatty acid substrate and blending. *Biotechnology Letters* 30: 1093–1100.

11 Develter, D.W.G. and Lauryssen, L.M.L. (2010). Properties and industrial applications of sophorolipids. *European Journal of Lipid Science and Technology* 112: 628–638.

12 Daverey, A. and Pakshirajan, K. (2010). Sophorolipids from *Candida bombicola* using mixed hydrophilic substrates: production, purification and characterization. *Colloids and Surfaces B* 79: 246–253.

13 Saci, F., Roelants, S.L.K.W., Soetaert, W. et al. (2022). Lyotropic liquid-crystalline phases of sophorolipids biosurfactants. *Langmuir* 38: 8564–8574.

14 Davidson, P. (2024). Liquid crystalline behavior of concentrated aqueous solutions of biosurfactants. *Current Opinion in Colloid and Interface Science* 69: 101777.

15 Hu, Y. and Ju, L.-K. (2001). Purification of lactonic sophorolipids by crystallization. *Journal of Biotechnology* 87: 263–272.

16 Dhasaiyan, P., Pandey, P.R., Visaveliya, N. et al. (2014). Vesicle structures from bolaamphiphilic biosurfactants: experimental and molecular dynamics simulation studies on the effect of unsaturation on sophorolipid self-assemblies. *Chemistry A European Journal* 20: 6246–6250.

17 Pal, S., Chatterjee, N., Das, A.K. et al. (2023). Sophorolipids: a comprehensive review on properties and applications. *Advances in Colloid and Interface Science* 313: 102856.

18 Hirata, Y., Ryu, M., Igarashi, K. et al. (2009). Natural synergism of acid and lactone type mixed sophorolipids in interfacial activities and cytotoxicities. *Journal of Oleo Science* 58: 565–572.

19 Hirata, Y., Ryu, M., Oda, Y. et al. (2009). Novel characteristics of sophorolipids, yeast glycolipid biosurfactants, as biodegradable low-foaming surfactants. *Journal of Bioscience and Bioengineering* 108: 142–146.

20 Asmer, H.-J., Lang, S., Wagner, F. et al. (1988). Microbial production, structure elucidation and bioconversion of sophorose lipids. *Journal of American Oil Chemists' Society* 65: 1460–1466.

21 Van Bogaert, I.N.A., Saerens, K., De Muynck, C. et al. (2007). Microbial production and application of sophorolipids. *Applied Microbiology and Biotechnology* 76: 23–34.

22 Claus, S., Sanchez, L.J., and Van Bogaert, I.N.A. (2021). The role of transport proteins in the production of microbial glycolipid biosurfactants. *Applied Microbiology and Biotechnology* 105: 1779–1793.

23 Wang, H., Tsang, C.-W., To, M.H. et al. (2020). Techno-economic evaluation of a biorefinery applying food waste for sophorolipid production – a case study for Hong Kong. *Bioresource Technology* 303: 122852.

24 Elshafie, A.E., Joshi, S.J., Al-Wahaibi, Y.M. et al. (2015). Sophorolipids production by *Candida bombicola* ATCC 22214 and its potential application in microbial enhanced oil recovery. *Frontiers in Microbiology* 6: 1324.

25 Lang, S., Brakemeier, A., Heckmann, R. et al. (2000). Production of native and modified sophorose lipids. *Chimica Oggi – Chemistry Today* 18: 76–79.

26 Davila, A.-M., Marchal, R., and Vandecasteele, J.-P. (1992). Kinetics and balance of a fermentation free from product inhibition – sophorose lipid production by *Candida bombicola*. *Applied Microbiology and Biotechnology* 38: 6–11.

27 Casas, J.A. and Garcia-Ochoa, F. (1999). Sophorolipid production by *Candida bombicola*: medium composition and culture methods. *Journal of Bioscience and Bioengineering* 88: 488–494.

28 Göbbert, U., Lang, S., and Wagner, F. (1984). Sophorose lipid formation by resting cells of *Torulopsis bombicola*. *Biotechnology Letters* 6: 225–230.

29 Guilmanov, V., Ballistreri, A., Impallomeni, G. et al. (2002). Oxygen transfer rate and sophorose lipid production by *Candida bombicola*. *Biotechnology and Bioengeering* 77: 489–494.

30 Van Bogaert, I., Fleurackers, S., Van Kerrebroeck, S. et al. (2011). Production of new-to-nature sophorolipids by cultivating the yeast *Candida bombicola* on unconventional hydrophobic substrates. *Biotechnology and Bioengineering* 108: 734–741.

31 Davila, A.-M., Marchal, R., and Vandecasteele, J.-P. (1994). Sophorose lipid production from lipidic precursors: predictive evaluation of industrial substrates. *Journal of Industrial Microbiology* 13: 249–257.

32 Felse, P.A., Shah, V., Chan, J. et al. (2007). Sophorolipid biosynthesis by *Candida bombicola* from industrial fatty acid residues. *Enzyme and Microbial Technology* 40: 316–323.

33 Ashby, R.D., Nunez, A., Solaiman, D.K.Y. et al. (2005). Sophorolipid biosynthesis from a biodiesel co-product stream. *Journal of the American Oil Chemists' Society* 82: 625–630.

34 Kim, H.-S., Kim, Y.-B., and Lee, B.-S. (2005). Sophorolipid production by *Candida bombicola* ATCC 22214 from a corn-oil processing byproduct. *Journal of Microbiology and Biotechnology* 15: 55–58.

35 Fleurackers, S.J.J. (2006). On the use of waste frying oil in the synthesis of sophorolipids. *European Journal of Lipid Science and Technology* 108: 5–12.

36 Shah, V., Jurjevic, M., and Badia, D. (2008). Utilization of restaurant waste oil as a precursor for sophorolipid production. *Biotechnology Progress* 23: 512–515.

37 Daniel, H.-J., Otto, R.T., Reuss, M., and Syldatk, C. (1998). Sophorolipid production with high yields on whey concentrate and rapeseed oil without consumption of lactose. *Biotechnology Letters* 20: 805–807.

38 Daniel, H.-J., Reuss, M., and Syldatk, C. (1998). Production of sophorolipids in high concentration from deproteinized whey and rapeseed oil in a two stage fed batch process using *Candida bombicola* ATCC 22214 and *Cryptococcus curvatus* ATCC 20509. *Biotechnology Letters* 20: 1153–1156.

39 Zhang, Y., Jia, D., Sun, W. et al. (2018). Semicontinuous sophorolipid fermentation using a novel bioreactor with dual ventilation pipes and dual sieve-plates coupled with a novel separation system. *Microbial Biotechnology* 11: 455–464.

40 Roelants, S.L.K.W., Ciesielska, K., De Maeseneire, S.L. et al. (2016). Towards the industrialization of new biosurfactants: biotechnological opportunities for the lactone esterase gene from *Starmerella bombicola*. *Biotechnology and Bioengineering* 113: 550–559.

41 Ashby, R.D., McAloon, A.J., Solaiman, D.K.Y. et al. (2013). A process model for approximating the production costs of the fermentative synthesis of sophorolipids. *Journal of Surfactants and Detergents* 16: 683–691.

42 Ellas, A.M., Longati, A.A., Ellamla, H.R. et al. (2021). Techno-economic-environmental analysis of sophorolipid biosurfactant production from sugarcane bagasse. *Industrial and Engineering Chemistry Research* 60: 9833–9850.

43 Martinez, M., Rodríguez, A., Gea, T. et al. (2022). A simplified techno-economic analysis for sophorolipid production in a solid-state fermentation process. *Energies* 15: 4077.

44 Renkin, M. (2003). Environmental profile of sophorolipid and rhamnolipid biosurfactants. *La Rivista Italiana delle Sostanze Grasse* 58: 249–252.

45 Hillion, G., Marchal, R., Stoltz, C., et al. (1995). Use of a sophorolipid to provide free radical formation inhibiting activity or elastase inhibiting activity. *US patent* 5,756,471, filed 12 June 1995 and issued 26 may 1998.

46 Lydon, H.L., Baccile, N., Callaghan, B. et al. (2017). Adjuvant antibiotic activity of acidic sophorolipids with potential for facilitating wound healing. *Antimicrobial Agents and Chemotherapy* 61: e02547-16.

47 Elshikh, M., Moya-Ramírez, I., Moens, H. et al. (2017). Rhamnolipids and lactonic sophorolipids: natural antimicrobial surfactants for oral hygiene. *Journal of Applied Microbiology* 123: 1111–1123.

48 Dhar, P., Havskjold, H., Thornhill, M. et al. (2021). Toward green flotation: interaction of a sophorolipid biosurfactant with a copper sulfide. *Journal of Colloid and Interface Science* 585: 386–399.

49 Helle, S.S., Duff, S.J.B., and Cooper, D.G. (1993). Effect of surfactants on cellulose hydrolysis. *Biotechnology and Bioengineering* 42: 611–617.

50 Menon, V., Prakash, G., Prabhune, A., and Rao, M. (2010). Biocatalytic approach for the utilization of hemicellulose for ethanol production from agricultural residue using thermostable xylanase and thermotolerant yeast. *Bioresource Technology* 101: 5366–5373.

51 Gaur, V.K., Regar, R.K., Dhiman, N. et al. (2019). Biosynthesis and characterization of sophorolipid biosurfactant by Candida spp: application as food emulsifier and antibacterial agent. *Bioresource Technology* 285: 121314.

52 Cho, W.Y., Ng, J.F., Yap, W.H., and Goh, B.H. (2022). Sophorolipids—bio-based antimicrobial formulating agents for applications in food and health. *Molecules* 27: 5556.

53 Ozdener, M.H., Ashby, R.D., Jyotaki, M. et al. (2019). Sophorolipid biosurfactants activate taste receptor type 1 member 3-mediated taste responses and block responses to bitter taste *in vitro* and *in vivo*. *Journal of Surfactants and Detergents* 22: 441–449.

54 Chuo, S.C., Ahmad, A., Mohd-Setapar, S.H. et al. (2019). Utilization of green sophorolipids biosurfactant in reverse micelle extraction of antibiotics: kinetic and mass transfer studies. *Journal of Molecular Liquids* 276: 225–232.

55 Vaughn, S.F., Behle, R.W., Skory, C.D. et al. (2014). Utilization of sophorolipids as biosurfactants for postemergence herbicides. *Crop Protection* 59: 29–34.

56 Sieverding, E. (2017). Increasing yields by the use of sophorolipids. US Patent 0094968 A1, filed 16 April 2015 and issued 4 April 2017.

57 Shah, V. and Daverey, A. (2021). Effects of sophorolipids augmentation on the plant growth and phytoremediation of heavy metal contaminated soil. *Journal of Cleaner Production* 280: 124406.

58 Sharma, N., Lavania, M., and Lal, B. (2023). Biosurfactants: an emerging tool for the petroleum industries. *Frontiers in Microbiology* 14: 1254557.

59 Liu, H., Chen, H., Huang, G. et al. (2023). Remediation of oily soil using acidic sophorolipids micro-emulsion. *Chinese Journal of Chemical Engineering* 59: 270–278.

7

Mannosylerythritol Lipids

7.1 Introduction

Mannosylerythritol lipids, sometimes referred to as MELs, are glycolipids produced by yeast, in particular by strains of the genus *Pseudozyma* (reclassified into *Moeszlomyces*) and the genus *Ustilago*. They are composed of the hydrophilic moiety 4-O-β-D-mannopyranosylmeso-erythritol and two acyl chains, typically with a length of 8–12 carbon atoms, as hydrophobic part. Thus, mannosylerythritol lipids can be regarded as twin-tailed nonionic surfactants.

Mannosylerythritol lipids were described already 50 years ago, but it is only during the last 15–20 years that they have become of commercial interest. Other than for the interesting surface-active properties, mannosylerythritol lipids are of particular interest as they are produced in relatively high yield (up to 165 g/L) by nonpathogenic yeast strains, such as *Pseudozyma antarctica*. In this they differ from rhamnolipids, which are synthesized in high yields by the pathogenic strain *Pseudomonas aeruginosa*.

7.2 Chemical Structures and Properties

The structure of mannosylerythritol lipids is shown in Figure 7.1. As can be seen, they have long acyl chains connected to carbon atoms 2 and 3 of the mannose ring. These chains are normally 8–12 carbon atoms long (i.e. $n = 6$–10) but may also be slightly longer. The length depends on the yeast strain used and on the carbon source. The hydroxyl groups in positions 4 and 6 may, or may not, be acetylated and that substitution pattern governs the classification of the biosurfactants.

MEL-A is acetylated at both positions, MEL-B is acetylated only at the 6-position, MEL-C is acetylated only at the 4-position, and MEL-D is nonacetylated [1, 2]. Mannosylerythritol lipids that have been treated with acid can also lose one of the long-chain acyl groups in addition to the acetyl group or groups. Such mannosylerythritol lipids have been given the name MEL-G [3]. Fermentation may give predominantly one type of mannosylerythritol lipid but may also result in a mixture of the different mannosylerythritol congeners. For instance, while *Pseudozyma hubeiensis* produces mainly MEL-C when cultured with soybean oil as carbon source, *Pseudozyma rugulosa* gives a mixture of MEL-A, MEL-B, and MEL-C when fermented with the same carbon source [4].

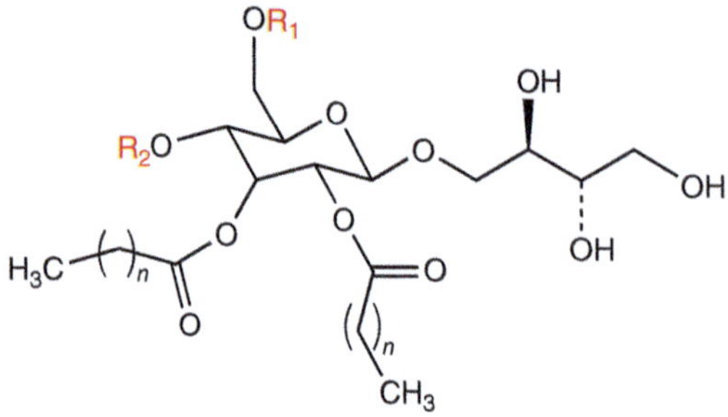

Figure 7.1 General structure of mannosylerythritol lipids. R_1 and R_2 are H or Ac. *n* is typically 6–10.

Most natural mannosylerythritol lipids are hydrophobic amphiphiles with very low solubility in water. Deacetylation, ultimately giving MEL-D, is a way to reduce the hydrophobicity, thus making the products more versatile as surfactants. Deacetylation can be performed enzymatically, using *Candida antarctica* as catalyst, but column chromatography is required in the work-up, which makes this procedure unsuitable for large-scale use [5]. Deacetylation of mannosylerythritol lipid can also be made by acid hydrolysis, as is shown in Figure 7.2. Also, under the conditions used (dilute sulfuric acid in methanol), free fatty acids, which are a byproduct in the fermentation process, will be transformed into methyl esters, as is also shown in Figure 7.2. Care must be taken not to use too high concentration of sulfuric acid because that will lead to hydrolysis not only of the acetyl groups at positions 4 and 6 but also of the ester bonds at positions 2 and 3, which will ultimately result in the very hydrophilic product mannose erythritol, which lacks surface activity. Partial hydrolysis of these esters yields MEL-G, as discussed above. The hydrophobic fatty acid methyl esters (and other hydrophobic impurities) can be removed by extraction with hexane [3].

The surface activity and the critical micelle concentration (CMC) of mannosylerythritol lipids obtained from different sources have been determined and some values are compiled

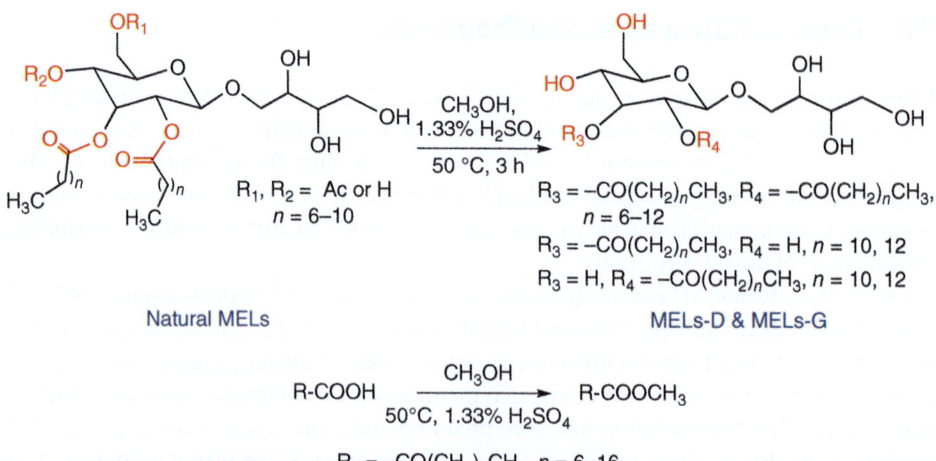

Figure 7.2 Top: Transformation of hydrophobic mannosylerythritol lipid to hydrophilic mannosylerythritol lipid through removal of the acetyl groups in positions 4 and 6; bottom: acid-catalyzed methylation of fatty acids present in the crude mannosylerythritol lipid. (Adapted from Yu et al. [3].)

Table 7.1 Critical micelle concentration (CMC) and surface tension at the CMC of mannosylerythritol lipids (MEL) from different inocula.

MEL type	CMC (M)	Surface tension (mN/m)	Reference
MEL A (from *Candida antarctica*)	2.7×10^{-6}	28	[6]
d-MEL-A[a)] (from *Pseudozyma crassa*)	5.2×10^{-6}	30[c)]	[7]
MEL-B (from *Candida antarctica*)	4.5×10^{-6}	28	[6]
d-MEL-B[b)] (from *Pseudozyma tsukubaensis*)	3.1×10^{-6}	26	[8]
MEL-C (from *Pseudozyma hubeiensis*)	6.0×10^{-6}	25	[9]
MEL-C (from *Pseudozyma graminicola*)	4.0×10^{-6}	24	[10]
MEL-C (from *Pseudozyma siamensis*)	4.5×10^{-6}	31	[11]
Monoacylated MEL (from *Pseudozyma antarctica*)	3.6×10^{-4}	34	[12]

a) Diastereomer of MEL-A having 4-O-β-D-mannopyranosyl-(2R, 3S)-erythritol as sugar moiety.
b) Diastereomer of MEL-B having 4-O-β-D-mannopyranosyl-(2R, 3S)-erythritol as sugar moiety.
c) The reference gives the value 26 mN/m, which is based on a surface tension dip at the CMC. Such a dip is typically related to the presence of an impurity [13], and a more realistic value, judging from the surface tension plot, is 30 mN/m.

(*Source:* Kitamoto et al. [14] / Glycolipid biosurfactants, mannosylerythritol lipids: Distinctive interfacial properties and applications in cosmetics and personal care products / Japan Oil Chemists' Society / CC BY 4.0.)

in Table 7.1. The values should be treated with caution, however, because the purity of the biosurfactant is usually unknown. It is well known that small amounts of hydrophobic impurities in a surfactant may drastically reduce the CMC, as well as the surface tension at the CMC.

Mannosylerythritol lipids self-assemble in aqueous solution into well-defined liquid crystalline structures [14–16]. Figure 7.3 shows the behavior of MEL-A and MEL-B. As can be seen, MEL-A gives a sponge phase in equilibrium with water up to a concentration of almost 60% mannosylerythritol lipid. This sponge phase is unusually temperature stable and remains intact to above 60 °C. Such a broad existence range with respect to both concentration and temperature is unusual. The sponge phases experienced with traditional nonionic surfactants having a polyoxyethylene chain as polar headgroup are much smaller with respect to both parameters [17]. It has been proposed that the high temperature stability of MEL-A's sponge phase is due to a stabilizing effect of the hydrogen bonding network of the sugar moieties [14].

As can also be seen from Figure 7.3, MEL-B shows a different pattern. It gives a lamellar liquid crystalline phase in equilibrium with water up to a concentration of 60% mannosylerythritol lipid; above that concentration, there is only the lamellar phase. Small-angle X-ray scattering has given a value for the lattice constant of 3.58 nm. The lamellar phase is also very temperature stable.

Both the sponge phase (sometimes referred to as L_3 phase) and the lamellar liquid crystalline phase are bicontinuous with surfactant bilayer films connected over macroscopic distances. The surfactant bilayers form zero or low mean curvature surfaces separating different water channels. Both phases require relatively hydrophobic amphiphiles. For instance, with the well-known nonionic surfactants of $C_{12}E_n$ type (dodecanol with *n*

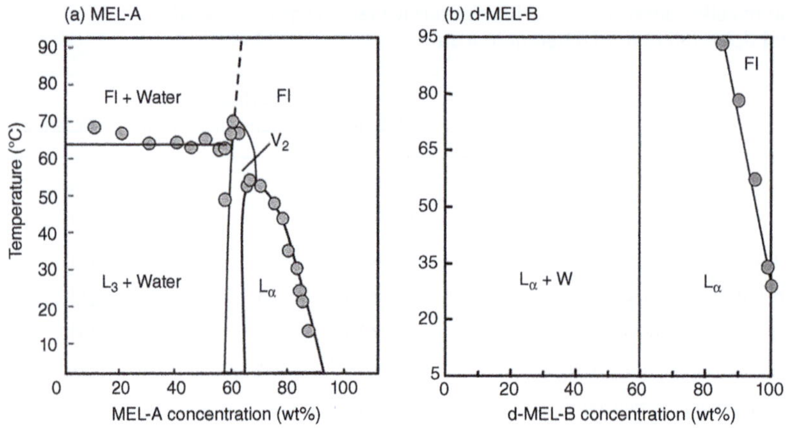

Figure 7.3 Binary phase diagrams of MEL-A and MEL-B. L_3 is sponge phase, V_2 is reverse bicontinuous cubic phase, L_α is lamellar liquid crystalline phase, and W is water. Fl stands for fluid isotropic phase. (From Reprinted with permission from Creative Commons with Kitamoto et al. [14] / Glycolipid biosurfactants, mannosylerythritol lipids: Distinctive interfacial properties and applications in cosmetics and personal care products / Japan Oil Chemists' Society / CC BY 4.0.)

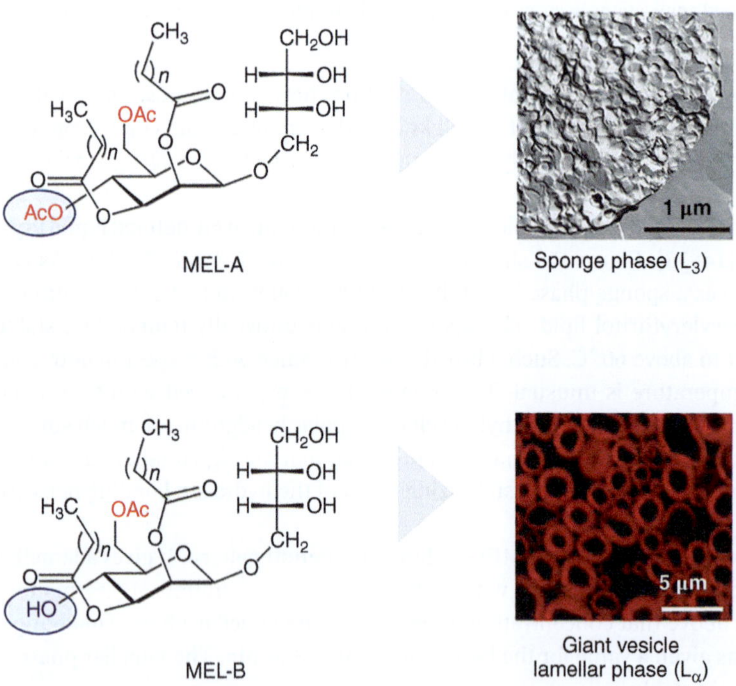

Figure 7.4 MEL-A self-assembles into a sponge, or L_3 phase (top) while MEL-B gives a lamellar phase, which on agitation may form giant vesicles (bottom). (From Reprinted with permission from Creative Commons with Kitamoto et al. [14] / Glycolipid biosurfactants, mannosylerythritol lipids: Distinctive interfacial properties and applications in cosmetics and personal care products / Japan Oil Chemists' Society / CC BY 4.0.)

oxyethylene groups attached), $C_{12}E_3$ gives a relatively large sponge phase in equilibrium with a lamellar phase, $C_{12}E_4$ gives a small sponge phase, $C_{12}E_5$ gives a very small sponge phase, and $C_{12}E_6$ and beyond do not give sponge phases [18].

An important distinction between the sponge phase and the lamellar liquid crystalline phase is that only the latter can form vesicles on agitation. The only difference between MEL-A and MEL-B is that the former contains two acetyl groups while the latter contains only one. It is interesting that such a rather small difference in the structure of these relatively large amphiphiles gives such a distinct effect on the self-assembly structures formed. This is illustrated in Figure 7.4.

The difference between MEL-A and MEL-B is also evident from the phase behavior in the presence of a hydrocarbon. This is important because many formulations within the cleaning and cosmetic sectors are emulsions or microemulsions. As can be seen from the phase diagrams in Figure 7.5, MEL-A gives a rather large water-in-oil microemulsion phase at low water concentrations, which is typical for very hydrophobic amphiphiles. At low MEL-A concentrations, there is a so-called Winsor III system, comprising a middle phase microemulsion, which is bicontinuous, in equilibrium with excess oil and water. This is the ideal phase behavior for chemical flooding, a procedure for enhanced oil recovery (EOR) [20], but probably not of any practical relevance because glycolipids are unlikely to survive the harsh conditions of EOR (stability for years at high temperature and often high salinity).

MEL-B, being a more balanced surfactant, gives a lamellar liquid crystalline phase at low water concentrations and a Winsor III system at low MEL-B concentrations. It is interesting that at approximately equal amounts of *n*-decane and water and around 20% MEL-B, a one-phase microemulsion is obtained. At higher MEL-B content, there is a phase transition from the bicontinuous microemulsion to a lamellar phase, which is a common behavior for balanced surfactants.

7.3 Production

Mannosylerythritol lipids are produced by fermentation, usually using yeast strains from the genus *Pseudozyma* or the genus *Ustilago* and with simple sugars, such as glucose or sucrose, or a triglyceride oil, such as soybean oil, castor oil, rapeseed oil, coconut oil, or olive oil, as carbon source. Mannosylerythritol lipids can also be obtained from fungi. The fact that they are produced from nonpathogenic organisms is important from a practical point of view because this may mean that requirements on the work-up and purification for products intended for the personal care and cosmetic sector may not be very severe. To reduce costs, inexpensive raw materials, such as agricultural wastes, distillery wastes, waste cooking oil, and animal fats, have also been explored as feedstock [21]. In fact, a cheap feedstock seems to be essential for mannosylerythritol lipids to be a commercially viable component in, for instance, cleaning formulations. The feedstock represents approximately 30% of the product cost for glycolipid biosurfactants [22]. Table 7.2 summarizes the microorganisms used and the yields obtained for different types of mannosylerythritol lipids.

The crude mannosylerythritol lipid is water insoluble and can be separated from the broth by centrifugation. After removal of the supernatant, the precipitate is usually dissolved

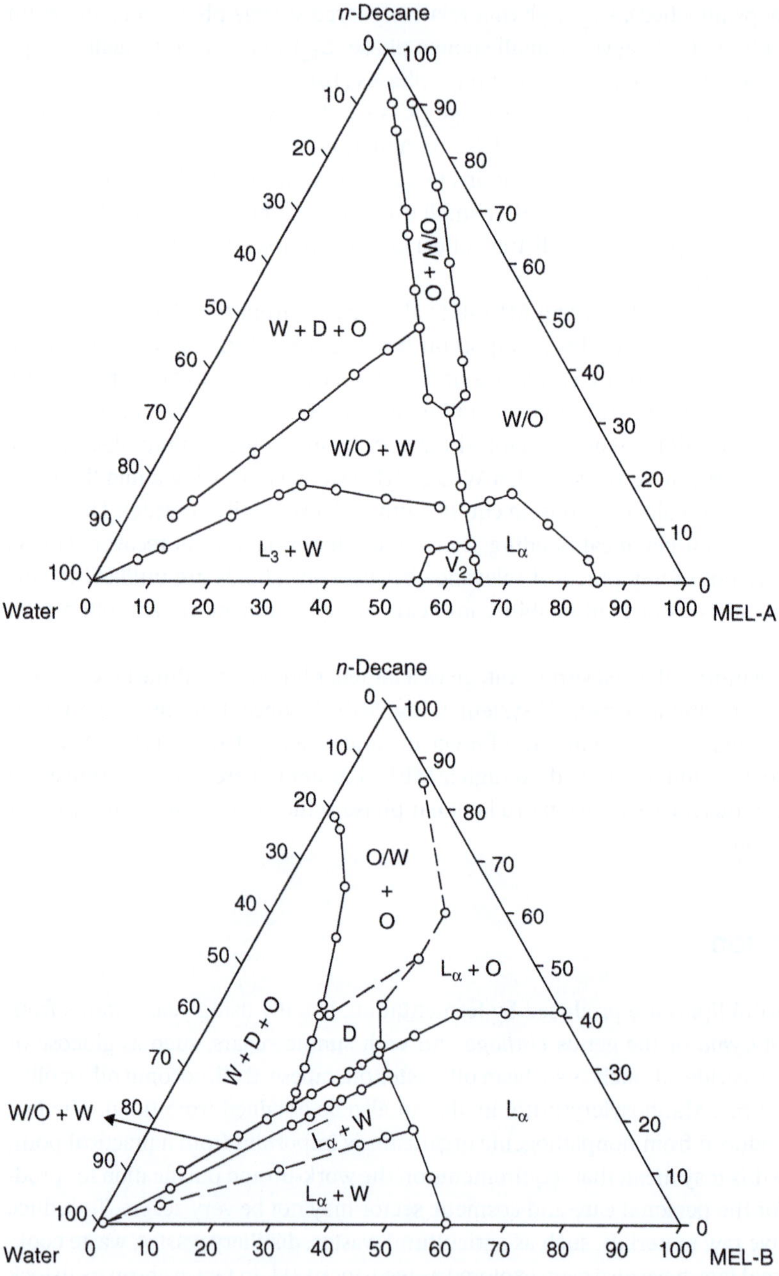

Figure 7.5 Phase diagram of the MEL-A/water/*n*-decane system (top) and the MEL-B/water/*n*-decane system (bottom), both at 25 °C. L_3 = sponge phase; V_2 = reverse bicontinuous cubic phase; L_α = lamellar phase; O/W = O/W-type microemulsion; W/O = W/O-type microemulsion; D = bicontinuous microemulsion; W = excess water; O = excess oil. (From Worakitkanchanakul et al. [19] / Phase behavior of ternary mannosylerythritol lipid/water/oil systems / with permission of Elsevier.)

Table 7.2 Yield of mannosylerythritol lipids produced from different microorganisms.

Microorganism	Main product	Yield (g/L)
Pseudozyma antarctica	MEL-A	140
Pseudozyma aphidis	MEL-A	165
Pseudozyma rugulosa	MEL-A	142
Pseudozyma parantarctica	MEL-A	107
Pseudozyma crassa	d-MEL-A[a)]	4.6
Pseudozyma tsukubaensis	d-MEL-B[b)]	73
Pseudozyma graminicola	MEL-C	9.6
Pseudozyma hubeiensis	MEL-C	129
Pseudozyma shanxiensis	MEL-C	2.7
Pseudozyma siamensis	MEL-C	19

a) Diasteromer of MEL-A bearing 4-O-β-D-mannopyranosyl-(2R,3S)-erythritol as the sugar moiety.
b) Diasteromer of MEL-B bearing 4-O-β-D-mannopyranosyl-(2R,3S)-erythritol as the sugar moiety.
(*Source:* Reprinted with permission from Creative Commons with Kitamoto et al. [14] / Glycolipid biosurfactants, mannosylerythritol lipids: Distinctive interfacial properties and applications in cosmetics and personal care products / Japan Oil Chemists' Society / CC BY 4.0.)

in ethanol and the ethanol solution is subsequently centrifuged to remove undissolved components. Further work-up is made depending on purity requirements.

Tailor-made mannosylerythritol lipids can be made by various gene alteration routes. With this type of engineering, specific side chains can be introduced in the structure [23].

The references above describe the production of mannosylerythritol lipid under laboratory conditions for small volumes. It is possible that some of these processes have been scaled up and used to produce the quantities required for the reported commercial sales. As discussed in the section "Producers," for the time being, these volumes are likely quite small.

Schonhoff et al. have extended laboratory and pilot plant data to industrial scale production of rhamnolipids and mannosylerythritol lipids from different substrates from the sugar industry (molasses and sugar beet pulp). Life cycle analysis, production cost, and market analysis were addressed. The results show that the use of molasses to produce mannosylerythritol lipids provides the lowest environmental burden and production cost. The market conditions considered and the production costs demonstrate the possibility of achieving competitive market positions [24].

The recent progress in bioprocess technology and advanced separation techniques have been encouraging and indicates that further developments are possible. However, for mannosylerythritol lipids to reach high-volume markets, a substantial improvement in the availability of larger volumes at accessible prices is required. This can probably only be achieved by using new strains, most likely recombinant strains.

7.4 Producers

Cayman Chemicals (Ann Arbor, MI, USA) is a biotechnology company founded in 1980. It supplies chemicals used principally by universities and pharmaceutical companies for the research and development of drugs. Mannosylerythritol lipid is on their product palette in 25, 50, and 100 mg doses.

Toyobo Co., Ltd. (Osaka, Japan) has 70 years of experience in fermentation. Their mannosylerythritol lipid is sold under the trade name Ceramela™ and was developed in cooperation with the National Institute of Advanced Industrial Science and Technology of Japan.

Biotopia Co., Ltd. (Cuncheon-si, Gangwon-do, Korea) is a biochemical company supplying mannosylerythritol lipid liquid.

Nippon Fine Chemicals Co., Ltd. (Osaka, Japan) is a manufacturer and seller of fine chemicals, cosmetics, and industrial chemical products. It provides a wide variety of cosmetic ingredients and personal care products for skin care, makeup, and hair care. They offer a premix of mannosylerythritol lipid and lecithin allowing a simple preparation of liposomes.

As of 2024, mannosylerythritol lipids do not appear in the product catalogues of specialty chemicals suppliers, such as Sigma-Aldrich, or of an important distributor, such as Carl Roth. Compared to other biosurfactants, such as lecithin, sophorolipids, and rhamnolipids, mannosylerythritol lipids are still at an early stage of development and exploitation. Many of the reports about the use of mannosylerythritol lipids as moisturizer in personal care products, as antimicrobial agent, and as wetting and spreading agent in crop protection formulations, seem to relate more to potential performance than documented commercial use. However, the interest in mannosylerythritol lipids is growing and it is likely that within a not-too-distant future they will have an established position on the market.

7.5 Existing and Potential Applications

As discussed in the section "Chemical structures and properties," mannosylerythritol lipids possess interesting and potentially useful physicochemical properties. From a performance point of view, they can be seen as viable alternatives to synthetic nonionic surfactants and the applications identified so far are outlined below. However, the cost and the availability of the individual grades and of the blends have so far limited the commercial use to biochemical research and to the use in a few top-of-the-range cosmetic products.

A breakthrough in production technology and a better knowledge of toxicology and environmental impact should open up a larger market in personal care (taking advantage of the self-assembling properties), pharmaceuticals (self-assembly and antimicrobial), crop protection (wetting), and many other industrial applications where superior wetting is essential, e.g. synthetic fibers spin finishing, printing inks, textiles dyeing, and many more.

7.5.1 Cosmetics

Mannosylerythritol lipids exhibit excellent moisturizing properties and are used as an active ingredient in cosmetics and personal care products. Their moisturizing activity has been compared to that of ceramide 3, an essential component of the stratum corneum. Ceramide 3 has two long chains and a polar headgroup composed of three hydroxyl groups and an amide bond and like MEL-A and MEL-B, ceramide 3 forms a lamellar liquid crystalline phase in water [25]. MELs have been found to be very efficient in maintaining the water barrier in the epidermis, thus keeping the skin hydrated [23].

It has been reported that mannosylerythritol lipids exhibit restorative properties for damaged hair. The lifted cuticle of damaged hair was significantly restored and repaired after treatment with MEL-A, an effect similar to that obtained with natural ceramide. Treatment with mannosylerythritol lipid was found to give smoother and stronger hair. Mannosylerythritol lipid also accelerated hair growth by improving the viability of fibroblasts and papilla cells of the epidermal layer of the hair [14].

MEL-B is sold commercially by the Japanese company **Toyobo Co., Ltd**. (Osaka, Japan) as an active ingredient (pseudo-ceramide) under the trade name Ceramela and is distributed around the world [14]. The product is advertised in the promotional leaflet as: "a natural moisturizing agent made from fermented olive oil which accomplishes penetration moisturizing and strengthens barrier function."

The Korean company **DKBIO** (Daesan, Korea) is reported to produce mannosylerythritol lipid for use in their cosmetic products (personal communication). Mineral cultural media have been claimed to be used by both Toyobo and DKBIO [26].

A premix product consisting of MEL-B, hydrogenated lecithin, and a polyhydric alcohol is commercially available from **Nippon Fine Chemicals Co., Ltd.** (Osaka, Japan) under the trade name Phytopresome™ mannosylerythritol lipid. A dispersion of multilamellar vesicles of the size range 50–100 nm is produced from this mixture by simply mixing it with water. This is more advantageous than the formation of vesicles from hydrogenated lecithin only, which requires stringent conditions, using sonication or high-pressure emulsification.

It is important to note that the safety of mannosylerythritol lipids has been officially certified to meet the requirements of the Japan Chemical Substances Control Law (CSCL). This fact, combined with increased product availability, is clearing the way for a wider use of mannosylerythritol lipids in a variety of skin care and makeup products, but also in hair and body shampoos, conditioners, and creams.

7.5.2 Disinfection

Mannosylerythritol lipids exhibit high antimicrobial activity against Gram-positive bacteria and the activity depends on the length of the acyl chains. It has been demonstrated that for both MEL-A and MEL-B, a chain length of 8–10 carbon atoms is optimal; shorter and longer chain length mannosylerythritol lipids are less potent [27, 28]. The activity against most Gram-positive bacteria is strong; the values of minimum inhibitory concentration (MIC) are smaller than those of biobased surfactants of glycolipid type, such as sucrose monocaprate and sorbitan monolaurate, as well as most rhamnolipids. Of particular interest for food processing is that mannosylerythritol lipids have been found to be effective antibacterial agents for foodborne pathogens, such as *Listeria monocytogenes* [29]. A particularly good effect was obtained by combining treatment with MEL-A with high hydrostatic pressure [30]. Mannosylerythritol lipids are not very active on Gram-negative bacteria, however. In this they resemble rhamnolipids, which are also active against Gram-positive but not

Gram-negative bacteria. Sophorolipids, on the other hand, exhibit antimicrobial activity against both Gram-positive and Gram-negative bacteria.

Mannosylerythritol lipids also function as microbial antiadhesives, i.e. agents that prevent microbial colonization. The combined antimicrobial and antiadhesive effect makes mannosylerythritol lipids suitable for use in the food processing industry to control the growth of harmful microorganisms [23].

7.5.3 Crop Protection

Mannosylerythritol lipids have been suggested as wetting agents for agricultural applications. Poorly wettable plant leaves need amphiphiles to reduce the contact angle of the formulated pesticide and to improve the adhesion. The wetting agents are usually synthetic nonionic surfactants, and they are generally used by mixing them with pesticides and/or insecticides. Natural amphiphiles such as glycolipids have an environmental advantage.

The wetting ability of mannosylerythritol lipids on hydrophobic plastic surfaces, as well as on plant leaf surfaces, has been evaluated based on contact angle measurements. It was found that contact angles of 22° or lower were reached, which was below what was obtained with conventional nonionics, such as fatty alcohol ethoxylates or ethoxylated sorbitan esters. Table 7.3 shows results from such comparative wetting tests.

Of particular interest was that solutions of MEL-A also spread on leaves of *Gramineae* plants (wheat and rice), surfaces which are known to be particularly difficult to wet because of a combination of hydrophobicity and irregularity [31]. The results indicate that mannosylerythritol lipids can be used as wetting/spreading agent in agrochemical formulations.

Mannosylerythritol lipids are not only efficient wetting and spreading agents on leaves but they are also effective in suppressing powdery mildew in wheat plants, presumably due to inhibition of conidial germination. The effect is dependent on the fungal species. For some species, MEL-A is most effective, whereas for other species, MEL-B or MEL-C is more effective [32].

Table 7.3 Contact angle of surfactant solutions on a hydrophobic plastic film and on plant leaves. Measurements were made 100 s after application of the drop.

Surfactant	Plastic film (GelBond)	Wheat	Rice	Mulberry	Strawberry
Only water (control)	78.5	ND	ND	76.9	94.2
MEL-A	8.4	14.1	17.1	10.3	18.8
MEL-B	10.2	17.2	13.8	8.4	22.0
MEL-C	13.1	10.5	17.5	6.5	29.4
Brij 35	42.2	113.1	121.1	43.0	49.3
Tween 20	43.1	113.2	125.8	51.6	55.4

ND: Not determined since the measurement failed.
Brij 35 is ethoxylated (23 EO) lauryl alcohol.
Tween 20 is ethoxylated (20 EO) sorbitan monolaurate.
(From Fukuoka et al. [31] / Application of yeast glycolipid biosurfactant, mannosylerythritol lipid, as agrospreaders / reproduced with permission of Japan Oil Chemists' Society.)

References

1 Arutchelvi, J.I., Bhaduri, S., Uppara, P.V., and Doble, M. (2008). Mannosylerythritol lipids: a review. *Journal of Industrial Microbiology & Biotechnology* 35: 1559–1570.

2 Yu, M., Kiu, Z., Zeng, G. et al. (2015). Characteristics of mannosylerythritol lipids and their environmental potential. *Carbohydrate Research* 407: 63–72.

3 Yu, F., Yang, Q., Cui, T. et al. (2024). Preparation and isolation of hydrophilic mannosylerythritol lipids via chemical modification and stepwise extraction. *Journal of Surfactants and Detergents* 27: 725–736.

4 Konishi, M., Morita, T., Fukuoka, T. et al. (2007). Production of different types of mannosylerythritol lipids as biosurfactants by the newly isolated yeast strains belonging to the genus Pseudozyma. *Applied Microbiology and Biotechnology* 75: 521–531.

5 Fukuoka, T., Yanagihara, T., Imura, T. et al. (2011). Enzymatic synthesis of a novel glycolipid biosurfactant, mannosylerythritol lipid-D and its aqueous phase behavior. *Carbohydrate Research* 346: 266–271.

6 Kitamoto, D., Yanagishita, H., Shinbo, T. et al. (1993). Surface active properties and antimicrobial activities of mannosylerythritol lipids as biosurfactants produced by *Candida antarctica*. *Journal of Biotechnology* 29: 91–96.

7 Fukuoka, T., Kawamura, M., Morita, T. et al. (2008). A basidiomycetous yeast, *Pseudozyma crassa*, produces novel diastereomers of conventional mannosylerythritol lipids as glycolipid biosurfactants. *Carbohydrate Research* 343: 2947–2955.

8 Fukuoka, T., Yanagihara, T., Imura, T. et al. (2012). The diastereomer of mannosylerythritol lipids have different interfacial properties and aqueous phase behavior, reflecting the erythritol configuration. *Carbohydrate Research* 351: 81–86.

9 Konishi, M., Morita, T., Fukuoka, T. et al. (2008). Efficient production of mannosylerythritol lipids with high hydrophilicity by *Pseudozyma hubeiensis* KM-59. *Applied Microbiology and Biotechnology* 78: 37–46.

10 Morita, T., Konishi, M., Fukuoka, T. et al. (2008). Identification of *Pseudozyma graminicola* CBS 10092 as a producer of glycolipid biosurfactants, mannosylerythritol lipids. *Journal of Oleo Science* 57: 123–131.

11 Morita, T., Konishi, M., Fukuoka, T. et al. (2008). Production of glycolipid biosurfactants, mannosylerythritol lipids, by *Pseudozyma siamensis* and their interfacial properties. *Journal of Bioscience and Bioengineering* 105: 493–502.

12 Fukuoka, T., Morita, T., Konishi, M. et al. (2007). Structural characterization and surface-active properties of a new glycolipid biosurfactant, mono-acylated mannosylerythritol lipid, produced from glucose by *Pseudozyma antarctica*. *Applied Microbiology and Biotechnology* 76: 801–810.

13 Kronberg, B., Holmberg, K., and Lindman, B. (2014). *Surface Chemistry of Surfactants and Polymers*, 242–243. Chichester, UK: Wiley.

14 Kitamoto, D., Fukuoka, T., Saika, A., and Morita, T. (2021). Glycolipid biosurfactants, mannosylerythritol lipids: distinctive interfacial properties and applications in cosmetics and personal care products. *Journal of Oleo Science* 71: 1–13.

15 Imura, T., Hikosaka, Y., Worakitkanchanakul, W. et al. (2007). Aqueous-phase behavior of natural glycolipid biosurfactant mannosylerythritol lipid A: sponge, cubic, and lamellar phases. *Langmuir* 23: 1659–1663.

16 Davidson, P. (2024). Liquid crystalline behavior of concentrated aqueous solutions of biosurfactants. *Current Opinion in Colloid and Interface Science* 69: 101777.

17 Kronberg, B., Holmberg, K., and Lindman, B. (2014). *Surface Chemistry of Surfactants and Polymers*, 143–145. Chichester, UK: Wiley.

18 Kronberg, B., Holmberg, K., and Lindman, B. (2014). *Surface Chemistry of Surfactants and Polymers*, 142. Chichester, UK: Wiley.

19 Worakitkanchanakul, W., Imura, T., Fukuoka, T. et al. (2009). Phase behavior of ternary mannosylerythritol lipid/water/oil systems. *Colloids and Surfaces B* 68: 207–212.

20 Han, X., Lu, M., Fan, Y. et al. (2021). Recent developments on surfactants for enhanced oil recovery. *Tenside Surfactants Detergents* 58: 164.

21 Banat, I.M., Satpute, S.K., Cameotra, S.S. et al. (2014). Cost effective technologies and renewable substrates for biosurfactants' production. *Frontiers in Microbiology* 697: 1–18.

22 Jiménez-Penalver, P., Rodríguez, A., Daverey, A. et al. (2019). Use of wastes for sophorolipids production as a transition to circular economy: state of the art and perspectives. *Reviews in Environmental Science and Biotechnology* 18: 413–435.

23 Zulkifli, W.N.F.W.M., Hayes, D.G., Maurad, Z.A., and Zan, A.M. (2024). A review on mannosylerythritol lipids and their properties, applications and roles in the circular economy. *Journal of Surfactants and Detergents* 27: 691–702.

24 Schonhoff, A., Stöckigt, G., Wulf, C. et al. (2023). Biosurfactants production with substrates from sugar industry - environmental, cost, market and social aspects. *RSC Sustainability* 1: 1798–1813.

25 Morita, T., Fukuoka, T., Imura, T., and Kitamoto, D. (2015). Mannosylerythritol lipids: production and applications. *Journal of Oleo Science* 64: 133–141.

26 Tran, Q.G., Ryu, A.J., Choi, Y.J. et al. (2022). Enhanced production of biosurfactants through genetic engineering of *Pseudozyma sp.* SY 16. *Korean Journal of Chemical Engineering* 39: 997–1003.

27 Nashida, J., Nishi, N., Takahashi, Y. et al. (2018). Systematic and stereoselective total synthesis of mannosylerythritol lipids and evaluation of their antibacterial activity. *Journal of Organic Chemistry* 83: 7281–7289.

28 Coelho, A.L.S., Feuser, P.E., Carciofi, B.A.M. et al. (2020). Mannosylerythritol lipids: antimicrobial and biomedical properties. *Applied Microbiology and Biotechnology* 104: 2297–2318.

29 Liu, X., Shu, Q., Chen, Q. et al. (2020). Antibacterial efficacy and mechanism of mannosylerythritol lipids-A on *Listeria monocytogenes. Molecules* 25: 4857.

30 Liu, X., Zhang, L., Pang, X. et al. (2022). Synergistic antibacterial effect and mechanism of high hydrostatic pressure and mannosylerythritol Lipid-A on *Listeria monocytogenes. Food Control* 135: 108797.

31 Fukuoka, T., Yoshida, S., Nakamura, J. et al. (2015). Application of yeast glycolipid biosurfactant, mannosylerythritol lipid, as agrospreaders. *Journal of Oleo Science* 64: 689–695.

32 Yoshida, S., Koitabashi, M., Nakamura, J. et al. (2015). Effects of biosurfactants, mannosylerythritol lipids, on the hydrophobicity of solid surfaces and infection behaviors of plant pathogenic fungi. *Journal of Applied Microbiology* 119: 215–224.

8

Trehalose Lipids

8.1 Introduction

Trehalose lipids are a family of biosurfactants based on the disaccharide trehalose, which is acylated with long-chain fatty acids [1]. Trehalose, with the IUPAC name α-D-glucopyranosyl-(1 → 1)-α-D-glucopyranoside, is a nonreducing disaccharide in which the two glucose residues are linked through the anomeric position to one another. Different trehalose-containing glycolipids are known to be produced by several microorganisms belonging to the mycolates group, such as *Arthrobacter*, *Nocardia*, *Rhodococcus*, and *Gordonia* [2]. The chemical structure of a trehalose lipid was first elucidated by Noll et al. [3].

The most common trehalose lipid is trehalose 6,6′-dimycolate, which is trehalose esterified in the 6-position of each glucose unit by mycolic acid. Mycolic acids are composed of a long β-hydroxy chain with a shorter α-alkyl side chain. Each molecule contains between 60 and 90 carbon atoms. The exact number of carbons varies between species and can be used as an identification aid. Most mycolic acids also contain functional groups: tri- and tetraesters of trehalose with one or two succinoyl moieties, respectively, are found in nature and are referred to as succinoyl trehalose lipids [4]. They are sometimes treated as a separate class of glycolipids but are included in this chapter.

One major challenge for the industrial use of trehalose lipids is the optimization of their production and downstream processing. This is especially true in the case of the *Rhodococcal* glycolipids, which are often bound to cellular envelopes and are usually produced along with other surface-active lipids.

8.2 Chemical Structures and Properties

Trehalose lipids can have one or two long acyl chains connected to the trehalose moiety by ester bonds to the 6-OH of the sugar ring(s). Figure 8.1 shows representative structures of trehalose mono- and diesters.

As mentioned above, trehalose lipids with succinoyl moieties attached to the sugar rings together with long-chain acyl substituents are also common and typically produced by *Rhodococcus* species. Figure 8.2 shows the structure of a typical succinoyl trehalose

Natural Surfactants: Science, Technology, and Industrial Aspects, First Edition. Guido Bognolo and Krister Holmberg.
 Published 2026 by John Wiley & Sons Ltd.

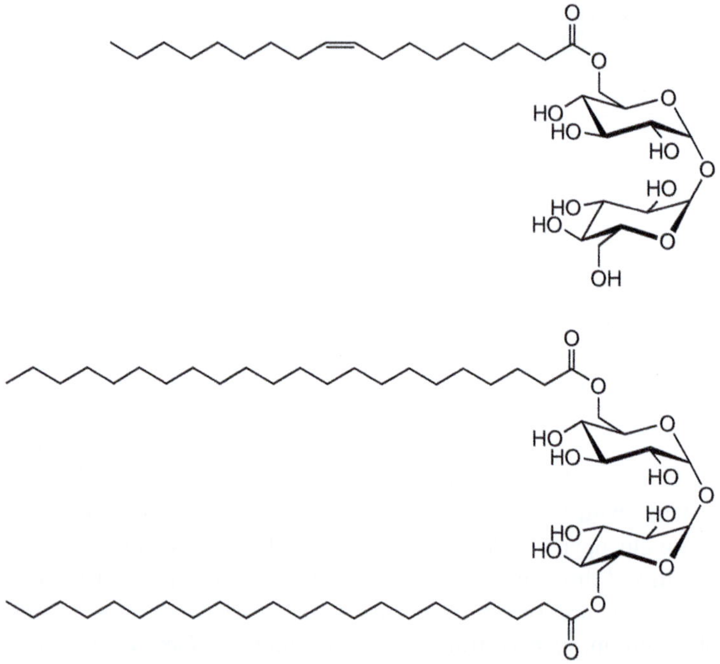

Figure 8.1 Top: trehalose 6-oleate (6-octadec-9-enoyl-α,α′-trehalose); bottom: trehalose 6,6′-dibehenate (6,6′-didocasanoyl-α,α′-trehalose).

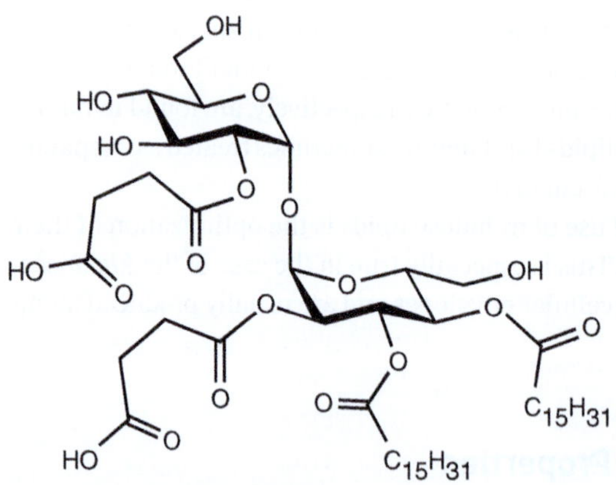

Figure 8.2 A succinoyl trehalose lipid: 3,4-di-O-palmitoyl-2,2′-di-O-succinoyl-α,α-trehalose. (Adapted from Jana et al. [5].)

lipid, trehalose substituted with two monoesters of succinic acid and two esters of hexadecanoic acid.

Trehalose monoesters of the type shown in Figure 8.1, top, have been synthesized and characterized. Table 8.1 shows CMC values for trehalose lipids of this type with the length of a saturated acyl chain ranging from C8 to C16. Values for alkyl maltosides with the same

Table 8.1 CMC values for trehalose lipids of the general structure shown in Figure 8.1, top, and with 8, 10, 12, 14, and 16 carbons in a saturated acyl chain. For comparison, CMC values for the corresponding alkyl maltosides, a more established type of disaccharide surfactant, are also given. (The values for the trehalose lipids are from Dojindo Laboratories Ltd., Kumamoto, Japan; those for the alkyl maltosides are from Anatrace Products LLC, Maumee, OH, USA.)

Number of carbons in hydrophobic tail	CMC of trehalose lipids (mM)	CMC of alkyl maltosides (mM)
8	5.6	19.5
10	3.0	1.8
12	0.15	0.17
14	0.012	0.01
16	0.0061	0.0006

number of carbon atoms in the hydrophobic tail are also given. As expected, the CMC values depend strongly on the length of the chain. An interesting observation that can be made from the table is that the chain length dependence differs between the two sugar-based surfactants. For the shortest derivatives, the C8 amphiphiles, the trehalose lipid has a CMC more than three times lower than that of the alkyl maltoside, for the C12 derivatives, the CMC values are about the same, and for the C16 amphiphiles, the trehalose lipid has a CMC 10 times higher than the alkyl maltoside. The reason for this difference in chain length dependence is not clear.

Marqués et al. studied the chemical composition and physicochemical properties of trehalose lipids from *Rhodococcus erythropolis* 51T7 grown on tetradecane as carbon source [6]. The surfactant produced after 72 hours of culture was identified as a mixture of at least six components, all tetraesters with hydrophobic moieties in the C9–C11 range. All the components contained the succinoyl residue; thus, the product was a succinoyl trehalose lipid. The CMC was pH dependent. As the pH increased from 4 to 7.4, the trehalose lipid became more anionic, resulting in a moderate increase in the CMC value (from 34 mg/L to 50 mg/L) as a result of increasing electrostatic repulsion between the negatively charged carboxylate groups on the trehalose molecules. The surface tension at the CMC was remarkably low, 27.9 mN/m, comparable to or even lower than what is obtained with the best petrochemical surfactants. The glycolipid gave a water-kerosene interfacial tension of 5 mN/m, which is not extremely low. However, it is important to keep in mind that the surface and interfacial tension measurements were made with the mixture of six or more glycolipids and maybe other lipids as well. This means that the interfaces will be covered by a mixture of species, and one cannot relate the values to any specific amphiphilic molecule. Conventional anionic surfactants can also give low values of surface and interfacial tension if they contain small amounts of amphiphilic impurities, such as fatty alcohols.

In succinoyl trehalose lipids, one of the carboxyl groups of the succinic acid is free. Thus, such glycolipids are weak acids with a pKa value of around 6. One would then expect the glycolipid to be sensitive to the ionic strength at higher pH, but that seems not to be the case. The CMC value, measured at pH 7.4 by the Wilhelmy plate method, was the same over the NaCl range 0–300 mM. This is surprising and indicates that the trehalose lipid behaves like a nonionic surfactant, also slightly above the pKa of the acid. The behavior can only be explained by a very high degree of counterion binding. It is interesting that the same behavior was found for another class of glycolipids, rhamnolipids, see paper by Chen et al. [7] and discussion in the section "Chemical structure and properties" in Chapter 5.

An interesting and potentially useful observation made by Marqués et al. was that the trehalose lipid gave rise to extremely stable emulsions both with paraffin oil and with isopropyl myristate as oil. The isopropyl myristate emulsion with the relative amounts of oil:water:glycolipid 11.3:81.8:7.5 was studied in some detail. The emulsion formed was water-in-oil despite the low ratio of oil to water (11.3:81.8). It was thick and remained intact for three months at 25 °C. The high degree of stability was probably due to the trehalose lipid forming lamellar liquid crystals in the lamellas separating the water drops, which were approximately 5 μm in size. Polarized light microscopy was used to study the liquid crystal phases. An L_a phase was observed at room temperature while at 70 °C a hexagonal phase was detected. According to a review from 2024, the paper by Marqués et al. is the only published study on liquid crystalline phases formed by trehalose lipids and since liquid crystals are known to be important for the stability of emulsions, the review states that "this research topic is certainly worth revisiting not only by texture observations but also by XRD". XRD stands for X-ray diffraction [8].

The observation that trehalose lipids can give low values of surface tension and of interfacial tension between a hydrocarbon and water had been observed before by Lang and Philp [9]. Table 8.2 shows values of surface and interfacial tension for whole broths as well as isolated trehalose lipids. This class of glycolipids obviously has a surface activity that well matches that of the conventional surfactants that dominate the market today.

Surface tension values below 30 mN/m have also been reported by dos Reis et al. for trehalose lipids produced by *Fusarium fujikuroi* UFSM-BAS-01 [10].

Vogt Singer et al. isolated and determined the chemical and physical properties of a biosurfactant synthesized by hexadecane-grown *Rhodococcus* species H13-A. The biosurfactant was an anionic glycolipid consisting of one major and ten minor components. The hydrophilic portion of the main component was trehalose, thus the biosurfactant was a trehalose lipid. The major species was identified as 2,3,4,6,2′,3′,4′,6′-octaacyltrehalose. The glycolipid exhibited a critical micelle concentration of 1.5 g/L and a minimum interfacial tension of 2×10^{-2} mN/m against decane, which is a very low value. Addition of pentanol, a common cosurfactant in such systems, reduced the interfacial tension value further, to 6×10^{-5} mN/m. Such an ultralow interfacial tension value indicates the formation of a Winsor III system with a middle-phase microemulsion. This is also in line with the phase behavior of the glycolipid [11].

Table 8.2 Surface tension, interfacial tension, and CMC values for trehalose lipids from *Rhodococcus* strains.

Strain	Product	Surface tension (mN/m)	Interfacial tension (mN/m)	CMC (mg/L)
R. aurantiacus 80001	Whole broth	28	0.35 (a)	n.d.
R. erithropolis DSM 43215	Trehalose dicorynomycolates	36	17 (a)	4
R. erithropolis DSM 43215	Trehalose monocorynomycolates	32	14 (a)	4
R. erithropolis DSM 43215	Trehalose 2,2′,3,4, tetraester	26	<1 (a)	15
R. erithropolis SD-74	Succinoyl trehalose lipid (Na salt)	26	8 (b)	200
R. rubropertinctus 60 003	Whole broth	31	2.0 (a)	n.d.
R. terrae 70006	Whole broth	30	3.0 (a)	n.d.
R. strain 51T7	Cell-free supernatant	30	n.d.	n.d.
R. strain H13-A	Octaacyl trehalose	n.d.	0.02 (c)	1.5

Remarks: (a) *n*-hexadecane/water, (b) kerosene/water, (c) *n*-decane/water.
(*Source:* Lang et al. [9] / Surface-active lipids in Rhodococci / reproduced with permission of Springer Nature.)

Table 8.3 Hydrodynamic diameter of a trehalose lipid produced by *Rhodococcus fascians* at different concentrations above the CMC.

Concentration (g/L)	Hydrodynamic diameter (nm)	Polydispersity index
0.15	60.8 ± 1.4	0.137 ± 0.04
0.20	71.7 ± 1.7	0.129 ± 0.11
0.30	81.8 ± 0.9	0.181 ± 0.09
0.40	150 ± 1.4	0.361 ± 0.12
0.50	157 ± 2.1	0.266 ± 0.08

(*Source:* Janek et al. [12] / Trehalose lipid biosurfactant reduces adhesion of microbial pathogens to polystyrene and silicone surfaces: An experimental and computational approach / Frontiers.)

Janek et al. isolated and purified a trehalose lipid produced by *Rhodococcus fascians* BD8 using hexadecane as carbon source. The purified glycolipid gave a surface tension of 34 mN/m and the CMC was 0.14 g/L. Table 8.3 shows the variation of the hydrodynamic diameter of the biosurfactant aggregates with the concentration as determined by dynamic light scattering at concentrations above the CMC. As can be seen, the aggregate size increased from 60 to 157 nm as the trehalose lipid concentration increased from 0.15 to 0.5 g/L. The values indicate that the micelles are large and nonspherical. The authors speculate that there may be a micelle to vesicle transition at high biosurfactant concentration but no evidence is put forward for that [12].

8.3 Production

In recent years, several strains producing extracellular trehalose lipids fed on both insoluble and soluble substrates have been identified. The cell-bound trehalose lipids seem to be produced only on hydrocarbon substrates while the extracellular glycolipids are also produced on soluble feedstuff. Table 8.4 gives examples of *Rhodococcus* strains producing trehalose lipids, the substrates used, and whether the glycolipid is cell-bound or extracellular.

Commercial success and efficiency of the entire biotechnological process are key points for microbial surfactants, as they are for most biotechnological products. This is especially true for trehalose lipids which are often cell-bound, thus increasing downstream costs and reducing production yield. The strategies that have been adopted involve the use of cheap substrates, including wastes, optimization of the fermentation conditions and of the recovery process, and the development of overproducing strains. However, comparatively little

Table 8.4 The main glycolipid-producing *Rhodococcus* species, the main trehalose lipids produced along with the carbon substrate used and their cellular deposition.

Strain	Product	Carbon substrate	Cellular deposition
R. erythropolis	Trehalose dicorynomycolates	*n*-Alkanes	Cell bound
R. erythropolis DSM 43215	Trehalose monocorynomycolates and trehalose dicorynomycolates	C14-C15 *n*-Alkanes or kerosene	Extracellular (70%)
R. H13-A	Glycolipids	*n*-Alkanes and fatty alcohol	Cell bound and extracellular
R. erythropolis S1	Glucose monomycolate, trehalose monomycolate, and trehalose dimycolate	*n*-Alkanes	Cell bound
R. ruber	Trehalose dicorynomycolates	Hydrocarbons	Cell bound
R. erythropolis EK-1	Trehalose monocorynomycolates and trehalose dicorynomycolates	Ethanol	Extracellular
R. erythropolis	Trehalose-2,2′,3,4-tetraester	*n*-Alkanes	Cell bound
R. erythropolis MS11	Trehalose esterified with succinic acid and decanoic acid	*n*-Hexadecane	Extracellular
R. wratislaviensis BN38	Trehalose tetraester	*n*-Hexadecane	Cell bound
R. erythropolis	Succinoyl trehalose lipids	*n*-Hexadecane	Extracellular
R. erythropolis ATCC 4277	Biosurfactant	Glycerol	Cell bound
R. erythropolis 3C-9	Trehalose lipids	*n*-Hexadecane	Cell bound
R. SD-74	Succinoyl trehalose lipids	*n*-Hexadecane	Extracellular

(*Source:* Franzetti et al. [1] / Production and applications of trehalose lipid biosurfactants / reproduced with permission of John Wiley & Sons.)

work has been made with the microbial strain *Rhodococcus*, as well as with the genera *Gordonia* and *Acinetobacter*.

Concerning the optimization of the fermentation conditions, *Rhodococcus erythropolis* SD-74 abundantly produces succinoyl trehalose lipids from n-hexadecane, the two main components of the exolipid being 2,3,4,2′-di-*O*-succinoyl-di-*O*-alkanoyl-(C16: 87%, C14: 13%)-trehalose and 2,3,4-mono-*O*-succinoyl-di-*O*-alkanoyl-(C16: 75%, C14: 25%)-trehalose. After incubation for 10 days, the exolipid concentration reached about 40 g/L, corresponding to about 50% yield on weight basis of the n-hexadecane consumed. The production was favored by potassium phosphate buffer over sodium phosphate buffer, and by iron(II) sulphate heptahydrate, yeast extracts, and potassium nitrate as nitrogen source. The addition of fructose, glucose, glycerol, citrate, or succinate did not improve the productivity [13].

To reduce the cost of production, the use of low-cost substrates is essential. *Rhodococcus erythropolis* 16 LM.USTHB was used to convert residual sunflower frying oil, a cheap renewable substrate, into extracellular glycolipids with good surface activity, documented by its ability to lower the surface tension of the crude broth down to 31.9 mN/m [14]. In a screening study of biosurfactant producers on renewable low-cost substrates, Ruggeri et al. isolated *Rhodococcus* sp. BS32, which was able to produce extracellular biosurfactants when grown on a variety of carbon sources, such as molasses, glucose, hexadecane, and rapeseed oil [15].

A recombinant *Gordonia amarae* was developed by insertion, stable maintenance, and expression of the *Vitreoscilla* hemoglobin gene (vgb), resulting in enhanced production of trehalose lipids in the engineered strain [16]. This seems to be the only study reported with a recombinant strain for production of trehalose lipids.

8.4 Skin Toxicity

The potential skin irritation of trehalose lipids produced by *Rhodococcus erythropolis* 51T7 was assayed with mouse fibroblast and human keratinocyte. In the keratinocyte test, the trehalose lipids had an IC50 of 90 μg/mL (NRU test) and 70 μg/mL (MTT test). In the fibroblast test, the IC50 concentration was 91.1 μg/mL (NRU test) and 117.4 μg/mL (MTT test). The values were higher than those obtained with the anionic surfactant sodium dodecyl sulfate (SDS), indicating that the trehalose lipid is less irritating than this commercial surfactant and could be a candidate for galenic and cosmetic formulations [6].

8.5 Existing Applications

Trehalose esters of saturated straight-chain fatty acids ranging from C8 to C16 with the structure shown in Figure 8.1, top have been commercialized by **Dojindo Laboratories** Ltd. (Kumamoto, Japan) for use as stabilizer and protective substance, for example as detergent in the field of proteomic research. The CMC values decrease from 5.6 mM for the C8 glycolipid to 0.0061 mM for the C16 product.

Mycobacterium flavescens strain EX-91 was used for the development of a product that was used for decontamination of an oil-polluted water body. It also proved effective for the treatment of engine oil-contaminated wastewater from a nuclear power station [1, 17].

8.6 Potential Applications

The structure of short-chain trehalose lipids presents analogies with the alkyl polyglucosides (APGs) produced via chemical synthesis of C8–C14 natural alcohols and glucose and which have reached a respectable market position. There are two main differences between the two classes:

- the glucose content (2 moieties in the trehalose lipid, against typically 1.3–1.5 moieties in the APG)
- an ester bond connecting the polar and the nonpolar part in the trehalose lipid versus a glycosidic (acetal) bond in the APG.

These differences are manifested in much higher CMC values for a trehalose lipid than for an APG with the same length of the hydrocarbon tail. However, by adjusting the chain lengths of the hydrophobic tails, amphiphiles with the same CMC can be obtained.

A practically important issue is that APGs higher than C14 are difficult to produce, and what is commercially available as C16–C18 is in reality a mixture of the intended APG and unreacted fatty alcohol. The trehalose lipids with a C14 or a C16 tail, having CMC values shown in Table 8.4, may represent interesting alternatives to long-chain APGs and function as emulsifiers not only in the personal care sector, but across the totality of the household and industrial applications, provided they can be produced in commercial volumes and at reasonable costs.

The use of trehalose lipids has been reported in the environmental field, as additives that can enhance the solubility of hydrophobic compounds and stimulate the biodegradation of hydrocarbons in contaminated soils [1]. In one study, the addition of *Rhodococcus ruber* strain IEGM AC219 and surfactant complexes from various *Rhodococcus* strains to remediate crude oil–contaminated agricultural soil enhanced hydrocarbon degradation over a 3-month period. Hydrocarbon degradation was further enhanced when biophile systems were set with increased ventilation and nutrient addition [18]. Trehalose lipids have also been proposed for enhanced oil recovery and for cleaning of oil storage tanks.

A trehalose tetraester has been suggested for the pretreatment of synthetic fibers both to remove lipophilic components and for enhanced dispersion of dyes, ensuring better and more uniform fiber penetration [19].

As mentioned above, trehalose lipids are less skin-irritating than many of the synthetic surfactants used in cosmetic products today [6]. This means that there is a potential for these glycolipids in the cosmetic sector.

Trehalose-6,6′-dimycolate, or cord factor (see Figure 8.1), has been extensively studied from a medical point of view due to the fact that it plays a central role in pathogenesis during infection. Despite promising pharmaceutical applications, the use of *Mycobacterial* cord factor is limited by the relatively high toxicity of the molecule and the potential pathogenicity of the producing strain. Cord factor produced by *Rhodococcus* sp. 4306 was demonstrated to exhibit lower toxicity *in vitro*, as well as *in vivo*, and the difference in toxicity is believed to be due to a difference in length of the hydrophobic chains [1, 20].

References

1 Franzetti, A., Gandolfi, I., Bestetti, G. et al. (2010). Production and applications of trehalose lipid biosurfactants. *European Journal of Lipid Science and Technology* 112: 617–627.

2 Henkel, M. and Hausmann, R. (2019). Diversity and classification of microbial surfactants. In: *Biobased Surfactants*, 2nd Ed (ed. D.G. Hayes, D.K.Y. Solaiman, and R.D. Ashby), 41–63. Champaign, IL: AOCS Press.

3 Noll, H., Bloch, H., Asselineau, J., and Lederer, E. (1956). The chemical structure of the cord factor of *Mycobacterium tuberculosis*. *Biochimica et Biophysica Acta* 20: 299–309.

4 Tokumoto, Y., Nomura, N., Uchiyama, H. et al. (2009). Structural characterization and surface-active properties of a succinoyl trehalose lipid produced by *Rhodococcus* sp. SD-74. *Journal of Oleo Science* 58: 97–102.

5 Jana, S., Mondal, S., and Kulkarni, S.S. (2017). Chemical synthesis of biosurfactant succinoyl trehalose lipids. *Organic Letters* 19: 1784–1787.

6 Marqués, A.M., Pinazo, A., Farfan, M. et al. (2009). The physicochemical properties and chemical composition of trehalose lipids produced by *Rhodococcus erythropolis* 51T7. *Chemistry and Physics of Lipids* 158: 110–117.

7 Chen, M.L., Penfold, J., Thomas, R.K. et al. (2010). Solution self-assembly and adsorption at the air–water interface of the monorhamnose and dirhamnose rhamnolipids and their mixtures. *Langmuir* 26: 18281–18292.

8 Davidson, P. (2024). Liquid crystalline behavior of concentrated aqueous solutions of biosurfactants. *Current Opinion in Colloid and Interface Science* 69: 101777.

9 Lang, S. and Philp, J.C. (1998). Surface-active lipids in *Rhodococci*. *Antonie van Leeuwenhoek* 74: 59–70.

10 dos Reis, C.B.L., Morandini, L.M.B., Bevilacqua, C.B. et al. (2018). First report of the production of a potent biosurfactant with α, β-trehalose by *Fusarium fujikuroi* under optimized conditions of submerged fermentation. *Brazilian Journal of Microbiology* 49: 185–192.

11 Vogt Singer, M.E., Finnerty, W.R., and Tunelid, A. (1990). Physical and chemical properties of a biosurfactant synthesized by *Rhodococcus* species H13-A. *Canadian Journal of Microbiology* 36: 746–750.

12 Janek, T., Krasowska, A., Czyżnikowska, Z. et al. (2018). Trehalose lipid biosurfactant reduces adhesion of microbial pathogens to polystyrene and silicone surfaces: an experimental and computational approach. *Frontiers in Microbiology* 9: 2441.

13 Uchida, Y., Misawa, S., Nakahara, T., and Tabuchi, T. (1989). Factors affecting the production of succinoyl trehalose lipids by *Rhodococcus erythropolis* SD-74 grown on n-alkanes. *Agricultural and Biological Chemistry* 53: 765–769.

14 Sadouk, Z., Hacene, H., and Tazerouti, A. (2008). Biosurfactants production from low cost substrate and degradation of diesel oil by a Rhodococcus strain. *Oil & Gas Science and Technology* 63: 747–753.

15 Ruggeri, C., Franzetti, A., Bestetti, G. et al. (2009). Isolation and characterization of surface active compound - producing bacteria from hydrocarbon-contaminated environments. *International Biodeterioration & Biodegradation* 63: 936–942.

16 Dogan, I., Pagilla, K.R., Webster, D.A., and Stark, B.C. (2006). Expression of *Vitreoscilla* hemoglobin in *Gordonia amarae* enhances biosurfactant production. *Journal of Industrial Microbiology and Biotechnology* 33: 693–700.

17 Ermolenko, Z.M., Kholodenko, V.P., Chugunov, V.A. et al. (1997). A mycobacterial strain isolated from the oil of the Ukhtinskoe oil field: identification and degradative properties. *Microbiology* 66: 542–545.

18 Christofi, N. and Ivshina, I.B. (2002). Microbial surfactants and their use in field studies of soil remediation. *Journal of Applied Microbiology* 93: 915–929.

19 Fracchia, L., Ceresa, C., Franzetti, A. et al. (2015). Industrial applications of biosurfactants. In: *Biosurfactants: Production and Utilization - Processes, Technologies, and Economics* (ed. N. Kosaric and F.V. Sukan), 245–267. Boca Raton, FL: CRC Press.

20 Ueda, S., Fujiwara, N., Naka, T. et al. (2001). Structure-activity relationship of mycoloyl glycolipids derived from *Rhodococcus* sp. 4306. *Microbial Pathogenesis* 30: 91–99.

9

Cellobiose Lipids

9.1 Introduction

Cellobiose lipids are a group of microbial biosurfactants secreted as secondary metabolites by many microorganisms from the family Ustilaginaceae, with *Ustilago maydis* being the most investigated producer of the biosurfactant. They were first discovered in 1950 by Haskins while screening a wide range of fungi for their ability to metabolize glucose and agricultural wastes [1].

Cellobiose lipids are based on the disaccharide cellobiose. Cellobiose is a reducing sugar, composed of two β-glucose units forming a β(1 → 4) bond. Its IUPAC name is 4-*O*-β-D-glucopyranosyl-β-D-glucopyranose.

9.2 Chemical Structures and Properties

Cellobiose lipids are usually produced as a mixture of different acylated low molecular weight D-glucolipids, linked to a hydroxypalmitic acid via an ω-hydroxyl group [2]. Depending on the producing microorganism, typical strain-associated structural variations can be observed.

When grown on vegetable oils and their derivatives, the smut fungus *Ustilago maydis* produces several different glycolipids under nitrogen-limiting conditions with mannosylerythritol lipids being the major product and cellobiose lipids only a smaller component. The yield of the more polar cellobiose lipids becomes the dominant product when glucose is used as carbon source. The cellobiose lipid obtained from *Ustilago maydis* DSM 4500 after cultivation on glucose had cellobiose glycosidically linked to the ω-hydroxyl group of either 2,15,16-trihydroxyhexadecanoic acid or 15,16-dihydroxyhexadecanoic acid [3]. Ustilagic acid is the name of a specific cellobiose lipid, produced by *Ustilago maydis*. Its structure is shown in Figure 9.1. As can be seen, there is an acetyl group in 6′-position and a longer acyl group in 2″-position [4, 5]. An additional variant where the carboxyl group has been transformed into a methyl ester is known as cellobiose lipid-C [3].

Several variants exist in the family of cellobiose lipids. The fungus *Pseudozyma flocculosa* produces flocculosin, a cellobiose lipid with the cellobiose moiety esterified in 2″-position with 2-hydroxyoctanoic acid and acetylated at positions 6′ and 3″. The structure is shown in Figure 9.2 [6].

Natural Surfactants: Science, Technology, and Industrial Aspects, First Edition. Guido Bognolo and Krister Holmberg.

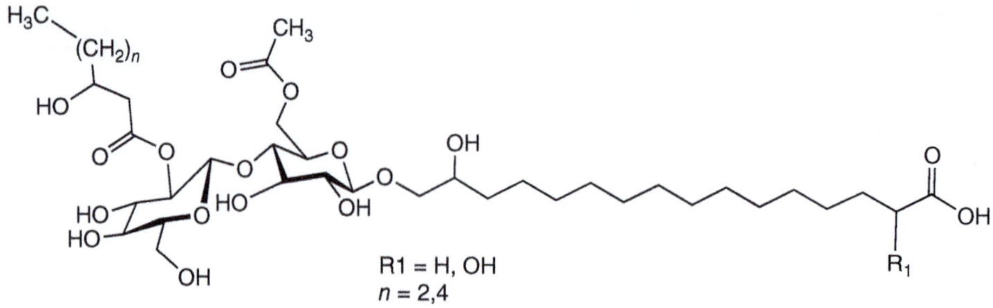

Figure 9.1 Structures of typical cellobiose lipids. In ustilagic acid, R_1 is OH and *n* is 2. (From Spoeckner et al. [3] / Glycolipids of the smut fungus Ustilago maydis from cultivation on renewable resources / reproduced with permission of Springer Nature.)

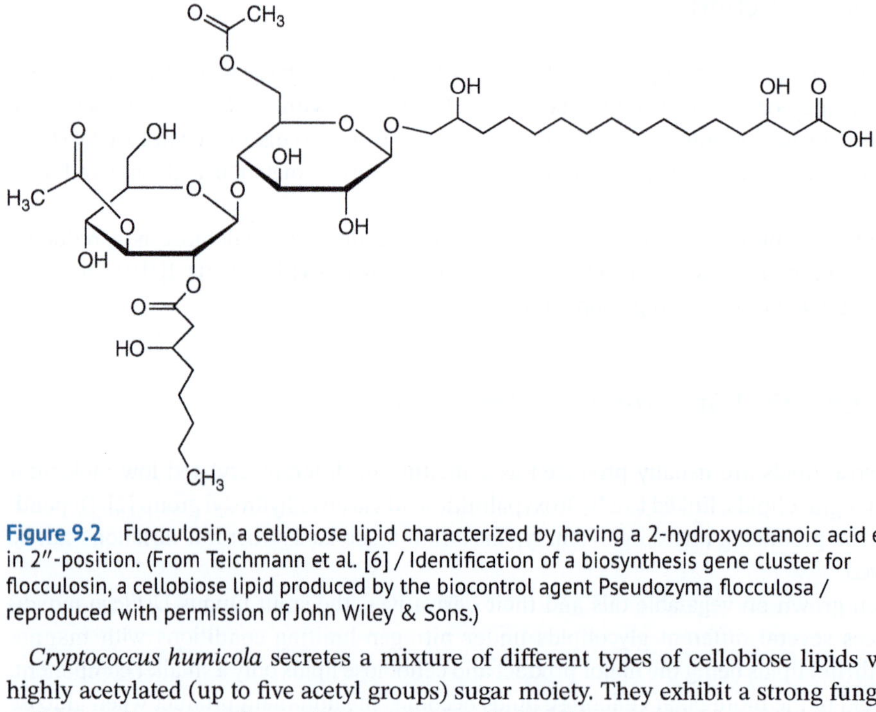

Figure 9.2 Flocculosin, a cellobiose lipid characterized by having a 2-hydroxyoctanoic acid ester in 2″-position. (From Teichmann et al. [6] / Identification of a biosynthesis gene cluster for flocculosin, a cellobiose lipid produced by the biocontrol agent Pseudozyma flocculosa / reproduced with permission of John Wiley & Sons.)

Cryptococcus humicola secretes a mixture of different types of cellobiose lipids with a highly acetylated (up to five acetyl groups) sugar moiety. They exhibit a strong fungicidal activity and some surface activity, reducing the surface tension of water to 37 mN/m. In the product mixture, there are not only cellobiose lipids with the normal hydroxyhexadecanoic (palmitic) chain, but there are also those with a hydroxyoctadecanoic (stearic) chain [7, 8]. In the cellobiose lipid produced by *Pseudozyma fusiformata,* a 2,15,16-trihydroxyhexadecanoic acid is glycosidically linked to the cellobiose and 3-hydroxycaproic acid and acetic acid are linked as O-acyl substituents, corresponding to the structure of cellobiose lipid-B produced by *Ustilago maydis* [9].

From a structural point of view, cellobiose lipids resemble di-rhamnolipids, as well as the acidic (open) form of sophorolipids, in that the molecule consists of a long hydrophobic chain with a carboxyl group at one end and a disaccharide at the other end. However, cellobiose lipids typically have longer acyl substituents on the glycone moiety than the two other biosurfactants.

Cellobiose lipids are relatively hydrophilic amphiphiles compared to most other glycolipids, but the values for the critical micelle concentration (CMC) and the surface tension at the CMC vary depending on the producing species [10]. For example, a cellobiose lipid derived from *Cryptococcus humicola* has a CMC of 4.1×10^{-4} M at pH 7. The value is pH dependent, however, because the carboxylate group will be increasingly more protonated as the pH is decreased. At pH 4, the CMC is 3.3×10^{-5}. The surface tension values at the CMC are 42 mN/m at pH 7 and 43 mN/m at pH 4, which are not very low values [11]. A cellobiose lipid with a completely deprotonated carboxyl group, such as the sodium salt of cellobiose lipid from *Cryptococcus humicola*, exhibits higher surface activity, giving a value of surface tension of 35 mN/m [12].

The sodium salt of cellobiose lipids forms soft gels at a concentration of around 5 g/L. The sol–gel temperature is approximately 50 °C and studies by freeze-fracture transmission electron microscopy showed that the gels consisted of an entangled fibrous network [12].

9.3 Production

Oraby et al. studied *Sporisorium scitamineum* as cellobiose lipid producer and analyzed factors potentially important for cellobiose lipid production [13]. It was hypothesized that since *Sporisorium scitamineum* is closely related to *Ustilago maydis*, the factors affecting the glycolipid production should be similar. The factors examined were the effect of pH value (2.5, 4.0, 6.7), the carbon source (glucose, fructose, sucrose), and the C/N ratio (i.e. mole carbon/mole nitrogen with urea as nitrogen source). The microbial growth and the cellobiose lipid production were monitored.

A pH of 2.5 resulted in the highest product titers. An increase of urea concentration from 0.6 to 1.2 g/L had a positive effect on biomass formation, and the glycolipid formation was optimal at a C/N ratio of 84, using 0.6 g/L of urea. Among the carbon sources examined, sucrose gave a higher yield of cellobiose lipids than glucose. Comparing different media compositions, vitamins were identified as not being necessary for cellobiose lipid synthesis.

In a subsequent study from the same research group, process modeling was performed as a means to develop more efficient processes to produce cellobiose lipids. Simulation of different process scenarios showed that the choice of the aeration unit was important, and the study also included optimization of the fermentation conditions and the downstream recovery processes. The economic consequences for industrial-scale production were also analyzed [13].

9.4 Potential Applications

There seems not to be any existing use of cellobiose lipids today. However, the increasing interest in glycolipids in general may trigger applications also of this class of biosurfactants in the future. Cellobiose lipids are reported to have antibacterial and antifungal properties. As mentioned above, they are also gel-forming with a minimum gelation concentration of 5 g/L and the gels are soft. Aqueous gels with a fungicidal activity may be of interest as soft materials in the food and cosmetics areas [12, 14, 15].

References

1 Haskins, R.H. (1950). Biochemistry of the ustilaginales: 1. Preliminary cultural studies of *Ustilago zeae*. *Canadian Journal of Research* 28c: 213–223.

2 Eveleigh, D.E., Dateo, G.P., and Reese, E.T. (1964). Fungal metabolism of complex glycosides: Ustilagic acid. *Journal of Biological Chemistry* 239: 839–844.

3 Spoeckner, S., Wray, V., Nimtz, M., and Lang, S. (1999). Glycolipids of the smut fungus *Ustilago maydis* from cultivation on renewable resources. *Applied Microbiology and Biotechnology* 51: 33–39.

4 Teichmann, B., Linne, U., Hewald, S. et al. (2007). A biosynthetic gene cluster for a secreted cellobiose lipid with antifungal activity from *Ustilago maydis*. *Molecular Microbiology* 66: 525–533.

5 Yang, X.-L., Awakawa, T., Wakimoto, T., and Abe, I. (2013). Induced production of the novel glycolipid ustilagic acid C in the plant pathogen *Ustilago maydis*. *Tetrahedron Letters* 54: 3655–3657.

6 Teichmann, B., Labbé, C., Lefebvre, F. et al. (2011). Identification of a biosynthesis gene cluster for flocculosin, a cellobiose lipid produced by the biocontrol agent *Pseudozyma flocculosa*. *Molecular Microbiology* 79: 1483–1495.

7 Puchkov, E.O., Zähringer, U., Lindner, B. et al. (2002). The mycocidal, membrane-active complex of *Cryptococcus humicola* is a new type of cellobiose lipid with detergent features. *Biochimica et Biophysica Acta - Biomembranes* 1558: 161–170.

8 Morita, T., Fukuoka, T., Imura, T., and Kitamoto, D. (2013). Accumulation of cellobiose lipids under nitrogen-limiting conditions by two ustilaginomycetous yeasts, *Pseudozyma aphidis* and *Pseudozyma hubeiensis*. *FEMS Yeast Research* 13: 44–49.

9 Kulakovskaya, T.V., Shashkov, A.S., Kulakovskaya, E.V. et al. (2005). Ustilagic acid secretion by *Pseudozyma fusiformata* strains. *FEMS Yeast Research* 5: 919–923.

10 Valkenburg, A.D., Ncube, M.Z., Teke, G.M. et al. (2025). Cellobiose lipids: applications, production, and downstream processing. *Trends in Biotechnology*, https://doi.org/10.1016/j.tibtech.2024.09.011 43.

11 Morita, T., Ishibashi, Y., Fukuoka, T. et al. (2011). Production of glycolipid biosurfactants, cellobiose lipids, by *Cryptococcus humicola* JCM 1461 and their interfacial properties. *Bioscience, Biotechnology, and Biochemistry* 75: 1597–1599.

12 Imura, T., Yamamoto, S., Yamashita, C. et al. (2014). Aqueous gel formation from sodium salts of cellobiose lipids. *Journal of Oleo Science* 63: 1005–1010.

13 Oraby, A., Werner, N., Sungur, Z., and Zibek, S. (2020). Factors affecting the synthesis of cellobiose lipids by *Sporisorium scitamineum*. *Frontiers in Bioengineering and Biotechnology* 8: 555647.

14 Teichmann, B. (2009). Das Cellobioselipid Ustilaginsäure aus Ustilago maydis: Biosynthese und Transkriptionelle Regulation. Ph. D. thesis. Universität Marburg. Germany.

15 Trilisenko, L.V., Kulakovskaya, E.V., Kulakovskaya, T.V. et al. (2012). The antifungal effect of cellobiose lipid on the cells of *Saccharomyces cerevisiae* depends on carbon source. *Springerplus* 1: 18.

10

Xylolipids

10.1 Introduction

Xylolipids are a relatively unexploited class of biosurfactants consisting of xylose glycosidically linked to a hydroxy alkanoic acid. Xylose is a monosaccharide of the aldopentose type; thus, it contains five carbon atoms. It contains an aldehyde group, which means that it is classified as a reducing sugar. Xylose was first isolated from wood and the name is derived from "xylon," Greek for wood.

Xylolipids are structurally similar to rhamnolipids but with another sugar moiety as nonionic polar headgroup. They are secreted by the yeast *Pichia caribbica* when grown in xylose-rich media [1]. They can also be obtained from bacteria, for instance *Lactococcus lactis* (formerly *Streptococcus lactis*), which is classified as GRAS (generally regarded as safe) within the European Union and with an established use in the food industry [2]. A crude xylolipid fraction has also been obtained by growing the soil bacterium *Serratia nematodiphila* on a xylose-rich hydrolysate from rice straw. The xylolipid was not isolated in pure form, so no proper physicochemical characterization was made. However, the product was found to have antibacterial properties and be a good emulsifier [3].

A synthesis method to produce single-tailed decyl- and tetradecyl-xylolipids, as well as their rhamnolipid analogues, from bacterially produced 3-hydroxy fatty acids has been described [4].

10.2 Chemical Structures and Properties

Figure 10.1 shows the structures of two synthetic xylolipids, one based on a 3-hydroxy alkanoic acid and the other on two 3-hydroxy alkanoic acids linked together with an ester bond. Xylose is always the hydrophilic moiety of xylolipids but the hydrophobic part, the "tail" in surfactant terminology, can vary but is always based on hydroxy alkanoic acids. The two structures in Figure 10.1 are based on 3-hydroxy alkanoic acids and are made by organic synthesis. Xylolipids based on ω- or (ω - 1)-hydroxy alkanoic acids occur naturally and the acyl chain may be saturated or unsaturated. The xylolipid produced by *Pichia caribbica* is based on 17-hydroxyoleic acid as hydrophobic moiety and with the hydroxyl group linked to xylose via a β-glycosidic bond [1].

Natural Surfactants: Science, Technology, and Industrial Aspects, First Edition. Guido Bognolo and Krister Holmberg.

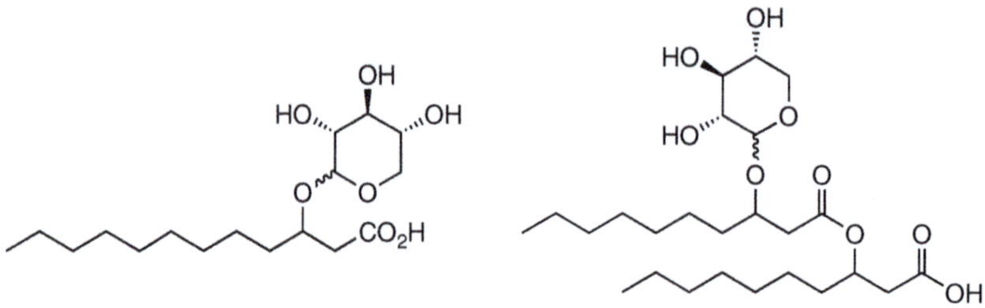

Figure 10.1 Xylolipids based on xylose glycosidically connected to (left) 3-hydroxydodecanoic acid and (right) two 3-hydroxydecanoic acid moieties linked together by an ester bond.

As shown in Figure 10.1, xylolipids contain a carboxyl group that may, or may not, be deprotonated. Thus, at low pH, the xylolipid contains only one polar group, the xylose ring, and can then be regarded as a nonionic surfactant. At higher pH, the carboxyl group becomes deprotonated, and the glycolipid will then contain two polar groups, one nonionic and one anionic. In this respect, xylolipids resemble several other glycolipids, e.g. rhamnolipids.

The chemical stability of the xylolipids is also very similar to that of rhamnolipids. The ester bond between the two hydroxy alkanoic acids is highly sensitive to alkali and moderately sensitive to acid and the acetal linkage (the glycosidic bond) between xylose and the hydroxy alkanoic acid is moderately sensitive to acid. Very low pH is not common in surfactant applications, but moderately high pH is not uncommon, for instance in cleaning. If the ester bond connecting the two hydroxy alkanoic acids in the right structure of Figure 10.1 is cleaved, then two new amphiphilic molecules will be generated, a xylolipid with a hydrophobic tail half the length of the original one and a hydroxy fatty acid salt. This mixture will obviously have physicochemical characteristics different from those of the original xylolipid.

The physicochemical properties of xylolipids seem not to have been much studied. Joshi-Navare et al. report a CMC of 1 mg/L and a surface tension at the CMC of 36 mN/m for a xylolipid based on 17-hydroxyoleic acid as hydrophobic tail [1]. The CMC value is low, which shows that the amphiphile is strongly hydrophobic, which, in turn, indicates that the carboxyl group at the end of the tail is protonated. There is no information on how the CMC value varies with pH. Compton et al. later reported CMC values of 0.15 mM and 0.11 mM for an enantiomer-pure and a racemic C14-xylolipid, respectively, at pH 4. At pH 8, the corresponding CMC values were much higher, 1.63 mM and 0.42 mM, respectively. These CMC values were similar to the values obtained for C14-rhamnolipids produced in a similar way [4].

10.3 Producers

There are no producers of naturally occurring xylolipids. The American start-up company **GlycoSurf** (Salt Lake City, UT, USA) has in its selling range several synthetic xylolipids, including those shown in Figure 10.1, spanning alkanoic acid chain lengths from C8 to C14.

The company also produces rhamnolipids, galactolipids, rhamnosides, and xylosides. Other sugar lipid derivatives are available for custom order.

10.4 Potential Applications

There seems not to be any existing industrial use of xylolipids. However, an antimicrobial effect against both Gram-positive bacteria (methicillin-resistant *Staphylococcus aureus*) and Gram-negative bacteria (*Pseudomonas aeruginosa*, as well as representative clinical pathogens of *Escherichia coli*) has been documented and use as a broad-spectrum antibacterial agent has been suggested [1, 2]. Xylopids may be regarded as an alternative to the more established sophorolipids as antibacterial ingredients in skin care products.

Xylolipids from *Lactococcus lactis* have been shown to have anti-oxidative properties [5]. In combination with their antimicrobial properties, they may have a future as natural preservatives in food and drinks.

References

1 Joshi-Navare, K., Singh, P.K., and Prabhune, A.A. (2014). New yeast isolate *Pichia caribbica* synthesizes xylolipid biosurfactant with enhanced functionality. *European Journal of Lipid Science and Technology* 116: 1070–1079.

2 Saravanakumari, P. and Mani, K. (2010). Structural characterization of a novel xylolipid biosurfactant from *Lactococcus lactis* and analysis of antibacterial activity against multi-drug resistant pathogens. *Bioresource Technology* 101: 8851–8854.

3 Panjiar, N., Mattam, A.J., Jose, S. et al. (2020). Valorization of xylose-rich hydrolysate from rice straw, an agroresidue, through biosurfactant production by the soil bacterium *Serratia nematodiphila*. *Science of the Total Environment* 729: 138933.

4 Compton, A., Deodhar, B., and Pemberton, J.E. (2022). Synthesis and characterization of rhamnolipids and xylolipids using biosynthetic and synthetic β-hydroxy acids. Abstract of presentation given at the American Chemical Society Spring Meeting, San Diego (March 20–24, 2022).

5 Nageshwar, L., Parameshwar, J., Rahman, P.K.S.M. et al. (2022). Anti-oxidative property of xylolipid produced by *Lactococcus lactis* LNH70 and its potential use as fruit juice preservative. *Brazilian Journal of Microbiology* 53: 2157–2172.

11

Lipopeptides

11.1 Introduction

Lipopeptides are a class of biosurfactants that consist of one or more lipid chains attached to a hydrophilic oligopeptide. The molecules are amphiphilic and self-assemble readily. The oligopeptide is usually cyclic, i.e. the amino acids form a loop, but lipoproteins with a linear oligopeptide moiety also exist. The ring is formed by a lactone bridge between a β-hydroxy group of the fatty acid and the carboxy-terminal function of the peptide. Cyclization of the peptide unit enhances the *in vivo* stability and can, thus, be expected to have evolved naturally as the favored form of bacterially expressed lipopeptides. The reduced proteolysis is due to the absence of free *C*- and *N*-termini. The cyclic lipopeptides are the ones that are of interest as surfactants and will be dealt with in this chapter [1, 2].

The most important cyclic lipopeptides, such as surfactin, iturin, and fengycin, are mainly produced by *Bacillus subtilis* [3]. Surfactin and iturin are very similar in structure, while fengycin is slightly different. They all have a peptide ring of 7 or 10 amino acids (including 2–4 D-amino acids) and a fatty acid chain of varying length: C13–C15 for surfactin, C14–C17 for iturin, and C15–C19 for fengycin. Surfactin, which is the most widely studied lipopeptide, is a secondary metabolite first discovered in the culture broth of *Bacillus subtilis* by Arima et al. [4]. Its structure was elucidated soon after [5]. It was given the name "surfactin" because of its pronounced surface activity.

Lipoproteins are complex assemblies of biomolecules such as apolipoproteins, triglycerides, cholesterol, and phospholipids. They orient their constituents so that the assembly becomes surface active, and their main biological function is to transport hydrophobic lipid molecules in aqueous media such as blood plasma or other extracellular fluids. Lipoproteins are biologically very important, but they are outside the scope of this book, which deals with relatively well-defined low-molecular-weight biosurfactants.

So-called antimicrobial peptides are a class of amphiphilic molecules closely related to lipopeptides. They do not have a hydrocarbon tail; instead, they attain their pronounced surface activity through the amino acid composition. Most antimicrobial peptides have a block rich in hydrophobic amino acids, such as tyrosine, and another block that contains cationic amino acids, such as lysine and arginine. Antimicrobial peptides are discussed in a separate chapter, Chapter 12, of the book.

Natural Surfactants: Science, Technology, and Industrial Aspects, First Edition. Guido Bognolo and Krister Holmberg.

11.2 Chemical Structures and Properties

Figure 11.1 shows the structure of the best-known lipopeptide, surfactin. Two out of the seven amino acids that constitute the loop are dicarboxylic acids, L-glutamic acid and L-aspartic acid. This means that the amphiphile becomes increasingly more anionic as the pH is raised. The seven-membered oligopeptide forms a closed structure by condensation with a β-hydroxy fatty acid.

The amino acid sequence in surfactin is typically L-glutamic acid – L-leucine – D-leucine – L-valine – L-aspartic acid – D-leucine – L-leucine, but other sequences also exist. The number of carbon atoms in the β-hydroxyacyl chain can vary between 13 and 15. Thus, surfactin has many congeners or isoforms, as has been determined by reversed-phase HPLC separation followed by MALDI-TOF-MS/MS (matrix-assisted laser desorption ionization time-of-flight tandem mass spectrometry) detection [6].

Surfactin is soluble in water above pH 5, with a maximum solubility at pH 8.0–8.5. It is also soluble in acetone, DMSO, lower alcohols, chloroform, and dichloromethane. It is stable at 100 °C for one hour, at 121 °C in an autoclave for 10 minutes, and in 6% aqueous NaCl at room temperature between pH 5 and 13 [7].

As mentioned above, the physicochemical properties of surfactin (and other cyclic lipopeptides) vary with pH since there are free carboxyl groups in the peptide ring. In plain water, surfactin is a very hydrophobic amphiphile with a reported CMC at room temperature of 7.5 μM [8], but somewhat higher CMC values have been obtained for other members of the surfactin family [9]. Other cyclic lipopeptides have given similar low CMC values [1]. In comparison, the very hydrophobic nonionic surfactant C12E3 has a much higher CMC, 50 μM [10]. A surface tension at the CMC of 27 mN/m has been reported for surfactin. This is a very low value, indicating a strong surface activity.

The aggregation of surfactin in water was studied by Shen et al. using small-angle neutron scattering (SANS) as the main technique. The aggregation number was unusually small, around 20, and the spherical micelles formed had a radius of 25 Å. As a comparison, a typical technical surfactant, such as sodium dodecyl sulphate (SDS), has an aggregation number of ca 60. The hydrocarbon chain and the four hydrophobic leucin residues formed the core of the micelle, while the shell was composed of the three other amino acid residues [11]. The same group also studied the size and structure of surfactin aggregates at the air-water and the hydrophobic solid-water interfaces. Neutron reflectivity measurements on perdeuterated and partly deuterated surfactin showed that the lipopeptide formed spherical aggregates on the surfaces with a thickness of 14 Å and an area per molecule of 147 $Å^2$. The value of area per molecule in the aggregate is unusually large, indicating that the polar headgroups are further apart than normal. Taken together, the SANS and neutron reflectivity experiments, supported by computer simulations, suggest that the hydrocarbon chain folds back into the structure, presumably to associate itself with the hydrophobic leucine and valine residues. According to the authors, surfactin at the surface behaves more like hydrophobic nanoparticles than as aggregates of a conventional surfactant. The small size of the surfactin aggregates in water was later confirmed by Hamley et al. Using cryo-transmission electron microscopy (cryo-TEM) and small-angle X-ray scattering (SAXS), spherical micelles with a radius of 20 Å were reported [12]. Not all cyclic lipopeptides form small spherical aggregates, however. For instance, mycosubtilin, another cyclic lipopeptide produced by *Bacillus subtilis* but with another amino acid sequence in the ring,

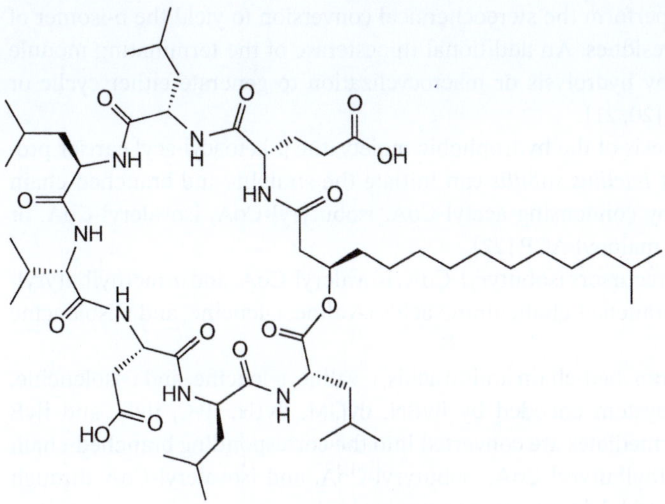

Figure 11.1 Structure of surfactin in protonated form with courtesy of Annija Stepulane.

was found to give extended nanotapes based on stacking of lipopeptide bilayers [12]. The difference in structure is obviously related to differences in surfactant packing parameters.

At high concentrations in water, above 50%, surfactin forms a lamellar liquid crystalline phase consisting of bilayers with interdigitated packing of the acyl chains, as demonstrated by polarized-light spectroscopy and SAXS [13, 14].

Treatment of surfactin with alkali, such as sodium methoxide in methanol, converts the cyclic structure to an open, linear one. The linear form is more hydrophilic than the normal, cyclic form and exhibits a different phase behavior [13]. Values of CMC and surface tension at the CMC have been determined for linear surfactin obtained by hydrolysis of natural, cyclic surfactin [15] and by organic synthesis [16]. The CMC values were 70–80 μM and 370 μM, respectively, i.e. considerably higher than the 7.5 μM reported for the cyclic lipopeptide.

Surfactin has got EU Nr. 607-428-2 and CAS Nr. 24730-31-2 and 252 023-70-4. The complete chemical name associated with the CAS Nr. 252023-70-4 is: 3-((3S,6R,9S,12S,15R,18S,21S,25R)-9-(carboxymethyl)-3,6,15,18-tetraisobutyl-12-isopropyl-25-(10-methylundecyl)-2,5,8,11,14,17,20,23-octaoxo-1-oxa-4,7,10,13,16,19,22-heptaazacyclopentacosan-21-yl) propanoic acid.

11.3 Biosynthesis

Surfactin is synthesized by a large multifunctional nonribosomal peptide synthetase (NRPS) that contains four modules, SrfAA, SrfAB, SrfAC, and SrfAD, and each module is in charge of adding one amino acid [17]. Furthermore, each of the modules contains at least three catalytic domains: an adenylation structural domain is responsible for the selection and activation of the amino acid substrate, a small peptidyl carrier protein (PCP) carries the aminoacyl adenosine substrates as enzyme-bound thioesters, and a condensation domain forms the peptide bonds between acyl-S-PCP intermediates [18, 19].

Epimerization domains perform the stereochemical conversion to yield the D-isomer of some of the incorporated residues. An additional thioesterase of the terminating module catalyzes product release by hydrolysis or macrocyclization to generate either cyclic or cyclic-branched molecules [20, 21].

Concerning the biosynthesis of the hydrophobic moiety, the β-ketoacyl-acyl carrier protein synthase III (FabH) of *Bacillus subtilis* can initiate the straight- and branched-chain fatty acid synthesis cycle by condensing acetyl-CoA, isobutyryl-CoA, isovaleryl-CoA, or α-methylbutyryl-CoA with malonyl-ACP [22].

The fatty acid synthesis precursors isobutyryl-CoA, isovaleryl-CoA, and α-methylbutyryl-CoA are derived from the branched-chain amino acids L-valine, L-leucine, and L-isoleucine [23, 24].

The biosynthesis of the branched-chain amino acids, L-valine, L-leucine, and L-isoleucine, shares the same enzyme system encoded by ilvBN, ilvGM, ilvIH, ilvC, ilvD, and ilvE [25, 26]. The resulting intermediates are converted into the corresponding branched-chain acyl-CoA precursors α-methylbutyryl-CoA, isobutyryl-CoA, and isovaleryl-CoA through the branched-chain α-keto acid dehydrogenase complex [27].

11.4 Production

In a 2023 article, Zhen et al. discussed earlier work on the influence of the fermentation parameters on the productivity under comparable conditions [17]. Comparing pH 6.0 and 8.0, the production was highest (1.90 g/L) at pH 8.0 for *Bacillus subtilis* BS5 [7]; however, *Bacillus amyloliquefaciens* produced 6.04 g/L surfactin at pH 7.0 [28]. For temperatures ranging from 25 to 45 °C and with *Bacillus subtilis* NLIM 0110, a maximum yield of 4.00 g/L was obtained at 37 °C [29]. The addition of a small amount of solid porous carrier, such as activated carbon or expanded clay, into the fermentation broth significantly increased surfactin production with *Bacillus subtilis* ATCC 21332. A culture medium containing 25 g/L of activated carbon gave a surfactin yield of 3.6 g/L, a value 36-fold higher than that obtained from carrier-free liquid culture [30].

Abdel-Mawgoud et al. found that cellular growth and surfactin production from *Bacillus subtilis* BS5 in a mineral salts medium were greatly affected by the environmental fermentation conditions and the medium components (carbon and nitrogen sources and minerals). Optimum conditions for high surfactin production with the shake flask method were found to be a slightly acidic initial pH (6.5–6.8), an incubation temperature of 30 °C, a 90% volumetric aeration percentage, and an inoculum size of 2% *v*/*v*. For media components, it was found that molasses (160 mL/L) and $NaNO_3$ (5 g/L) were the optimum carbon and nitrogen sources, respectively. The optimum trace elements were $ZnSO_4{\cdot}7H_2O$ (0.16 g/L), $FeCl_3{\cdot}6H_2O$ (0.27 g/L), and $MnSO_4{\cdot}H_2O$ (0.017 g/L) [7].

11.5 Production Costs

It has been estimated that lipopeptide biosurfactants synthesized in small amounts for applications in the pharmaceutical industry cost USD 20–130/mg. The specific costs for

Table 11.1 Yield of surfactin obtained with different genetically engineered *Bacillus subtilis* strains.

Bacterial strain	Genetic modification	Yield (g/L)	References
B. subtilis 168S23	Promoter substitution of key genes for precursor material	8.5	[35]
B. subtilis TS1726ΔspolVB	Deleting regulator of sporulation	9.6	[36]
B. subtilis THY-7/Pg3-srfA	Replacement of PsrfA with Pg3	9.7	[37]
B. subtilis 168S35	Overexpression of positive regulators and deletion of negative regulators	12.8	[38]
B. subtilis TS1726	Overexpression of key genes	13.4	[39]
B. subtilis TS1726ΔspolVB	Overexpression of key genes and medium optimization	16.7	[36]

surfactin and iturin were assessed at USD 92/mg and USD 20/mg, respectively [31]. The high production cost of lipopeptides compared with glycolipids is due to low yields and expensive substrates. As raw materials constitute the major share of the production costs, the use of agro-based and industrial wastes and residues can significantly decrease the operational costs [32].

In common with all other biosurfactants, reduction of the production cost for surfactin is an important issue, and the possible use of cheap waste biomass as a substitute for refined carbon and nitrogen sources has been investigated. For example, in a study of surfactin production using low-cost brewery waste as carbon source, *Bacillus subtilis* produced surfactin in an amount of 210 mg/L after 28 hours. Proper antimicrobial activity was demonstrated against all tested bacteria, including complete inactivation of *Pseudomonas aeruginosa*. Although the yield was totally unsatisfactory, the study indicated that low-cost substrates could be an alternative [33]. Attempts have also been made to produce surfactin from other wastes, such as distillers' grains, an important waste by-product from the liquor industry, using *Bacillus amyloliquefaciens* MT45 as the strain. The yield was low but could be somewhat improved when MT45 was co-cultured with other *Bacillus* strains [34].

The ultimate solution to obtain cheap surfactin, available in large quantities, will require strains with high long-term productivity. Such strains may either be natural ones or genetically engineered. Neither of the two options has materialized thus far. Even with targeted molecular engineering, the yields remain low, as shown in Table 11.1 below.

11.6 Toxicology, Environmental Properties

Some toxicological characteristics of the sodium salt of surfactin are given in Table 11.2.

According to Kaneca Corporation, the product Kaneka Surfactin is easily biodegradable by microorganisms (MITI method). It is hazardous to the aquatic environment (acute and chronic hazard) and to the ozone layer.

Table 11.2 Toxicological profile of sodium surfactin.

Acute toxicity (oral)	Not classified
Acute toxicity (dust/mist)	Category 4 (practically nontoxic; A/N)
Skin corrosion/irritation	Not classified
Eye damage/eye irritation	Category 2 (moderately irritant; A/N)
Skin sensitization	Not classified
Germ cell mutagenicity	Not classified

(*Source:* Adapted from Kaneka Corporation, CAS Nr. 302933-83-1, Safety data sheet.)

The ECHA registration dossier for sodium surfactin concludes that "A single dose oral administration of sodium surfactin to rats at the dose level of 2000 mg/kg body weight resulted in transient diarrhea and a suppression of body weight gain in both of males and females. No deaths occurred. The LD_{50} value was considered to be above 2000 mg/kg body weight in males and females." According to the classification provided by companies to ECHA for the REACH registration, surfactin causes serious eye irritation, is harmful if inhaled, and is suspected of damaging fertility of the unborn child.

11.7 Suppliers and Products

Kaneka Corporation (Osaka, Japan) has three surfactin-containing products in the selling range:

- *Kaneka Surfactin* (INCI name: Sodium Surfactin), which is presented as "a natural surfactant with high biodegradability and low skin irritation." It is a highly efficient amphiphile with a critical micelle concentration (CMC) of 0.0003%. The product is able to create a bicontinuous phase that incorporates both oil and water and which is claimed to be capable of making semi-transparent and very stable emulsions, probably so-called nanoemulsions, using very low dosage.
- *Kaneka Premix SQ401* (INCI name: Squalane (and) Glycerin (and) Sodium Surfactin), which is a gel that can act as an oil thickening agent and emulsifier. It has moisturizing properties and is suitable for the formulation of fine particle size emulsions. Typical applications include make-up removers, sun care, baby hair care, and skin care products.
- *Kaneka Premix T301* (INCI name: Caprylic/Capric Triglyceride (and) Glycerin (and) Sodium Surfactin), which is used in applications similar to those of Kaneka Premix SQ401.

Sabo S.p.A. (Levate, Italy) offers Sabosol RF-MB (INCI name: Coco-Glucoside (and) Squalane (and) Sodium Surfactin), a re-fattening agent for the formulation of toiletries.

Evonik Personal Care (Essen, Germany) promotes Symbio® Solv Clear Plus MB (INCI name: Caprylyl/Capryl Glucoside (and) Sodium Cocoyl Glutamate (and) Glyceryl Caprylate (and) Citric Acid (and) Polyglyceryl-6-Oleate (and) Sodium Surfactin), a natural solubilizer, comparable in performance to PEG-40 hydrogenated castor oil or Polysorbate 20.

Cayman Chemicals (Ann Arbor, MI, USA) provides chemicals used primarily by universities and pharmaceutical companies for the research and development of medicines and has Surfactin CAS Nr. 252023-70-4 on its selling range.

Bloomberg Biotechnology Corp., Ltd. (Fairfield, NJ, USA), a company specializing in biomaterials, offers Bloomsurfact™ sodium surfactin.

Alfa Chemistry (Holbrook, NY, USA), a supplier of chemicals, custom synthesis, and testing services, is a vendor of sodium surfactin.

Iturin A from *Bacillus subtilis* is available in mg quantities from **Sigma-Aldrich** (St. Louis, MO, USA).

11.8 Existing and Potential Applications

Surfactin has been extensively studied for its antibacterial, antiviral, antitumor, and hemolytic effects. Due to its high production costs and low yields, for the time being, surfactin has only found limited use. This applies both to the protonated form of surfactin and to its sodium salt. Both are used in some top of the range personal care products, like eye and lip contour serums, antiaging skin care cleaners, cleansing gels, and hair care formulations. These are today accessible in Japan, and in a few countries outside Japan, but are always produced by Japanese companies. Among the formulators, **Kanebo Cosmetics Inc**. (Tokyo, Japan) is in the front line, followed by **Nippon Menard Cosmetics Co.,** Ltd. (Nagoya, Japan) and **Ozeki Corp**. (Nishinomiya City, Japan).

At the time of writing these notes (2025), surfactin is available in the market in mg quantities, at prices of around 100 USD/mg. It is obvious that the potential uses, suggested by many authors, outside the pharmaceutical/therapeutic and high-premium cosmetics domains are totally unrealistic. If in the future, surfactin should become available in industrial volumes and at reasonable prices, these applications will surely be reconsidered, but in the foreseeable future, they will remain unfulfilled dreams.

11.9 Regulatory

Surfactin is TSCA registered in the United States.

Sodium surfactin is listed in the European List of Notified Chemical Substances (ELINCS). It is registered under the REACH Regulation and is manufactured in and/or imported to the European Economic Area, at ≥1 to <10 tpa.

References

1 Hamley, I.W. (2015). Lipopeptides: from self-assembly to bioactivity. *Chemical Communication* 51: 8574–8583.

2 Xue, Y., Wang, M., Zhao, P. et al. (2018). Gram-negative bacilli-derived peptide antibiotics developed since 2000. *Biotechnology Letters* 40: 1271–1287.

3 Zhao, H., Shao, D., Jiang, C. et al. (2017). Biological activity of lipopeptides from *Bacillus*. *Applied Microbiology and Biotechnology* 101: 5951–5960.

4 Arima, K., Kakinuma, A., and Tamura, G. (1968). Surfactin, a crystalline peptidelipid surfactant produced by *Bacillus subtilis*: isolation, characterization and its inhibition of fibrin clot formation. *Biochemical and Biophysical Research Communications* 31: 488–494.

5 Kakinuma, A., Hori, M., Isono, M. et al. (1969). Determination of amino acid sequence in surfactin, a crystalline peptidolipid surfactant produced by Bacillus subtilis. *Agricultural and Biological Chemistry* 33: 971.

6 Yang, H., Li, X., Li, X. et al. (2015). Identification of lipopeptide isoforms by MALDI-TOF-MS/MS based on the simultaneous purification of iturin, fengycin, and surfactin by RP-HPLC. *Analytical and Bioanalytical Chemistry* 407: 2529–2542.

7 Abdel-Mawgoud, A.M., Aboulwafa, M.M., and Hassouna, N.A.-H. (2008). Optimization of surfactin production by Bacillus subtilis isolate BS5. *Applied Biochemistry and Biotechnology* 150: 305–325.

8 Heerklotz, H. and Seeling, J. (2001). Detergent-like action of the antibiotic peptide surfactin on lipid membranes. *Biophysical Journal* 81: 1547–1554.

9 Peypoux, F., Bonmatin, J.M., and Wallach, J. (1999). Recent trends in the biochemistry of surfactin. *Applied Microbiology and Biotechnology* 51: 553–563.

10 Rosen, M.J. and Kunjappu, J.T. (2012). *Surfactants and Interfacial Phenomena*, 4the, 153. Hoboken, N.J.: Wiley.

11 Shen, H.-H., Thomas, R.K., Chen, C.-Y. et al. (2009). Aggregation of the naturally occurring lipopeptide, surfactin, at interfaces and in solution: an unusual type of surfactant. *Langmuir* 25: 4211–4218.

12 Hamley, I.W., Dehsorkhi, A., Jauregi, P. et al. (2013). Self-assembly of three bacterially-derived bioactive lipopeptides. *Soft Matter* 9: 9572–9578.

13 Imura, T., Ikeda, S., Aburai, K. et al. (2013). Interdigitated lamella and bicontinuous cubic phases formation from natural cyclic surfactin and its linear derivative. *Journal of Oleo Science* 62: 499–503.

14 Davidson, P. (2024). Liquid crystalline behavior of concentrated aqueous solutions of biosurfactants. *Current Opinion in Colloid and Interface Science* 69: 101777.

15 Morikawa, M., Hirata, Y., and Imanaka, T. (2000). A study on the structure-function relationship of lipopeptide biosurfactant. *Biochimica et Biophysica Acta* 1488: 211–218.

16 Dufour, S., Deleu, M., Nott, K. et al. (2005). Hemolytic activity of new linear surfactin analogs in relation to their physico-chemical properties. *Biochimica et Biophysica Acta* 1726: 87–95.

17 Zhen, C., Ge, X.-F., Lu, Y.-T. et al. (2024). Chemical structure, properties and potential applications of surfactin, as well as advanced strategies for improving its microbial production. *AIMS Microbiology* 9: 195–217.

18 Ongena, M. and Jacques, P. (2008). *Bacillus* lipopeptides: versatile weapons for plant disease biocontrol. *Trends in Microbiology* 16: 115–125.

19 Mootz, H.D. and Marahiel, M.A. (1997). Biosynthetic systems for nonribosomal peptide antibiotic assembly. *Current Opinion in Chemical Biology* 1: 543–551.

20 Yeh, E., Kohli, R.M., Bruner, S.D., and Walsh, C.T. (2004). Type II thioesterase restores activity of a NRPS module stalled with an aminoacyl-S-enzyme that cannot be elongated. *ChemBioChem* 5: 1290–1293.

21 Kopp, F. and Marahiel, M.A. (2007). Macrocyclization strategies in polyketide and nonribosomal peptide biosynthesis. *Natural Product Reports* 24: 735–749.

22 Hu, F., Liu, Y., and Li, S. (2019). Rational strain improvement for surfactin production: enhancing the yield and generating novel structures. *Microbial Cell Factories* 18: 42.

23 Singh, A.K., Zhang, Y.-M., Zhu, K. et al. (2009). FabH selectivity for anteiso branched-chain fatty acid precursors in low-temperature adaptation in *Listeria monocytogenes*. *FEMS Microbiology Letters* 301: 188–192.

24 Choi, K.H., Heath, R.J., and Rock, C.O. (2000). Beta-ketoacyl-acyl carrier protein synthase III (FabH) is a determining factor in branched-chain fatty acid biosynthesis. *Journal of Bacteriology* 182: 365–370.

25 Park, J.H., Lee, K.H., Kim, T.Y., and Lee, S.Y. (2007). Metabolic engineering of *Escherichia coli* for the production of L-valine based on transcriptome analysis and *in silico* gene knockout simulation. *Proceedings of the National Academy of Sciences* 104: 7797–7802.

26 Park, J.H. and Lee, S.Y. (2010). Metabolic pathways and fermentative production of L-aspartate family amino acids. *Biotechnology Journal* 5: 560–577.

27 Debarbouille, M., Gardan, R., Arnaud, M., and Rapoport, G. (1999). Role of BkdR, a transcriptional activator of the SigL-dependent isoleucine and valine degradation pathway in *Bacillus subtilis*. *Journal of Bacteriology* 181: 2059–2066.

28 Yang, N., Wu, Q., and Xu, Y. (2020). Fermentation optimization for the production of surfactin by *Bacillus amyloliquefaciens*. *China Biotechnology* 40: 51–58.

29 Amani, H., Haghighi, M., and Keshtkar, M.J. (2013). Production and optimization of microbial surfactin by *Bacillus subtilis* for *Ex Situ* enhanced oil recovery. *Petroleum Science and Technology* 31: 1249–1258.

30 Yeh, M.-S., Wei, Y.-H., and Chang, J.-S. (2005). Enhanced production of surfactin from *Bacillus subtilis* by addition of solid carriers. *Biotechnology Progress* 21: 1329–1334.

31 Dhanarajan, G. and Sen, R. (2014). Cost analysis of biosurfactant production from a scientist's perspective. In: *Biosurfactants. Production and Utilization – Processes, Technologies, and Economics* (ed. N. Kosaric and F.V. Sukan), 153–162. Boca Raton, Fl: CRC Press.

32 Shahid, M.K., Kashif, A., Choi, Y. et al. (2022). Circular bioeconomy perspective of agro-waste-based biochar. In: *Biomass, Biofuels, Biochemical. Circular Bioeconomy: Technologies for Waste Remediation* (ed. S. Varjani, A. Pandey, M.J. Taherzadeh, et al.), 223–243. Amsterdam: Elsevier.

33 Nazareth, T.C., Zanutto, C.P., Tripathi, L. et al. (2020). The use of low-cost brewery waste product for the production of surfactin as a natural microbial biocide. *Biotechnology Reports* 28: e00537.

34 Zhi, Y., Wu, Q., and Xu, Y. (2017). Production of surfactin from waste distillers' grains by co-culture fermentation of two *Bacillus amyloliquefaciens* strains. *Bioresource Technology* 235: 96–103.

35 Dhali, D., Coutte, F., Arias, A.A. et al. (2017). Genetic engineering of the branched fatty acid metabolic pathway of *Bacillus subtilis* for the overproduction of surfactin C_{14} isoform. *Biotechnology Journal* 12: 1600574.

36 Wang, M., Yu, H., Li, X., and Shen, Z. (2020). Single-gene regulated non-spore-forming *Bacillus subtilis*: construction, transcriptome responses, and applications for producing enzymes and surfactin. *Metabolic Engineering* 62: 235–248.
37 Jiao, S., Li, X., Yu, H. et al. (2017). In situ enhancement of surfactin biosynthesis in *Bacillus subtilis* using novel artificial inducible promoters. *Biotechnology and Bioengineering* 114: 832–842.
38 Wu, Q., Zhi, Y., and Xu, Y. (2019). Systematically engineering the biosynthesis of a green biosurfactant surfactin by *Bacillus subtilis* 168. *Metabolic Engineering* 52: 87–97.
39 Wang, M., Yu, H., and Shen, Z. (2019). Antisense RNA-based strategy for enhancing surfactin production in *Bacillus subtilis* TS1726 via overexpression of the unconventional biotin carboxylase II to enhance ACCase activity. *ACS Synthetic Biology* 8: 251–256.

12

Antimicrobial Peptides

12.1 Introduction

Antimicrobial peptides (AMPs), often referred to as host defence peptides, are a class of small peptides that widely exist in nature. They are an important part of the innate immune system of different organisms. AMPs have a wide range of inhibitory effects against bacteria, fungi, parasites, and viruses. The emergence of antibiotic-resistant microorganisms and the increasing concerns about the extensive use of antibiotics have resulted in the development of new AMPs with good application prospects in medicine, food, animal husbandry, agriculture, and aquaculture [1].

AMPs have several advantages over antibiotics, such as broader range of antimicrobial activities, including antifungal, antiviral, and antibacterial, and they have not been reported to contribute to resistance development [2].

In 1922, Alexander Fleming discovered the first antimicrobial protein, lysozyme, a key component of the innate immune system against foreign pathogens, which marked the birth of modern innate immunology. Since then, many antibiotics and AMPs have been discovered. As of January 2025, the Antimicrobial Peptide Database (APD) contains 5099 peptides, including 3300 natural AMPs from the different kingdoms (2580 from animals, 410 from bacteria, 268 from plants, 29 from fungi, 8 from protists [i.e. eukaryotic organisms that are not an animal, plant, or fungus], and 5 from archaea). There are also 1299 synthetic and 231 predicted AMPs [3].

According to the statistics of the APD, AMPs can be divided into different categories that can be summarized as follows [1]:

- Antibacterial (60%)
- Antifungal (26%)
- Anticancer (5%)
- Antiviral (4%)
- Antiparasitic (3%)
- Antihuman immunodeficiency virus (HIV) (2%)

AMPs are also divided into several subgroups on the basis of amino acid sequence, the net charge of the peptide, the protein structure, and the source. These are reported in Table S1 of the work of Huan et al. [1]. Most of them have a net charge of + 2 to + 9 and contain up to 50 amino acids [4], but anionic AMPs also exist.

Natural Surfactants: Science, Technology, and Industrial Aspects, First Edition. Guido Bognolo and Krister Holmberg.

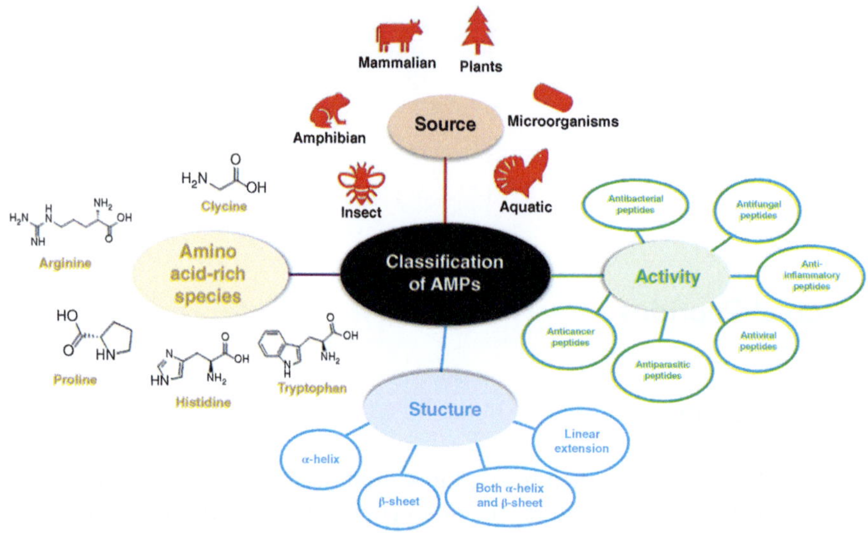

Figure 12.1 Different ways to classify antimicrobial peptides (AMPs). (Reprinted from Huan et al. [1] / Frontiers Media S.A / CC BY.)

Figure 12.1 illustrates different ways to classify antimicrobial peptides.

12.2 Chemical Structures and Properties

AMPs generally contain between 12 and 50 amino acids. Most AMPs are positively charged, and the charge originates from arginine, lysine, and/or histidine residues. (The positive charge of the histidine residue requires relatively acidic conditions.) The majority of the amino acids in the polypeptide are uncharged and hydrophobic, however, this means that many AMPs have a distinct amphiphilic character because of a structure with one or more blocks of hydrophobic amino acids and blocks of positively charged residues. Thus, they contain both polar and nonpolar domains and can be regarded as a kind of biosurfactant; however, values usually used to characterize amphiphilic compounds, such as surface tension, interfacial tension, and critical micelle concentration, are seldom reported. One may assume that the reduction in surface and interfacial tension that can be obtained with most AMPs is modest compared to what can be obtained with the biosurfactants described in previous chapters in this book, such as sophorolipids, rhamnolipids, and lipopeptides. The amphiphilic character is, however, sufficient for enabling the peptides to partition into biological membranes, thus exerting their antimicrobial action. Figure 12.2 shows the structure of a small AMP intended for immobilization to a support and providing antibacterial effect to a medical implant, see Section 6.2 below.

The secondary structure of the peptides can take various forms [6]. The AMPs may be α-helical or contain β sheets. Peptides containing one or more α-helices are the most common [7]. Peptides that contain β-sheets have few or no helical domains. AMPs may also be extended, or they may have a structure that is a mixture of these different motifs. Figure 12.3

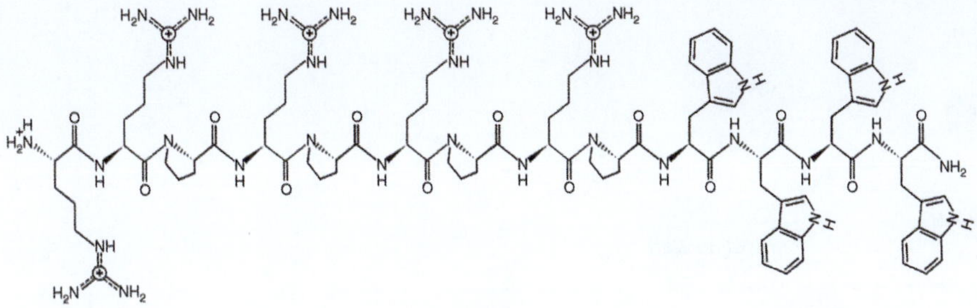

Figure 12.2 Primary structure of a cationic antimicrobial peptide intended for immobilization to a support. (Stepulane [5] / with permission of Elsevier.)

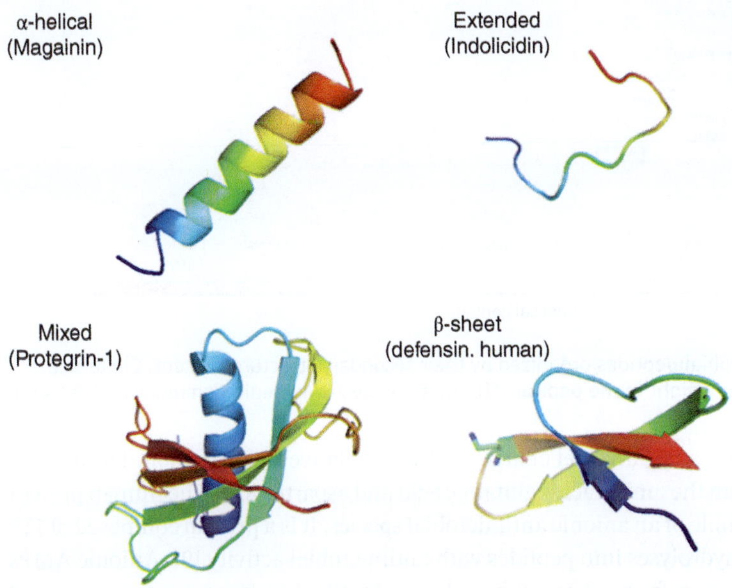

Figure 12.3 Common natural motifs in antimicrobial peptides. (Ymahn / Wikimedia Commons / CC BY-SA 3.0.)

illustrates different secondary structures. It is common that the helices are built up such that one side is mainly composed of charged amino acids while the other side largely contains hydrophobic residues. Such helices are distinctly amphiphilic [8].

The presence of two or more disulfide bonds in the peptide gives rise to β-stranded structures. So-called β-hairpins are also common. In these two strands, oriented in an antiparallel direction, are connected by a short loop of two to five amino acids.

The structure of AMPs varies widely. Some are rich in α-helices and some contain mostly β-sheets. Some are very small while others are relatively large. Figure 12.4 illustrates the broad variation in structures.

An interesting observation is that many of the peptides do not have a distinct secondary structure in bulk solution; it is only when they partition into the biological membrane that they fold into a distinct configuration.

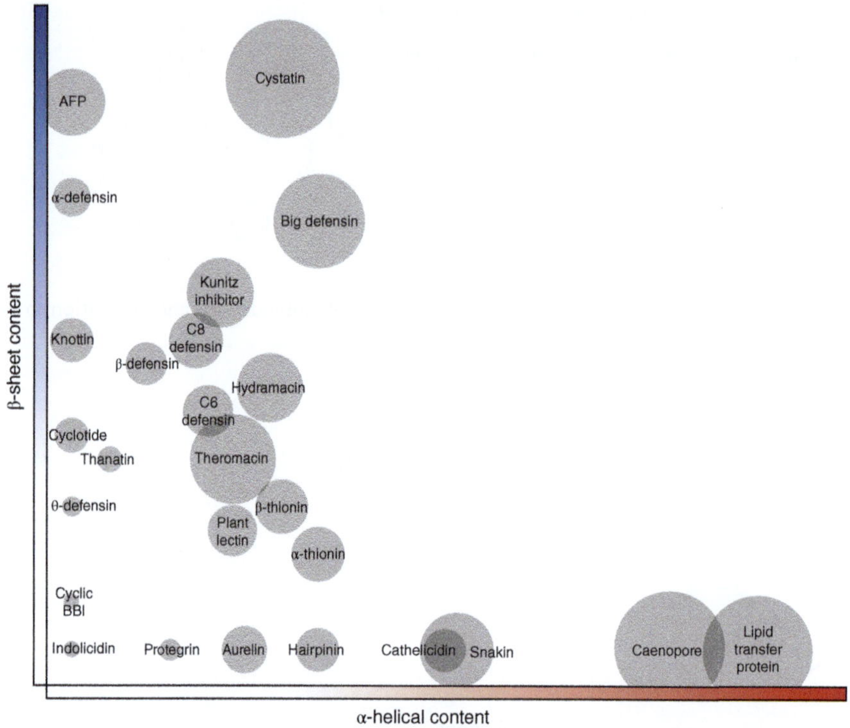

Figure 12.4 Antimicrobial peptides organized by their secondary structure content. Circle size indicates the molecular weight of the peptide. (Thomas Shafee / Wikimedia Commons / CC BY 4.0.)

AMPs may also have a negative net charge but that is relatively uncommon. The anionic charges originate from the amino acids glutamic acid and aspartic acid. The human protein dermcidin is an example of an anionic antimicrobial species. It is a protein composed of 110 amino acids, and it hydrolyses into peptides with antimicrobial activity [9]. Anionic AMPs with net charges ranging from −1 to −8 have been identified in frogs, toads, newts, and salamanders. Most of them show antimicrobial activity but their mode of action is obviously different from that of their more common cationic relatives [10].

12.3 Sources

AMPs can be obtained from many sources, the most important being:

Mammals and other vertebrates. Cathelicidins and defensins are the main families of AMPs in the human body and exhibit different expressions depending on the stage of body development. AMPs are found in many parts of the body such as skin, eyes, ears, mouth, respiratory tract, lung, intestine, and urethra. In human breast milk, they also play an important role because they can decrease the morbidity and mortality of breast-feeding infants [11]. Dairy is an important source of AMPs, which are generated through enzymatic hydrolysis in milk. Several AMPs, notably lactoferricin B (LfcinB), have been isolated from α-lactalbumin, β-lactoglobulin, lactoferrin, and casein fractions [12].

Amphibians. Frogs are the main source of amphibian antimicrobial peptides, the most famous one being magainin; the skin secretions of frogs from the genera *Xenopus*, *Silurana*, *Hymenochirus*, and *Pseudohymenochirus* [13]. Cancrin has been reported as the first antimicrobial peptide from the sea amphibian *Rana cancrivora* [14].

Insects. AMPs are mainly synthesized in fat bodies and blood cells of insects, which is one of the main reasons for insects' strong adaptability to survival [15]. Cecropin is the best-known family of insect AMPs, and it can be found in guppy silkworm, bees, and *Drosophila*. The number of AMPs varies considerably between species, for example, invasive harlequin ladybird (*Harmonia axyridis*) and black soldier fly (*Hermetia illucens*) have up to 50 AMPs, while pea aphid (*Acyrthosiphon pisum*) has none [16]. Jellein, a peptide derived from bee royal jelly, shows promising effects on several bacteria and fungi, and its lauric acid-conjugated form can inhibit the parasite *Leishmania major* [17].

Microorganisms. AMPs can be obtained from microorganisms like bacteria and fungi. Some well-known peptides are nisin and gramicidin from *Lactococcus lactis*, *Bacillus subtilis*, and *Bacillus brevis* [18]. Due to the high price of chemical synthesis of AMPs, the biological expression has attracted increased attention. Specific yeast species like *Pichia pastoris* and *Saccharomyces cerevisiae*, and bacteria like *Escherichia coli* and *Bacillus. subtilis*, as well as plants have been used as expression systems [19], although special attention must be given to expressions from pathogenic microorganisms.

On top of these sources, many antimicrobial peptides have been extracted and isolated from the stems, seeds, and leaves of plants. They are classified into several groups, including thionins, defensins, and snakins [20]. The number of marine-derived antimicrobial peptides is also constantly increasing, and many have shown promising results in *in vitro* and *in vivo* tests. For example, As-CATH4 shows an immunity-stimulating effect *in vivo* and can enhance the anti-infective capability of drugs used in combination with it [21]. Myticusin-beta is an immune-related antimicrobial peptide from *Mytilus coruscus* and a promising alternative to antibiotics [22].

12.4 Production

AMPs for commercial purposes can be produced by liquid-phase or solid-phase synthesis, a hybrid of both solid-phase and solution-phase synthesis, or recombinant DNA technologies [23]. When obtained from foods, peptides must be released from the parent protein to be activated. This can be achieved by hydrolysis and the processing of foods by curing, fermentation, or ripening, or by digestive enzymes, including commercial proteolytic enzymes [24–26].

Suda et al. have proposed a protocol for the production and purification of nisin, a widely used antimicrobial peptide produced by the food-grade microorganism *Lactococcus lactis*, that can be easily applied with minor modifications to the production of other AMPs [27].

Plant and animal proteins are mainly used to obtain AMPs [28]. Conventional methods of AMPs recovery are fermentation, chemical, and enzymatic hydrolysis. The most effective methods for the biotechnological production of peptides are enzymatic hydrolysis and microbial fermentation of proteins [29].

Enzymatic hydrolysis uses commercial enzymes for the cleavage of peptide bonds in proteins. This method may provide advantages in obtaining AMPs that are not provided by other methods such as fermentation, chemical hydrolysis, or digestion, including major improvement in the reaction rate [30]. However, the cost of enzymes is quite high, and there is a need for special maintenance of their activity. The yield and productivity of the process are usually low, and the formation of undesirable secondary metabolites is also possible [31].

Greater selectivity and process speed can be achieved by using ultrafiltration membranes. The method involves continuous pumping of the enzyme and substrate mixture from the reaction vessel into a membrane filter, through which only small and hydrolyzed species pass, while large species (polypeptides, nonhydrolyzed substrate, enzymes) are returned back to the hydrolysis tank [32]. The hydrolysis process can also be improved by pretreating the parent protein, which ensures higher hydrolysis rates while minimizing the amount of enzyme used. Microwaves, ultrasound, high-voltage pulsed electric field, and high-pressure hydrostatic enzymatic hydrolysis are used for this purpose [33].

Some novel technologies are currently applied in combination with enzymatic hydrolysis. These are eco-friendly, innovative, and sustainable techniques like hydrostatic pressure processing, ultrasound-assisted and microwave-assisted extraction, pulsed electric field processing, ohmic heating, and subcritical water hydrolysis [34].

Microbial fermentation can generate peptides of different sizes and sequences with different biological activities, as occurs during gastrointestinal digestion [35]. This is a more economical and cost-effective method for AMPs production. The microbial conversion of various protein-containing substrates occurs via solid-state fermentation or submerged fermentation depending on the available free water volume [36].

Peptidase-producing microorganisms such as filamentous fungi, yeasts, *Bacillus* species, and lactic acid bacteria (LAB) are used to obtain AMPs [37]. They are one of the most valuable microorganisms for the production of AMPs due to their high adaptability to various environments and to animal and plant substrates, their safety (they are generally recognized as safe [GRAS]) and most importantly, the presence of an effective and diverse proteolytic system [38, 39].

Some yeast organisms, which also possess a set of proteases and peptidases responsible for protein degradation and belong to the GRAS category, are also widely used in dairy products to break down milk proteins and enrich them with bioactive peptides and amino acids necessary for growth [40].

The most widely used microbial fermentation processes are submerged fermentation (microorganism culture in a liquid medium containing nutrients) and solid-state fermentation (microorganism growth on nutrient-rich solid substrates) [39, 41].

The recent years have seen the emergence of plant systems as production platforms for AMPs. Such production systems have been found to give good yields of well-defined peptides in combination with a cost-effective process [42].

Compared to bacterial and yeast expression systems, plant-based molecular farming is safer since there is essentially no chance of product contamination with infections that affect humans and animals, or of endotoxins because plants do not have these pollutants.

Substantial progress has been achieved in the production of AMPs in plant systems through synthetic biology [43].

Although the production of AMPs in plants is considered a prospective and emerging field of research, crucial challenges in terms of stability, function, and yield of the product remain [44].

There are two main methods for producing AMPs in plants [45]:

- Steady integration of antimicrobial peptide genes into the plant genome
- Temporary transformation techniques

Figure 12.5 illustrates molecular farming of AMPs in plants.

12.5 Mechanism of Action

The first and key question in research on AMPs is how they specifically target the invading pathogen while sparing the host cells. This targeting specificity has been attributed to differences in the composition of the cell membranes between the pathogens and the host cells.

In general, lipids and proteins are the main components of the cellular membrane and form the phospholipid bilayer as the basic scaffold for the cell membrane. Phosphatidylcholine (PC) and phosphatidyl ethanolamine (PE) are normally uncharged, while hydroxylated phospholipids such as phosphatidylserine (PS), cardiolipin (CL), and phosphatidylglycerol (PG) are negatively charged. PS, PG, and CL are found in bacterial pathogens but have little or no presence in the mammalian cytoplasmic membrane [47]. This is contrary to PE and PC, which are commonly found in mammalian cell membranes [48].

Moreover, the lipopolysaccharide of Gram-negative bacteria and the lipoteichoic acid of Gram-positive bacteria carry a large number of anionic groups, which increase the amount of negative charges in the membrane. The negative charges in the fungal membranes, on the other hand, mainly result from the phosphomannan and other related components, such as phosphatidylinositol (PI), PS, and diphosphatidylglycerol (DPG) [49]. Consequently, the cationic AMPs selectively interact with the negatively charged membrane through electrostatic interaction which partially explains the targeting specificity of AMPs.

There are two modes of action for the AMPs:

- Extracellular via membrane targeting mode
- Intracellular mode

12.5.1 Extracellular Via Membrane Targeting Mode

The membrane targeting may follow any of three models:

- The Carpet-Like
- The Barrel-Stave
- The Toroidal Pore

In the **Carpet model**, the AMPs cover the cell membrane with the hydrophilic end facing the solution and the hydrophobic end facing the phospholipid bilayer. They then

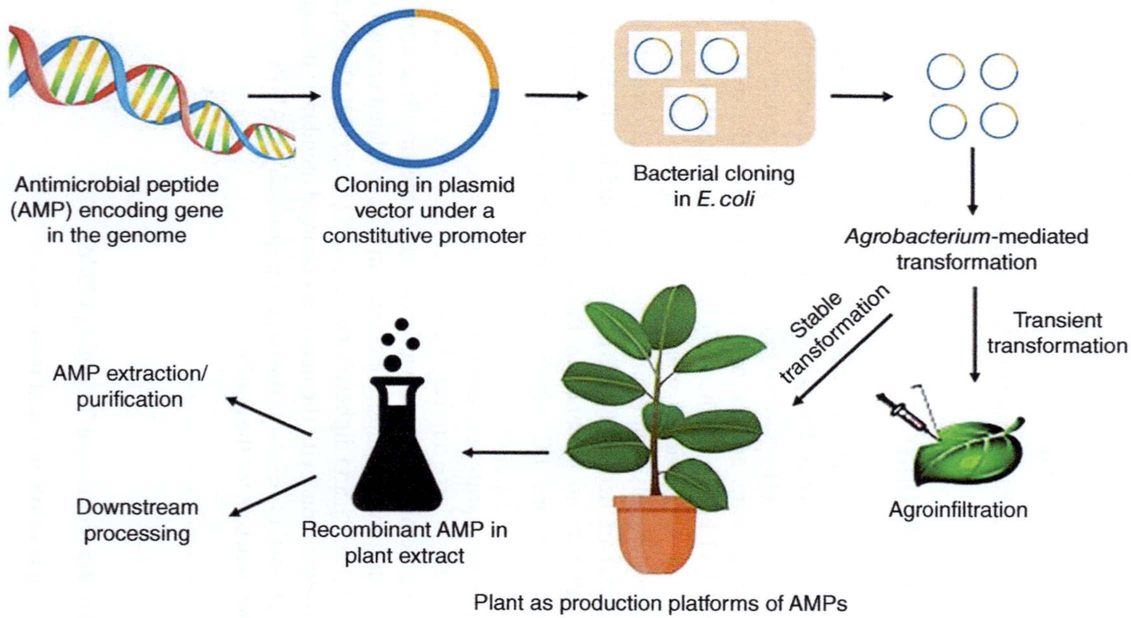

Figure 12.5 Plant systems as "production platforms" for molecular farming of antimicrobial peptides. (Reprinted from Tiwari et al. [46] / MDPI / CC BY 4.0.)

destroy the cell membrane in a "detergent-like" manner [50]. Human cathelicidin LL-37 works through this mechanism, which requires a high concentration threshold. Polarized light-attenuated total reflection Fourier transform infrared spectroscopy (ATR-FTIR) confirmed that cecropin P1 deposits flat on the bacterial cell membranes, destabilizing and eventually destroying them [51]. Other AMPs with β-sheet structure also work according to this model [52, 53]

In the **Barrel-stave model**, the AMPs undergo self-aggregation, penetrate the bilayer of the cell membrane in the form of multimers, and form channels that result in cytoplasmic outflow. In severe cases, AMPs can induce cell membrane collapse and lead to cell death [54]. Alamethicin works according to this model.

In the **Toroidal pore model**, also known as the Wormhole model, AMPs, vertically embedded in the cell membrane, accumulate and then bend to form a ring hole with a diameter of 1–2 nm [55, 56]. Typical examples of AMPs that work according to this model are magainin 2, lacticin Q, and arenicin. Furthermore, cationic peptides, including TC19, TC84, and BP2, disrupt the membrane barrier by creating fluid domains [57].

The three models are illustrated in Figure 12.6.

The membrane targeting mechanisms can be further refined to address the large differences in the lipid composition of the cell membranes of bacteria, fungi, and mammals. Many AMPs take advantage of the differences in membrane composition to exert their effects.

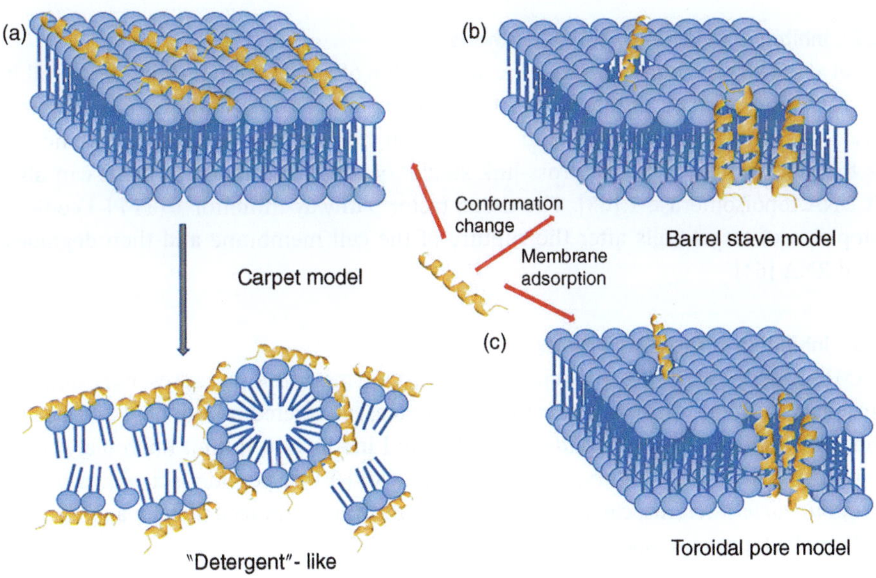

Figure 12.6 Models of action of extracellular antimicrobial peptides. (a) Carpet model: antimicrobial peptides cover the surface and then destroy the cell membrane in a "detergent-like" mode. (b) Barrel stave model: antimicrobial peptides self-aggregate and penetrate the bilayer of the cell membrane in the form of multimers forming channels in the bilayer. (c) Toroidal pore model: antimicrobial peptides, vertically embedded in the cell membrane, accumulate and then bend to form a ring hole. (Huan et al. [1] / Frontiers Media S.A / CC BY 4.0.)

12.5.2 Intracellular Mode

In the nonmembrane targeting mechanism, the AMPs pass the cell membranes, enter the cytoplasm, and identify and act on the target by one of the following mechanisms:

- Inhibition of protein biosynthesis
- Inhibition of nucleic acid biosynthesis
- Inhibition of protease activity
- Inhibition of cell division

12.5.2.1 Inhibition of Protein Biosynthesis

AMPs affect transcription, translation, and assembly into functional proteins through molecular chaperone folding by interfering with related enzymes and effector molecules. For example, Bac71–35 targets ribosomes to inhibit protein translation [58], whereas Tur1A inhibits protein synthesis in *Escherichia coli* and *Thermus thermophilus* by inhibiting the transition from the initial phase to the extension phase. Some AMPs have different targets. For instance, genome-wide transcription shows that the antimicrobial peptide DM3 can affect many important intracellular pathways of protein biosynthesis [59].

Chaperones are key proteins for the correct folding and assembling of newly synthesized proteins, inducing stereoisomerism, thus exposing them to the action of AMPs. Both pyrrhocoricin and drosocin can prevent DnaK from refolding misfolded proteins by inducing a permanent closure of the DnaK peptide-binding cavity [60–62].

12.5.2.2 Inhibition of Nucleic Acid Biosynthesis

AMPs can affect key enzymes or induce the degradation of nucleic acid molecules to inhibit nucleic acid biosynthesis. Indolicidin, a tryptophan-rich cationic antimicrobial peptide composed of only 13 amino acids and amidated at the carboxyl terminus, specifically targets the abasic site of DNA to cross-link single- or double-stranded DNA. It can also inhibit DNA topoisomerase 1 [63]. The tissue factor pathway inhibitor 1 (TFPI-1) enters the cytoplasm of target cells after the rupture of the cell membrane and then degrades DNA and RNA [64].

12.5.2.3 Inhibition of Protease Activity

Many AMPs can inhibit protease activity, thus disrupting metabolic processes. For example, histatin 5 has a strong inhibitory effect on the proteases secreted by the host and bacteria. The neutrophil-activating peptide-2 (NAP-2) and indolicidin inhibit microbial serine proteases, elastase, and chymotrypsin [61]. Cathelicidin-BF, a peptide isolated from the venom of *Bungarus fasciatus*, can effectively inhibit thrombin-induced platelet aggregation and further block protease-activated receptor 4 [65].

12.5.2.4 Inhibition of Cell Division

AMPs inhibit cell division by inhibiting DNA replication and DNA damage response (SOS response), blocking the cell cycle, or causing the failure of chromosome separation [66]. For instance, APP can efficiently kill *Candida albicans* because of its cell-penetrating efficiency, strong DNA-binding affinity, and ability to induce S-phase arrest in intracellular environments [67]. The developmental regulator MciZ, a 40-amino acid peptide endogenously produced during *Bacillus subtilis* sporulation is an effective inhibitor of bacterial cell division, Z-ring formation, and localization [68].

12.6 Existing and Potential Applications

12.6.1 Medicine

The application of AMPs in medicine is in development. At the end of 2022, over 60 AMPs had been approved by the FDA and more than 400 were in clinical trials [69]. Only seven of those had reached the market, mainly as topical medications and, in some cases involving serious infections, as injectables [70].

Despite their favorable features, AMPs are generally unstable in the gastrointestinal tract and in other body fluids. They have poor absorption and distribution, as well as fast metabolic degradation and excretion, which result in low bioavailability. Their biological activity is affected by salt concentration and pH [71]. For this reason, the most promising applications for AMPs are those involving topical applications [72].

12.6.1.1 Wound Healing and Skin Infections

Skin and soft tissue infections are one of the most common microbial infections in humans, and antimicrobial peptides can be a new therapeutic option, thanks to their broad spectrum of biological activities, because skin pathogens include bacteria but also protozoa, fungi, and viruses [73].

AMPs can promote wound healing after skin injury by encouraging the keratinocyte and fibroblast migration, thereby dramatically speeding up wound healing. Even though they have clear advantages over traditional antibiotics, their wide use in therapeutic interventions is constrained by concerns about toxicity, stability, and costs. Current breakthroughs in nanotechnology have sparked interest in unlocking the enormous potential of AMPs as biofilm-inhibiting agents and broad-spectrum antibiotics to counter the risk of bacterial resistance. Using nanotechnological techniques allows a controlled delivery of antimicrobials while retaining a high level of activity [74].

Some AMPs play a vital role in both the skin barrier and skin function [75] and thus have the potential to treat multiple skin diseases, like melanoma, acne, diabetic foot ulcer, and psoriasis [76].

New strategies are being investigated to extend the applications of antimicrobial peptides in medicine. These include:

- Constructing precursors to reduce cytotoxicity and improve protease stability
- Using antimicrobial peptides in combination with existing antibacterial agents
- Inducing the correct expression of antimicrobial peptides with appropriate drugs and using engineered probiotics as vectors to express antimicrobial peptides.

For example, in the field of wound repair, different formulation strategies, such as loading AMPs in nanoparticles, hydrogels, creams, gels, ointments, or glutinous rice paper capsules, have been developed to effectively deliver AMPs to the wound [77, 78].

12.6.1.2 Dental Care

Dental caries is a prevalent oral disease and some acidogenic bacteria like *Streptococcus* species are the main caries-associated pathogens [79]. Several antimicrobial peptides have shown promise for treating oral candidiasis, for instance peptide PAC-113, which has been sold over the counter in Taiwan [80].

12.6.1.3 Ophthalmology

In ophthalmology, AMPs like Lactoferricin B and Protegrin-1 exhibit antimicrobial activity against eye-infecting pathogenic bacteria, such as *Staphylococcus aureus*, *Aspergillus* species, and *Candida albicans*; however, their application is still at the theoretical stage. With the popularity of contact lenses and the increase in cases of related eye infections, AMPs have good application prospects [81].

12.6.1.4 Antibacterial Peptides

Antibacterial peptides are the largest segment of AMPs and have a broad inhibitory effect on common pathogenic bacteria, such as VRE, *Acinetobacter baumannii*, and MRSA in clinical medicine, *Staphylococcus aureus*, *Listeria monocytogenes*, and *Escherichia coli* in food, and *Salmonella* and *Vibrio parahaemolyticus* in aquatic products. Many natural and synthetic AMPs like nisin, cecropins, and defensins have shown good inhibition activity to Gram-positive and Gram-negative bacteria [1].

12.6.1.5 Antifungal Peptides

Antifungal peptides address fungal infections with enhanced drug resistance. Many have shown excellent antifungal activities against common pathogenic fungi, such as *Aspergillus* and *Candida albicans*, yeast, filamentous fungi (e.g. *Aspergillus flavus*), mold in food and agriculture. Many synthetic peptides also show good antifungal activity [82]. A total of 37 antifungal peptides isolated from *Lactobacillus plantarum* TE10 and their mixtures were found to reduce *Aspergillus flavus* spore formation in fresh maize seeds [83].

12.6.1.6 Anticancer Peptides

Anticancer peptides work by [84, 85]:

- Inducing immune cells (such as dendritic cells) to kill tumor cells.
- Inducing the necrosis or apoptosis of cancer cells.
- Inhibiting angiogenesis to eliminate tumor nutrition and prevent metastasis.
- Activating certain regulatory functional proteins to interfere with the gene transcription and translation of tumor cells.

Tritrpticin and its analogues induce toxicity toward Jurkat cells *in vitro*, whereas indolicidin and puroindoline A can also act as anticancer peptides [86].

Both net charge and hydrophobicity play important roles in optimizing the anticancer activity of anticancer peptides and they can constrain and influence each other. Thus, achieving a balance between net charge and hydrophobicity is important for strong anticancer activity.

12.6.1.7 Antiviral Peptides

Viruses cause serious harm to human life and huge economic losses in animal husbandry. The COVID-19 has caused great loss of lives and property. Furthermore, foot-and-mouth disease virus, avian influenza virus (AIV), and HIV are long-term threats to human life.

Antiviral peptides show a strong killing effect on viruses mainly by [87]:

- Inhibiting virus attachment and virus cell membrane fusion.
- Destroying the virus envelope.
- Inhibiting virus replication.

12.6.1.8 Antiparasitic Peptides

Antiparasitic peptides are effective in killing a wide range of parasites that can transmit through a variety of routes. In recent research, Epi-1, a marine synthetic antimicrobial peptide, has been shown to remarkably inhibit *Trichomonas vaginalis* by destroying its membrane [88]. The four-amino acid antimicrobial peptide AMP-KDEL (lysine, aspartic acid, glutamic acid, and leucine) has been found to have a significant effect on the *Leishmania* parasite [89].

12.6.1.9 Anti-HIV Peptides

Anti-HIV peptides are a subclass of antiviral peptides. The most important examples of these peptides include defensins, LL-37, gramicidin D, caerin 1, maximin 3, magainin 2, dermaseptin-S1, dermaseptin-S4, siamycin-I, siamycin-II, and RP 71955 [82], and the antiviral peptide Fuzeon™ (Enfuvirtide) has been commercialized as an anti-HIV drug [90].

12.6.1.10 Endogenous Antimicrobial Peptides

The endogenous production of antimicrobial peptides is also relevant. For example, the administration of sodium butyrate has been shown to induce production of intestinal AMPs, beneficial for the treatment of infectious or inflammatory diseases [91].

12.6.1.11 Biofilm Inhibitors

Biofilms are ubiquitous in nature, having the ability to adhere to virtually any surface, either biotic or abiotic, including medical devices, causing chronic infections that are difficult to eradicate. The biofilm matrix protects bacteria from hostile environments, thus contributing to bacterial resistance to antimicrobial agents. Biofilms are very difficult to treat, with options restricted to the use of large doses of antibiotics or the removal of the infected device. AMPs represent good candidates to develop new antibiofilm drugs as they can act at different stages of biofilm formation, on disparate molecular targets, and with various mechanisms of action. These include inhibition of biofilm formation and adhesion, downregulation of quorum-sensing factors, and disruption of the preformed biofilm [4].

12.6.2 Medical Devices

The rapid increase of bacteria resistant to traditional antibiotics, e.g. methicillin-resistant *Staphylococcus aureus* (MRSA), is a growing concern with respect to the use of medical implants. In order to combat infections, antibiotics are frequently used together with implants and other medical devices. The combined use of AMPs and conventional antibiotics has shown promise as a way to more efficiently prevent infections by *Staphylococcus aureus* and, in particular, MRSA [92]. To increase the stability of an antimicrobial peptide, which had the amino acid sequence RRPRPRPRPWWWW-NH2, it was covalently

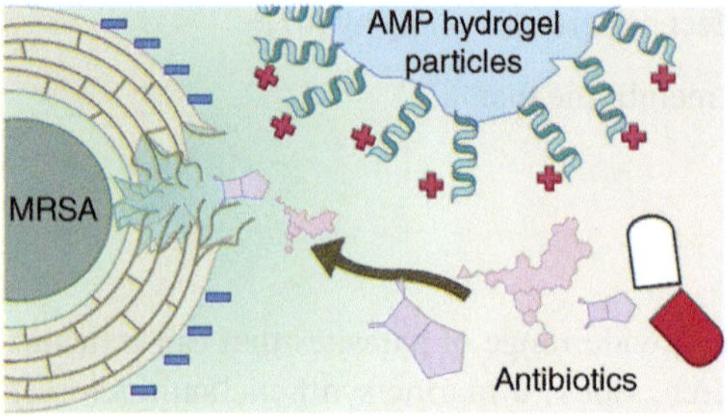

Figure 12.7 A combination of a cationic antibacterial peptide (AMP) immobilized on hydrogel particles and a conventional antibiotic destroys the cell wall of methicillin-resistant *Staphylococcus aureus* (MRSA). (Stepulane et al. [92] / Elsevier / CC BY 4.0.)

attached to hydrogel particles by coupling the terminal amino group to carboxyl groups in the hydrogel surface. Immobilization of AMPs to hydrogels or other surfaces has been shown to drastically reduce the rate of hydrolysis while retaining high antibacterial effect; rapid hydrolysis *in vivo* is a well-known shortcoming of these antimicrobial agents [93]. The conventional antibiotics were vancomycin and oxacillin. The combination of the antimicrobial peptide and the traditional antibiotics showed a strong synergistic effect. For example, the presence of the antimicrobial peptide gave a 64-fold reduction of the minimum inhibitory concentration of oxacillin against MRSA. Figure 12.7 shows a schematic of the combined action of the immobilized antimicrobial peptide and the conventional antibiotic on the cell wall of MRSA.

Biomaterials functionalized with AMPs have been developed for a range of medical devices, from biomedical implants to wound care products [94]. The approach can be regarded as an alternative to the traditional antibiotic therapy in the management of biomaterial-associated infections but, as discussed above, the two approaches can also be combined.

12.6.3 Pharmaceutical Industry

Peptide drugs account for 5–6% of the global pharmaceutical market [95]. The global peptide therapeutics market is led by international companies, including **Takeda Pharmaceutical** (Tokyo, Japan), **Pfizer** (New York, NY, USA), **Merck & Co.** (Rahway, NJ, USA), **Eli Lilly and Company** (Indianapolis, IN, USA), **Sanofi S.A.** (Paris, France), **AstraZeneca PLC** (Cambridge, UK), and **GlaxoSmithKline** (Brentford, UK). Over the years, the demand for peptide therapeutics has increased significantly: 114 peptides have already been approved by the regulatory authorities as therapeutic agents [96]. Their therapeutic indications include cancer, inflammatory, autoimmune, and metabolic diseases, and microbial infections. The latter may be a promising area for new antibiotic therapy [97–99]. Another interesting and very promising aspect is their cell-penetrating properties. Cell-penetrating peptides are now under study as drug delivery tools for anticancer, antibacterial, and antiviral therapies [100].

12.6.3.1 Therapeutic Antimicrobial Peptides on the Market

Bacitracin is a group of cyclic polypeptides produced by organisms of the licheniformis group of *Bacillus subtilis*. It has antimicrobial activity against many Gram-positive bacteria, including *Staphylococci*, *Streptococci*, and *Clostridia*, and was approved by the FDA in 1948 [101]. It is applied topically to treat local infections, mainly infections of the skin, ear, and eye.

Colistin (Polymyxin E) and **polymyxin B** (PMB) are lipopeptide antibiotics with activity against many Gram-negative bacteria. The polymyxins were approved for clinical use in the late 1950s but fell out of favor in the mid-1970s due to concerns about their potential nephrotoxicity and neurotoxicity [102]. They have since undergone many interesting modifications, not least conjugation with already known antibiotics, e.g. a clinical trial of colistin–rifampicin administered intravenous with new approvals by the FDA [103]. (For instance, colistimethate sodium, a form of colistin, was approved in 1999.)

Daptomycin is a lipopeptide isolated in the 1980s with impressive activity against Gram-positive, but not Gram-negative, bacteria [104]. It received approval by the FDA in 2003 and is currently widely used to treat *Staphylococcus* and *Enterococcus* infections [105].

Vancomycin is a tricyclic glycopeptide which acts against Gram-positive bacteria, including methicillin-resistant strains of *Staphylococcus aureus*, by inhibiting the synthesis of the peptidoglycan layer of the bacterial cell wall. It was first approved for use in the United States in 1958 and continues to be widely used, particularly with the recent rise in the incidence of serious MRSA infections [106].

Dalbavancin, **oritavancin**, and **telavancin** are small lipoglycopeptides derived from vancomycin. They are more potent and bactericidal than their prototype and are effective against vancomycin-resistant bacteria. They inhibit bacterial cell wall formation, and telavancin and oritavancin also disrupt bacterial cell membranes and affect membrane permeability. They were approved by the FDA between 2009 and 2014 [107].

12.6.3.2 Nanoparticle Formulations of Antimicrobial Peptides

Toxicity, together with instability and short half-life *in vivo*, is a major challenge for the clinical application of AMPs and many strategies have been tried to overcome the hurdle. One of the most promising is based on the encapsulation of AMPs in biocompatible nanocarriers in order to reduce local toxicity while maintaining efficacy [108]. Gold nanoparticles (AuNPs) are considered among the most biocompatible inorganic nanosystems and are suitable for coupling active molecules, such as AMPs [109]. Casciaro et al. designed a system named AuNPs@Esc(1–21) where gold nanoparticles are coated with Esculentin-1a(1–21), an antimicrobial peptide derived from frog skin, for use as a drug against infections caused by *Pseudomonas aeruginosa* [110]. Subaer et al. reported the powerful antimicrobial potential of the LL-37 peptide conjugated with AuNPs (LL-37@AuNPs) [111]. Various organic nanoparticle systems, like liposomes, lipid-based nanoparticles, polymeric micelles, and polymeric nanoparticles, have also been reported in the literature [112]. Today, poly(lactic-co-glycolic acid) (PLGA), an aliphatic polymer with a polyester structure that is formed by copolymerisation of poly-lactic acid (PLA) and poly-glycolic acid (PGA), is the most widely used polymer. PLGA has interesting properties such as controlled and sustained release, low cytotoxicity, tissue and cell biocompatibility, and prolonged residence time [113]. On this basis, it has been approved by the FDA and by the European Medicines Agency (EMA) as a superior drug carrier [114].

12.6.4 Food

AMPs have a good inhibitory effect on bacteria and fungi commonly found in food. Many AMPs are resistant to acids, alkali, and high temperatures and are easily hydrolyzed by proteases in the human body. They are therefore promising alternatives to conventional preservatives.

Nisin is a bacteriocin produced by *Lactococcus lactis* subspecies. It is classified as GRAS by the FDA and is used as a food preservative in a number of countries [115]. Polylysine is the other antimicrobial peptide that is FDA approved as a food additive [116].

Pediocin PA-1 is used as a food additive to inhibit the growth of *Listeriaactive monocytogenes*, which can cause meat deterioration [117]. Enterocin AS-48 is used to preserve cider, fruit, and vegetable juices, and enterocin CCM4231 to preserve soy milk [116, 118].

Encapsulating bacteriocins into liposomes is an original method used to overcome the problems of AMPs in food applications, such as proteolytic degradation or interaction with food ingredients [119].

12.6.5 Animal Husbandry and Aquaculture

Many AMPs have the potential to be used in poultry, swine, and ruminant breeding, as well as aquaculture, because they can improve production performance [120, 121], provide immunity, and promote intestinal health. Some of them have a stronger inhibitory effect on bacterial inflammation if used together with antibiotics [122, 123].

Aquaculture is a fast-growing industry focused on producing high-quality, protein-rich food. Aquaculture farms in fresh- and saltwater environments face challenges in controlling infectious diseases among their animals. Antibiotic usage can lead to antimicrobial resistance. Alternatives to antibiotics have been developed in the past few decades to tackle the threat of antimicrobial resistance. These include AMPs, phage therapy, nanocomposites, vaccines, gene therapy, quorum-sensing prevention, and biofilm inhibition [124].

12.6.6 Agriculture

Plant pathogenic infections of bacteria and fungi cause significant economic losses [125]. The rapid increase of antimicrobial resistance in plant pathogens has urged researchers to develop both new pesticides and management strategies for plant protection. The antimicrobial peptides show potential to eliminate plant pathogenic fungi and bacteria. Tang et al. have published an extensive overview of the status of the work on products and cultures [126].

12.7 Producers

There are scores of companies producing and selling AMPs and related services and expertise around the world. It is not the purpose of this book to provide a complete list; we just report a few to show the diversity of products and services and the geographic delocalization.

Ambio Pharm Inc., Shanghai, China. Produces biologically active peptides, angiotensin, defensin, cell penetrating peptides, glucagon-like peptides, enzyme inhibitors, hepcidins, venom and toxin peptides, receptor agonists and antagonists, linear and cyclic peptides. Has production facilities in Shanghai and North Augusta, CA, USA.

AMP Biotech, San Salvatore Telesino, Italy. Offers AMPs for cosmetic/dermocosmetic, nutraceutical, and pharmaceutical applications.

AMSBIO, Cambridge, MA, USA. Offers peptide modifications, analysis, and purification.

Biosynth, Staad, Switzerland. Manufacturer of GMP-compliant peptides, diagnostic peptides, lipidated peptides, and AMPs in general.

BOC Sciences, Shirley, NY, USA, and London, UK. Supplies functional Active Pharmaceutical Ingredients (APIs), cosmetic and cell penetrating peptides, peptide inhibitors, PNA monomers, resins, and amino acids.

Celdara Medical LLC, Lebanon, NH, USA. Develops products up to the clinical trials which they organize.

Flagship Biotech International Pvt Ltd., Mumbai, India. Produces peptides and modified peptides and offers consultation services and GMP facilities for bulk production of peptides.

Fujifilm Wako Pure Chemical Corporation, Osaka, Japan. Offers hepcidin, defensin, and other AMPs, for example, candidalysin, magainin, and plectasin.

GenScript, Tokyo, Japan. Provides diverse services, from custom peptide synthesis (including long chain peptides, highly hydrophobic peptides, and cyclic peptides) to peptide libraries.

Hycult Biotech, Uden, The Netherlands. Operates in the medical, pharmaceutical, and food industries.

Jiuzhou Pharmaceutical, Taizhou, China. Has GMP peptide production, analytical, and preparative systems.

KS-V Peptides, Hefei, China. Offers peptides and oral peptidomimetic drugs, peptide-targeted conjugated drugs, precision nucleic acid drug delivery with cyclic peptides, and conjugated lipid nanoparticles.

MedChemExpress LLC, Monmouth Junction, NJ, USA. Offers a wide range of structurally and synthetically diverse biologically active compounds for the pharmaceutical industry, as well as customized order volumes.

NovaBiotics Ltd., Aberdeen, UK. Produces amino acids for the cosmetic (nail care and skin care) and pharmaceutical (treatment of pulmonary exacerbations and maintenance of ventilatory functions) industries.

Nuritas, Dublin, Ireland. Active in the food sector. Uses artificial intelligence and DNA analysis to develop novel peptide ingredients.

PH Japan Co., Ltd., Hiroshima, Japan. Designs physiologically active peptides.

Piramal Pharma Solutions, Mumbai, India. Offers commercial supply of API peptides and finished dosage forms, and a range of services including peptide discovery and process development services.

QYABIO (China Peptide Co. Ltd.), Shanghai, China. Specialized in peptide synthesis, including isotope-labeled peptides, fluorescent-labeled peptides, branched peptides, cyclic peptides, and phosphorylated peptides.

SB-Peptides, Saint Égrève, France. Specialized in polypeptide synthesis, peptide pools, fluorescent-labeled peptides, isotope-labeled peptides, branched peptides, cyclic peptides, phosphorylated peptides, disulfide bridge determination, and peptide libraries.

WuXi Biologics, Wuxi, China. Manufactures microbial fermentation-derived products. Has production facilities in China, Singapore, Germany, Ireland, and the United States.

References

1 Huan, Y., Kong, Q., Mou, H., and Yi, H. (2020). Antimicrobial peptides: classification, design, application and research progress in multiple fields. *Frontiers in Microbiology* 11: 582779.

2 Deo, S., Turton, K.L., Kainth, T. et al. (2022). Strategies for improving antimicrobial peptide production. *Biotechnology Advances* 59: 107968.

3 University of Nebraske Medical Center, online September 10, 2025, Antimicrobial Peptide Database. http://aps.unmc.edu/AP.

4 Di Somma, A., Moretta, A., Canè, C. et al. (2020). Antimicrobial and antibiofilm peptides. *Biomolecules* 10: 652.

5 Stepulane, A. (2024). Soft amphiphilic biomaterials for antibacterial applications. PhD Thesis. Chalmers University of Technology, Gothenburg, Sweden. Courtesy of Annija Stepulane.

6 Teixeira, V., Feio, M.J., and Bastos, M. (2012). Role of lipids in the interaction of antimicrobial peptides with membranes. *Progress in Lipid Research* 51: 149–177.

7 Gennaro, R. and Zanetti, M. (2000). Structural features and biological activities of the cathelicidin-derived antimicrobial peptides. *Peptide Science* 55: 31–49.

8 Yeaman, M.R. and Yount, N.Y. (2003). Mechanisms of antimicrobial peptide action and resistance. *Pharmacological Reviews* 55: 27–55.

9 Schittek, B., Hipfel, R., Sauer, B., and et al (2001). Dermcidin: a novel human antibiotic peptide secreted by sweat glands. *Nature Immunology* 2: 1133–1137.

10 Dennison, S.R., Harris, F., Mura, M., and Phoenix, D.A. (2018). An atlas of anionic antimicrobial peptides from amphibians. *Current Protein & Peptide Science* 19: 823–838.

11 Field, C.J. (2005). The immunological components of human milk and their effect on immune development in infants. *The Journal of Nutrition* 135: 1–4.

12 Akalın, A.S. (2014). Dairy-derived antimicrobial peptides: action mechanisms, pharmaceutical uses and production proposals. *Trends in Food Science and Technology* 36: 79–95.

13 Conlon, J.M. and Mechkarska, M. (2014). Host-defense peptides with therapeutic potential from skin secretions of frogs from the family Pipidae. *Pharmaceuticals* 7: 58–77.

14 Lu, Y., Ma, Y., Wang, X. et al. (2008). The first antimicrobial peptide from sea amphibian. *Molecular Immunology* 45: 678–681.

15 Vilcinskas, A. (2013). Evolutionary plasticity of insect immunity. *Journal of Insect Physiology* 59: 123–129.

16 Shelomi, M., Jacobs, C., Vilcinskas, A., and Vogel, H. (2020). The unique antimicrobial peptide repertoire of stick insects. *Developmental & Comparative Immunology* 103: 103471.

17 Zahedifard, F., Lee, H., No, J.H. et al. (2020). Comparative study of different forms of jellein antimicrobial peptide on Leishmania parasite. *Experimental Parasitology* 209: 107823.

18 Cao, J., de la Fuente-Nunez, C., Ou, R.W., and et al. (2018). Yeast-based synthetic biology platform for antimicrobial peptide production. *ACS Synthetic Biology* 7: 896–902.

19 Parachin, N.S., Mulder, K.C., Viana, A.A.B. et al. (2012). Expression systems for heterologous production of antimicrobial peptides. *Peptides* 38: 446–456.

20 Tang, S.-S., Prodhan, Z.H., Biswas, S.K. et al. (2018). Antimicrobial peptides from different plant sources: isolation, characterisation, and purification. *Phytochemistry* 154: 94–105.

21 Semreen, M.H., El-Gamal, M.I., Abdin, S. et al. (2018). Recent updates of marine antimicrobial peptides. *Saudi Pharmaceutical Journal* 26: 396–409.

22 Oh, R., Lee, M.J., Kim, Y.-O. et al. (2020). Myticusin-beta, antimicrobial peptide from the marine bivalve, *Mytilus coruscus. Fish and Shellfish Immunology* 99: 342–352.

23 Kota, S. (2019). Peptide manufacturing methods and challenges. In: *Peptide Therapeutics* (ed. V. Strivastava), 111–150. London, UK: Royal Society of Chemistry.

24 Bersi, G., Barberis, S.E., Origone, A.L. et al. (2018). Bioactive peptides as functional food ingredients. In: *Role of Materials Science in Food Bioengineering* (ed. A.M. Grumezescu and A.M. Holban), 147–186. London, UK: Academic Press.

25 Chauhan, V. and Kanwar, S.S. (2020). Bioactive peptides: synthesis, functions and biotechnological applications. In: *Biotechnological Production of Bioactive Compounds* (ed. M.L. Verma and A.K. Chandel), 107–137. Amsterdam, The Netherlands: Elsevier.

26 Daliri, E.B.-M., Oh, D.H., and Lee, B.H. (2017). Bioactive peptides. *Foods* 6: 32.

27 Suda, S., Field, D., and Barron, N. (2017). Antimicrobial peptide production and purification. *Methods in Molecular Biology* 1485: 401–410.

28 Cruz-Casas, D.E., Aguilar, C.N., Ascacio-Valdés, J.A., and et al. (2021). Enzymatic hydrolysis and microbial fermentation: the most favorable biotechnological methods for the release of bioactive peptides. *Food Chemistry* 3: 100047.

29 Saubenova, M., Oleinikova, Y., Rapoport, A. et al. (2024). Bioactive peptides derived from whey proteins for health and functional beverages. *Fermentation* 10: 359.

30 Martínez-Medina, G.A., Barragán, A.P., Ruiz, H.A. et al. (2019). Fungal proteases and production of bioactive peptides for the food industry. In: *Enzymes in Food Biotechnology* (ed. M. Kuddus), 221–246. London, UK: Academic Press.

31 Cruz-Casas, D.E., Chávez-García, S.N., García-Flores, L.A. et al. (2023). Bioactive peptides from fermented milk products. In: *Enzymes Beyond Traditional Applications in Dairy Science and Technology* (ed. Y.S. Rajput and R. Sharma), 289–311. London, UK: Academic Press.

32 Rizzello, C.G., Cassone, A., Coda, R., and Gobbetti, M. (2011). Antifungal activity of sourdough fermented wheat germ used as an ingredient for bread making. *Food Chemistry* 127: 952–959.

33 Mikhaylin, S., Boussetta, N., Vorobiev, E., and Bazinet, L. (2017). High voltage electrical treatments to improve the protein susceptibility to enzymatic hydrolysis. *ACS Sustainable Chemistry and Engineering* 5: 11706–11714.

34 Ulug, S.K., Jahandideh, F., and Wu, J. (2021). Novel technologies for the production of bioactive peptides. *Trends in Food Science & Technology* 108: 27–39.

35 Park, Y.W. and Nam, M.S. (2015). Bioactive peptides in milk and dairy products: a review. *Korean Journal for Food Science and Animal Resources* 35: 831–840.

36 Chai, K.F., Voo, A.Y.H., and Chen, W.N. (2020). Bioactive peptides from food fermentation: a comprehensive review of their sources, bioactivities, applications, and future development. *Comprehensive Reviews in Food Science and Food Safety* 19: 3825–3885.

37 Kumari, R., Sanjukta, S., Sahoo, D., and Rai, A.K. (2022). Functional peptides in Asian protein rich fermented foods: production and health benefits. *Systems Microbiology and Biomanufacturing* 2: 1–13.

38 Lafarga, T. and Hayes, M. (2017). Effect of pre-treatment on the generation of dipeptidyl peptidase-IV-and prolyl endopeptidase-inhibitory hydrolysates from bovine lung. *Irish Journal of Agricultural and Food Research* 56: 12–24.

39 Melini, F., Melini, V., Luziatelli, F. et al. (2019). Health-promoting components in fermented foods: an up-to-date systematic review. *Nutrients* 11: 1189.

40 Melville, P.A., Benites, N.R., Ruz-Peres, M., and Yokoya, E. (2011). Proteinase and phospholipase activities and development at different temperatures of yeasts isolated from bovine milk. *The Journal of Dairy Research* 78: 385–390.

41 Hayes, M. and García-Vaquero, M. (2016). Bioactive compounds from fermented food products. In: *Novel Food Fermentation Technologies* (ed. K.S. Ojha and B.K. Tiwari), 293–310. Cham, Switzerland: Springer.

42 Vriens, K., Cammue, B.P.A., and Thevissen, K. (2014). Antifungal plant defensins: mechanisms of action and production. *Molecules* 19: 12280–12303.

43 Chahardoli, M., Fazeli, A., and Ghabooli, M. (2018). Recombinant production of bovine Lactoferrin-derived antimicrobial peptide in tobacco hairy roots expression system. *Plant Physiology and Biochemistry* 123: 414–421.

44 Shanmugaraj, B., Bulaon, C.J.I., Malla, A., and Phoolcharoen, W. (2021). Biotechnological insights on the expression and production of antimicrobial peptides in plants. *Molecules* 26: 4032.

45 Sinha, R. and Shukla, P. (2019). Antimicrobial peptides: recent insights on biotechnological interventions and future perspectives. *Protein and Peptide Letters* 26: 79–87.

46 Tiwari, P., Srivastava, Y., Sharma, A. et al. (2023). Antimicrobial peptides: the production of novel peptide-based therapeutics in plant systems. *Life* 13: 1875.

47 Lin, T.-Y. and Weibel, D.B. (2016). Organization and function of anionic phospholipids in bacteria. *Applied Microbiology and Biotechnology* 100: 4255–4267.

48 Enoki, T.A., Moreira-Silva, I., Lorenzon, E.N. et al. (2018). Antimicrobial peptide K^0-W^6-Hya1 induces stable structurally modified lipid domains in anionic membranes. *Langmuir* 34: 2014–2025.

49 Renne, M.F. and de Kroon, A.I.P.M. (2018). The role of phospholipid molecular species in determining the physical properties of yeast membranes. *FEBS Letters* 592: 1330–1345.

50 Oren, Z. and Shai, Y. (1998). Mode of action of linear amphipathic α-helical antimicrobial peptides. *Biopolymers* 47: 451–463.

51 Lyu, Y., Fitriyanti, M., and Narsimhan, G. (2019). Nucleation and growth of pores in 1,2-dimyristoyl-sn-glycero-3-phosphocholine (DMPC)/cholesterol bilayer by antimicrobial peptides melittin, its mutants and cecropin P1. *Colloids and Surfaces B* 173: 121–127.

52 Shenkarev, Z.O., Balandin, S.V., Trunov, K.I. et al. (2011). Molecular mechanism of action of β-hairpin antimicrobial peptide arenicin: oligomeric structure in dodecylphosphocholine micelles and pore formation in planar lipid bilayers. *Biochemistry* 50: 6255–6265.

53 Corrêa, J.A.F., Evangelista, A.G., de Melo Nazareth, T. et al. (2019). Fundamentals on the molecular mechanism of action of antimicrobial peptides. *Materialia* 8: 100494.

54 Lohner, K. and Prossnigg, F. (2009). Biological activity and structural aspects of PGLa interaction with membrane mimetic systems. *Biochimica et Biophysica Acta* 1788: 1656–1666.

55 Matsuzaki, K., Murase, O., Fujii, N., and Miyajima, K. (1995). Translocation of a channel-forming antimicrobial peptide, magainin 2, across lipid bilayers by forming a pore. *Biochemistry* 34: 6521–6526.

56 Matsuzaki, K., Murase, O., Fujii, N., and Miyajima, K. (1996). An antimicrobial peptide, magainin 2, induced rapid flip-flop of phospholipids coupled with pore formation and peptide translocation. *Biochemistry* 35: 11361–11368.

57 Omardien, S., Drijfhout, J.W., Vaz, F.M. et al. (2018). Bactericidal activity of amphipathic cationic antimicrobial peptides involves altering the membrane fluidity when interacting with the phospholipid bilayer. *Biochimica Et Biophysica Acta - Biomembranes* 1860: 2404–2415.

58 Mardirossian, M., Grzela, R., Giglione, C. et al. (2014). The host antimicrobial peptide Bac71-35 binds to bacterial ribosomal proteins and inhibits protein synthesis. *Chemistry & Biology* 21: 1639–1647.

59 Le, C.F., Gudimella, R., Razali, R. et al. (2016). Transcriptome analysis of *Streptococcus pneumoniae* treated with the designed antimicrobial peptides, DM3. *Scientific Reports* 6: 26828.

60 Kragol, G., Lovas, S., Varadi, G. et al. (2001). The antibacterial peptide pyrrhocoricin inhibits the ATPase actions of DnaK and prevents chaperone-assisted protein folding. *Biochemistry* 40: 3016–3026.

61 Le, C.-F., Fang, C.-M., and Sekaran, S.D. (2017). Intracellular targeting mechanisms by antimicrobial peptides. *Antimicrobial Agents and Chemotherapy* 61: e02340-16.

62 Wrońska, A.K. and Boguś, M.I. (2020). Heat shock proteins (HSP 90, 70, 60, and 27) in galleria mellonella (lepidoptera) hemolymph are affected by infection with *Conidiobolus coronatus* (entomophthorales). *PLoS One* 15: e0228556.

63 Subbalakshmi, C. and Sitaram, N. (1998). Mechanism of antimicrobial action of indolicidin. *FEMS Microbiology Letters* 160: 91–96.

64 He, S.-W., Zhang, J., Li, N.-Q. et al. (2017). A TFPI-1 peptide that induces degradation of bacterial nucleic acids and inhibits bacterial and viral infection in half-smooth tongue sole, *Cynoglossus semilaevis*. *Fish and Shellfish Immunology* 60: 466–473.

65 Shu, G., Chen, Y., Liu, T. et al. (2019). Antimicrobial peptide cathelicidin-BF inhibits platelet aggregation by blocking protease-activated receptor 4. *International Journal of Peptide Research and Therapeutics* 25: 349–358.

66 Lutkenhaus, J. (1990). Regulation of cell division in *E. Coli*. *Trends in Genetics* 6: 22–25.

67 Li, L., Sun, J., Xia, S. et al. (2016). Mechanism of antifungal activity of antimicrobial peptide APP, a cell-penetrating peptide derivative, against *Candida albicans*:

intracellular DNA binding and cell cycle arrest. *Applied Microbiology and Biotechnology* 100: 3245–3253.

68 Cruz, G.F., de Araujo, I., Torres, M.D.T. et al. (2020). Photochemically-generated silver chloride nanoparticles stabilized by a peptide inhibitor of cell division and its antimicrobial properties. *Journal of Inorganic and Organometallic Polymers and Materials* 30: 2464–2474.

69 Büyükkiraz, M.E. and Kesmen, Z. (2022). Antimicrobial peptides (AMPs): a promising class of antimicrobial compounds. *Journal of Applied Microbiology* 132: 1573–1596.

70 Chen, C.H. and Lu, T.K. (2020). Development and challenges of antimicrobial peptides for therapeutic applications. *Antibiotics* 9: 24.

71 Moretta, A., Scieuzo, C., Petrone, A.M. et al. (2021). Antimicrobial peptides: a new hope in biomedical and pharmaceutical fields. *Frontiers in Cellular and Infection Microbiology* 11: 668632.

72 Hancock, R.E.W. and Sahl, H.G. (2006). Antimicrobial and host-defense peptides as new anti-infective therapeutic strategies. *Nature Biotechnology* 24: 1551–1557.

73 Sunderkötter, C. and Becker, K. (2015). Frequent bacterial skin and soft tissue infections: diagnostic signs and treatment. *Journal der Deutschen Dermatological Gesellschaft* 13: 495–616.

74 Kanaujia, K.A., Mishra, N., Rajinikanth, P.S., and Saraf, S.A. (2024). Antimicrobial peptides as antimicrobials for wound care management: a comprehensive review. *Journal of Drug Delivery Science and Technology* 95: 105570.

75 Moravvej, H., Memariani, M., Memariani, H. et al. (2021). Can antimicrobial peptides be repurposed as a novel therapy for keloids? *Dermatology* 237: 293–295.

76 Alencar-Silva, T., Braga, M.C., Santana, G.O.S. et al. (2018). Breaking the frontiers of cosmetology with antimicrobial peptides. *Biotechnology Advances* 36: 2019–2031.

77 Borro, B.C., Nordström, R., and Malmsten, M. (2020). Microgels and hydrogels as delivery systems for antimicrobial peptides. *Colloids & Surfaces B* 187: 110835.

78 Torres, M.D.T., Sothiselvam, S., Lu, T.K., and de la Fuente-Nunez, C. (2019). Peptide design principles for antimicrobial applications. *Journal of Molecular Biology* 431: 3547–3567.

79 Izadi, N., Keikha, M., Ghazvini, K., and Karbalaei, M. (2020). Oral antimicrobial peptides and new therapeutic strategies for plaque-mediated diseases. *Gene Reports* 21: 100811.

80 Chen, L., Jia, L., Zhang, Q. et al. (2017). A novel antimicrobial peptide against dental-caries-associated bacteria. *Anaerobe* 47: 165–172.

81 Khan, S.A. and Lee, C.-S. (2020). Recent progress and strategies to develop antimicrobial contact lenses and lens cases for different types of microbial keratitis. *Acta Biomaterialia* 113: 101–118.

82 Madanchi, H., Shoushtari, M., Kashani, H.H. et al. (2019). Antimicrobial peptides of the vaginal innate immunity and their role in the fight against sexually transmitted diseases. *New Microbes and New Infections* 34: 100627.

83 Muhialdin, B.J., Algboory, H.L., Kadum, H. et al. (2020). Antifungal activity determination for the peptides generated by Lactobacillus plantarum TE10 against Aspergillus flavus in maize seeds. *Food Control* 109: 106898.

84 Wu, D., Gao, Y., Qi, Y. et al. (2014). Peptide-based cancer therapy: opportunity and challenge. *Cancer Letters* 351: 13–22.

85 Ma, R., Wong, S.W., Ge, L. et al. (2019). in vitro and MD simulation study to explore physicochemical parameters for antibacterial peptide to become potent anticancer peptide. *Molecular Therapy Oncolytics* 16: 7–19.

86 Arias, M., Haney, E.F., Hilchie, A.L. et al. (2020). Selective anticancer activity of synthetic peptides derived from the host defence peptide tritrpticin. *Biochimica et Biophysica Acta – Biomembranes* 1862: 183228.

87 Jung, Y., Kong, B., Moon, S. et al. (2019). Envelope-deforming antiviral peptide derived from influenza virus M2 protein. *Biochemical and Biophysical Research Communications* 517: 507–512.

88 Neshani, A., Zare, H., Eidgahi, M.R.A. et al. (2019). Epinecidin-1, a highly potent marine antimicrobial peptide with anticancer and immunomodulatory activities. *BMC Pharmacology and Toxicology* 20: 33.

89 Cao, L., Jiang, W., Cao, S. et al. (2019). in vitro leishmanicidal activity of antimicrobial peptide KDEL against *Leishmania tarentolae. Acta Biochimica et Biophysica Sinica* 51: 1286–1292.

90 Ashkenazi, A., Wexler-Cohen, Y., and Shai, Y. (2011). Multifaceted action of Fuzeon as virus-cell membrane fusion inhibitor. *Biochimica et Biophysica Acta – Biomembranes* 1808: 2352–2358.

91 Guaní-Guerra, E., Santos-Mendoza, T., Lugo-Reyes, S.O., and Terán, L.M. (2010). Antimicrobial peptides: general overview and clinical implications in human health and disease. *Clinical Immunology* 135: 1–11.

92 Stepulane, A., Rajasekharan, A.K., and Andersson, M. (2024). Antibacterial efficacy of antimicrobial peptide-functionalized hydrogel particles combined with vancomycin and oxacillin antibiotics. *International Journal of Pharmaceutics* 664: 124630.

93 Blomstrand, E., Rajasekharan, A.K., Atefyekta, S., and Andersson, M. (2022). Cross-linked lyotropic liquid crystal particles functionalized with antimicrobial peptides. *International Journal of Pharmaceutics* 627: 122215.

94 Kazemzadeh-Narbat, M., Cheng, H., Chabok, R. et al. (2021). Strategies for antimicrobial peptide coatings on medical devices: a review and regulatory science perspective. *Critical Reviews in Biotechnology* 41: 94–120.

95 Research and Markets. 21 October 2024. https://www.researchandmarkets.com (accessed 1 March 2024).

96 PepTherDia. http://peptherdia.herokuapp.com (accessed 1 March 2024).

97 Sharma, K., Sharma, K.K., Sharma, A., and Jain, R. (2023). Peptide-based drug discovery: current status and recent advances. *Drug Discovery Today* 28: 103464.

98 Muttenthaler, M., King, G.F., Adams, D.J., and Alewood, P.F. (2021). Trends in peptide drug discovery. *Nature Reviews Drug Discovery* 20: 309–325.

99 Anand, U., Bandyopadhyay, A., Jha, N.K. et al. (2023). Translational aspect in peptide drug discovery and development: an emerging therapeutic candidate. *BioFactors* 49: 251–269.

100 Rossino, G., Marchese, E., Galli, G. et al. (2023). Peptides as therapeutic agents: challenges and opportunities in the green transition era. *Molecules* 28: 7165.

101 U.S Food and Drug Administration (1948). FDA Approved Drug Products. http://www.accessdata.fda.gov/scripts/cder/drugsatfda/index.cfm?fuseaction=Search.DrugDetails (accessed 1 March 2024).

102 Tran, T.B., Velkov, T., Nation, R.L. et al. (2016). Pharmacokinetics/pharmacodynamics of colistin and polymyxin B: are we there yet? *International Journal of Antimicrobial Agents* 48: 592–597.

103 Roscia, G., Falciani, C., Bracci, L., and Pini, A. (2013). The development of antimicrobial peptides as new antibacterial drugs. *Current Protein & Peptide Science* 14: 641–649.

104 Gregoire, N., Chauzy, A., Buyck, J. et al. (2021). Clinical pharmacokinetics of daptomycin. *Clinical Pharmacokinetics* 60: 271–281.

105 Tally, F.P. and DeBruin, M.F. (2000). Development of daptomycin for gram-positive infections. *The Journal of Antimicrobial Chemotherapy* 46: 523–526.

106 LiverTox: Clinical and Research Information on Drug-Induced Liver Injury [Internet]. Bethesda (MD): National Institute of Diabetes and Digestive and Kidney Diseases; 2012-. Vancomycin. [Updated 2020 Aug 20]. https://www.ncbi.nlm.nih.gov/books/NBK548881 (accessed 1 March 2024).

107 Zhanel, G.G., Calic, D., Schweizer, F. et al. (2010). New lipoglycopeptides: a comparative review of dalbavancin, oritavancin and telavancin. *Drugs* 70: 859–886.

108 Cresti, L., Cappello, G., and Pini, A. (2024). Antimicrobial peptides towards clinical application – a long history to be concluded. *International Journal of Molecular Sciences* 25: 4870.

109 Casciaro, B., Ghirga, F., Quaglio, D., and Mangoni, M.L. (2020). Inorganic gold and polymeric poly(lactide-co-glycolide) nanoparticles as novel strategies to ameliorate the biological properties of antimicrobial peptides. *Current Protein & Peptide Science* 21: 429–438.

110 Casciaro, B., Moros, M., Rivera-Fernández, S. et al. (2017). Gold-nanoparticles coated with the antimicrobial peptide esculentin-1a(1-21)NH_2 as a reliable strategy for antipseudomonal drugs. *Acta Biomaterialia* 47: 170–181.

111 Subaer, S., Hartati, H., Ramadhan, I. et al. (2023). A simple synthesis and microstructure analysis of human peptide LL-37@Gold nanoparticles (known as LL-37@AuNPs) conjugates as antimicrobials and substances for wound healing. *Materials* 16: 7675.

112 Eleraky, N.E., Allam, A., Hassan, S.B., and Omar, M.M. (2020). Nanomedicine fight against antibacterial resistance: an overview of the recent pharmaceutical innovations. *Pharmaceutics* 12: 142.

113 Lu, Y., Cheng, D., Niu, B. et al. (2023). Properties of poly(lactic-co-glycolic acid) and progress of poly(lactic-co-glycolic acid)-based biodegradable materials in biomedical research. *Pharmaceuticals* 16: 454.

114 Sharma, S., Parmar, A., Kori, S., and Sandhir, R. (2016). PLGA-based nanoparticles: a new paradigm in biomedical applications. *Trends in Analytical Chemistry* 80: 30–40.

115 Khan, I. and Oh, D.-H. (2016). Integration of nisin into nanoparticles for application in foods. *Innovative Food Science & Emerging Technologies* 34: 376–384.

116 Santos, J.C.P., Sousa, R.C.S., Otoni, C.G. et al. (2018). Nisin and other antimicrobial peptides: production, mechanisms of action, and application in active food packaging. *Innovative Food Science & Emerging Technologies* 48: 179–194.

117 Settanni, L. and Corsetti, A. (2008). Application of bacteriocins in vegetable food biopreservation. *International Journal of Food Microbiology* 121: 123–138.

118 Rai, M., Pandit, R., Gaikwad, S., and Kövics, G. (2016). Antimicrobial peptides as natural bio-preservative to enhance the shelf-life of food. *Journal of. Food Science and Technology* 53: 3381–3394.

119 da Silva Malheiros, P., Daroit, D.J., and Brandelli, A. (2010). Food applications of liposome-encapsulated antimicrobial peptides. *Trends in Food Science and Technologies* 21: 284–292.

120 Liu, T., She, R., Wang, K. et al. (2008). Effects of rabbit sacculus rotundus antimicrobial peptides on the intestinal mucosal immunity in chickens. *Poultry Science.* 87: 250–254.

121 Bao, H., She, R., Liu, T. et al. (2009). Effects of pig antibacterial peptides on growth performance and intestine mucosal immune of broiler chickens. *Poultry Science* 88: 291–297.

122 Wang, J., Dou, X., Song, J. et al. (2019). Antimicrobial peptides: promising alternatives in the post feeding antibiotic era. *Medicinal Research Reviews* 39: 831–859.

123 Cote, C.K., Blanco, I.I., Hunter, M. et al. (2020). Combinations of early generation antibiotics and antimicrobial peptides are effective against a broad spectrum of bacterial biothreat agents. *Microbial Pathogenesis* 142: 104050.

124 Mohaideen, N.S.M.H., Vaani, S., and Hemalatha, S. (2023). Antimicrobial peptides. *Current Pharmacology Reports* 9: 433–454.

125 Liu, S., Wang, W., Deng, L. et al. (2019). Control of sour rot in citrus fruit by three insect antimicrobial peptides. *Postharvest Biology and Technology* 149: 200–208.

126 Tang, R., Tan, H., Dai, Y. et al. (2023). Application of antimicrobial peptides in plant protection: making use of the overlooked merits. *Frontiers in Plant Science* 14: 1139539.

13

Lipopolysaccharides

13.1 Introduction

Polysaccharides with hydrophobic side chains can be very surface active. This has been exploited in a large number of semisynthetic polymers, such as hydrophobically modified cellulose and starch derivatives. Different hydrophobic modifications of pullulan, a maltotriose-based polysaccharide, have also been investigated for the solubilization of membrane proteins [1]. The side chains of these polymeric amphiphiles are typically hydrocarbon chains of 12–18 carbon atoms covalently attached to the polysaccharide backbone. Hydrophobically modified ethylhydroxyethylcellulose (HM-EHEC), see Figure 13.1, is a representative example. In a typical product, every 20th anhydroglucose ring carries the hydrophobic substituent. HM-EHEC and similar products are commercially important and widely used to control the rheology of coating formulations. They are also used as stabilizers of emulsions and suspensions [2].

HM-EHEC is a nonionic amphiphilic polymer. The side chains are typically attached to the backbone polymer by reacting ethylhydroxyethylcellulose with a glycidyl ether of a long-chain fatty alcohol. Sometimes the fatty alcohol is first ethoxylated before it is reacted with epichlorohydrin to give the glycidyl ether, as is the case for the molecule shown in Figure 13.1. This procedure results in a hydrolytically stable ether linkage. There are also anionic amphiphilic cellulose derivatives, e.g. hydrophobically modified carboxymethylcellulose (HM-CMC). HM-CMC is typically prepared by reacting carboxymethylcellulose with a long-chain amine to form an amide bond to some of the carboxyl groups, leaving the majority of the carboxyl groups unreacted to give an anionic character to the surface-active polymer [3].

Even if the backbone polysaccharide in such substances is a natural polymer, such as cellulose and starch, the product is a man-made, not a natural, amphiphile. Such compounds are therefore outside the scope of this book. However, as will be discussed further down in this chapter, nature provides similar products.

Before providing examples of natural hydrophobically modified polysaccharides, often referred to as lipopolysaccharides (LPS), it is interesting to note that cellulose, the most abundant polymer on earth, also exhibits surface activity. This may sound surprising because one normally does not associate amphiphilicity with homopolymers like a plain polysaccharide. However, the glucose ring can attain a conformation that makes one side hydrophobic and the other side hydrophilic. A well-known example of this behavior is

Natural Surfactants: Science, Technology, and Industrial Aspects, First Edition. Guido Bognolo and Krister Holmberg.

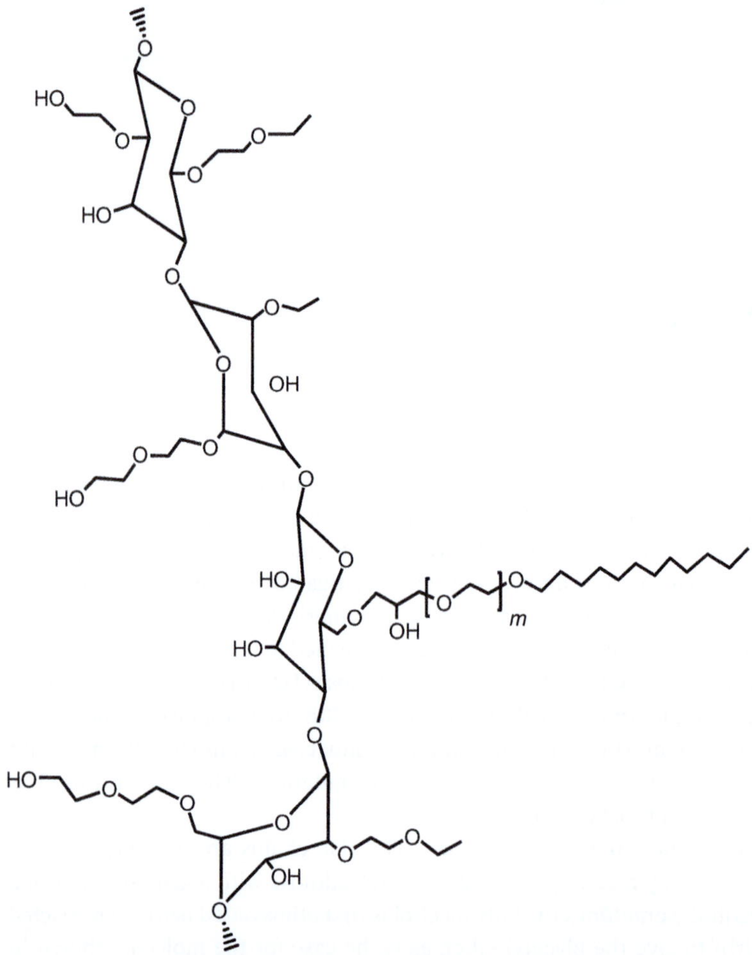

Figure 13.1 Hydrophobically modified ethylhydroxyethylcellulose (HM-EHEC). The glycidyl ether of an ethoxylated fatty alcohol is reacted at the 6-position of an anhydroglucose ring. (Courtesy of Dr. Leif Karlson, Nouryon, Stenungsund. Sweden.)

cyclodextrin, also a homopolymer composed of anhydroglucose units. Cyclodextrin is water soluble while at the same time it can solubilize hydrophobic molecules in its interior. This behavior makes cyclodextrin of interest for the delivery of hydrophobic drugs [4].

As pointed out by Lindman, the insolubility of cellulose is due to the specific spatial arrangement of the polysaccharide chains [5, 6]. A multitude of inter- and intramolecular hydrogen bonding results in the formation of flat ribbons with sides that may either be polar or nonpolar. In an aqueous environment, hydrophobic interactions will cause the nonpolar sides to adhere to each other, which results in extreme water insolubility. The polar sides of the ribbons will repel hydrocarbons and other nonpolar solvents. Thus, cellulose is an example of a substance that is both hydrophobic and oleophobic. This behavior is very different from that of starch, where the anhydroglucose rings do not form ribbons with distinct polar and nonpolar characteristics. The difference in the 1,4-glycosidic linkage, α (for starch) or β (for cellulose), is evidently decisive for the orientation, which in turn governs many of the properties. Starch is a water-soluble polymer (although the dissolution can be tedious).

Even simpler homopolymers than polysaccharides can exhibit amphiphilicity. Poly(ethylene glycol) (PEG) is surface active and reduces the surface tension of water considerably. The amphiphilicity is explained by the conformation. The conformation of the oxyethylene group that has the lowest energy is the one that is *gauche* around the C—C bond and *anti* around the C—O bond. This conformation has a large dipole moment because the ethylene moieties and the ether oxygen atoms are oriented in different directions. This dipole moment gives rise to strong hydration of the polyoxyethylene chains, which, in turn, leads to high water solubility of PEGs and of nonionic surfactants with reasonably long polyoxyethylene chains [7].

13.2 Chemical Structures and Properties

Figure 13.2 shows a schematic of a lipopolysaccharide. Hydrophobic chains are attached to a polysaccharide backbone, usually by ester bonds but also by amide bonds if the polysaccharide contains glucosamine or some other amino sugar.

The degree of substitution (DS), i.e. the average number of hydroxyl groups per repeating unit that are replaced by a substituent, can vary widely. A low DS will give a polymer with high water solubility but low surface activity. The surface activity will increase with increasing DS because the higher the concentration of hydrophobic side chains, the larger will be the driving force for the polymer to go to the air–water interface where it will attain a configuration such that the hydrophobic chains will protrude into the air and the polysaccharide backbone will reside in the aqueous phase. This is analogous to the behavior of low-molecular-weight amphiphiles, which go to the air–water interface, where they form a closely packed monolayer if the bulk concentration is high enough. Increasing the concentration of the low-molecular-weight amphiphile further will lead to the formation of micelles and the concentration where the first micelles are formed is called the critical micelle concentration, or the CMC. Polymeric surfactants, such as LPS, can also form micelles, but the micelles formed will often be intramolecular, not intermolecular, if the

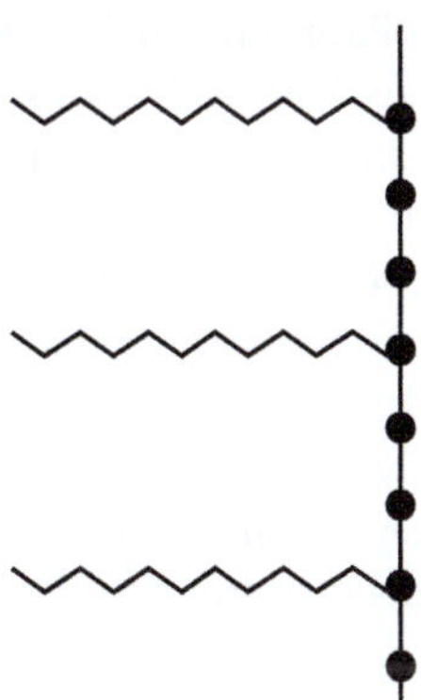

Figure 13.2 An illustration of a lipopolysaccharide. The structure is very simplified. The polysaccharide chain (the connected black dots) need not be linear. As will be discussed below, the biological LPS often have a branched backbone. The lipophilic side chains may also be branched.

Figure 13.3 Schematic structure of a lipopolysaccharide of the type present in the outer membrane of Gram-negative bacteria. (Mike Jones / Wikimedia Commons / CC BY-SA 3.0.)

DS is high. Such intramolecular micelles are closed aggregates with the lipid chains in the center and the polysaccharide backbone at the periphery. Such structures lack surface activity and are usually not very useful. Thus, there is usually an optimum in DS, which will depend on variables such as the length of the polysaccharide chain and of the lipophilic side chains, as well as on whether the polysaccharide chain contains charged sugar units.

In biology, the term lipopolysaccharide is intimately associated with a component of the outer membrane of Gram-negative bacteria. They often go under the name LPS and they are the most abundant antigen on the cell surface of most Gram-negative bacteria. They contribute to the structural integrity of the bacteria, and they provide the membrane with steric stabilization.

These membrane LPS typically consist of three distinct parts: an outer polysaccharide called the O-antigen, an inner core oligosaccharide, and finally Lipid A. The structure is schematically shown in Figure 13.3. As can be seen, the structure of the membrane LPS is much more complex than that of the idealized molecule shown in Figure 13.2, as well as of the semisynthetic polymer shown in Figure 13.1. Bacterial LPS are large molecules.

These LPS are endotoxins, and the toxicity of Gram-negative bacteria is caused by Lipid A. It is the six long acyl chains of lipid A that constitute the hydrophobic part of LPS, the part that anchors the large molecule to the membrane. The endotoxicity is governed by the acyl chain length and by the substitution pattern [8]. The structure of Lipid A can vary but Figure 13.4, which shows *Escherichia coli* Lipid A, is a representative example. As can be seen, it is a bisphosphorylated diglucosamine with six long acyl chains attached by ester and amide bonds.

The polysaccharide, the O-antigen, protrudes from the cell surface. These chains are not linear, they contain branches, which are not shown on the figure. It is the long polysaccharide chains that provide the steric stabilization, which prevents certain types of chemical attack.

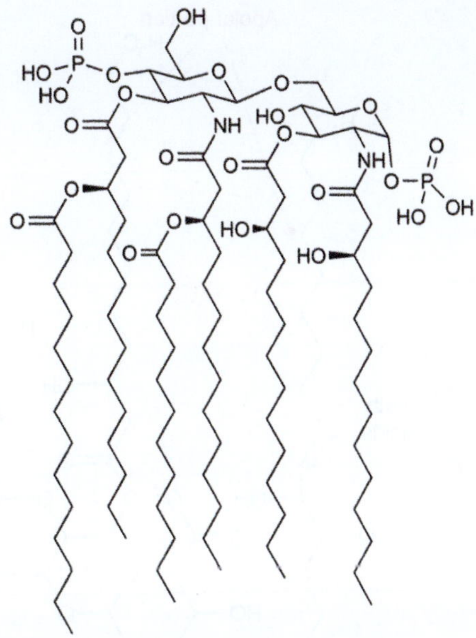

Figure 13.4 Structure of Lipid A. (Edgar181 / Wikimedia Commons / CC BY 4.0.)

13.3 The Emulsan Story

Emulsan is probably the best-known example of a lipopolysaccharide with pronounced emulsifying properties. Such biopolymers are sometimes referred to as bioemulsifiers. They are usually not very effective in reducing the surface tension of water, but they are effective in reducing the interfacial tension between oil and water. They can give remarkably stable oil-in-water emulsions. In the literature, there is often a distinction between bioemulsifiers and biosurfactants. Bioemulsifiers are high-molecular-weight molecules, often a complex between heterolipopolysaccharides and proteins while biosurfactants are low-molecular-weight amphiphiles, such as sophorolipids, rhamnolipids, and lipopeptides [9, 10]. Bioemulsifiers are also effective in solubilizing poorly water-soluble substances.

Emulsan is a product produced by *Acinetobacter* strains, in particular *Acinetobacter calcoaceticus*. It was initially believed to be an anionic lipoheteropolysaccharide composed of the three amino sugars D-galactosamine, D-galactosamineuronic acid, and diamino-6-deoxy-D-glucose, present in a ratio of 1:1:1. The molecular weight was reported to be around 1000 kDa [11]. It has later been found that Emulsan is a mixture of two main components, a lipopolysaccharide and a high-molecular-weight cationic exopolysaccharide. The ratio between the two is 80:20 [12]. Both components are important for the physical chemical behavior of the product. The lipopolysaccharide is very heterogeneous and contains a protein fraction in addition to the polysaccharide and the long alkyl chains. Figure 13.5 shows a partial structure of the lipopolysaccharide.

Emulsan was found to be an extremely efficient stabilizer for emulsions of crude oil in water, an observation that triggered an interest in using it for the transport of heavy oil. Pumping heavy, high viscous oil in pipelines requires heating, which is difficult and costly, particularly in cold climates. Since the viscosity of an emulsion is mainly governed

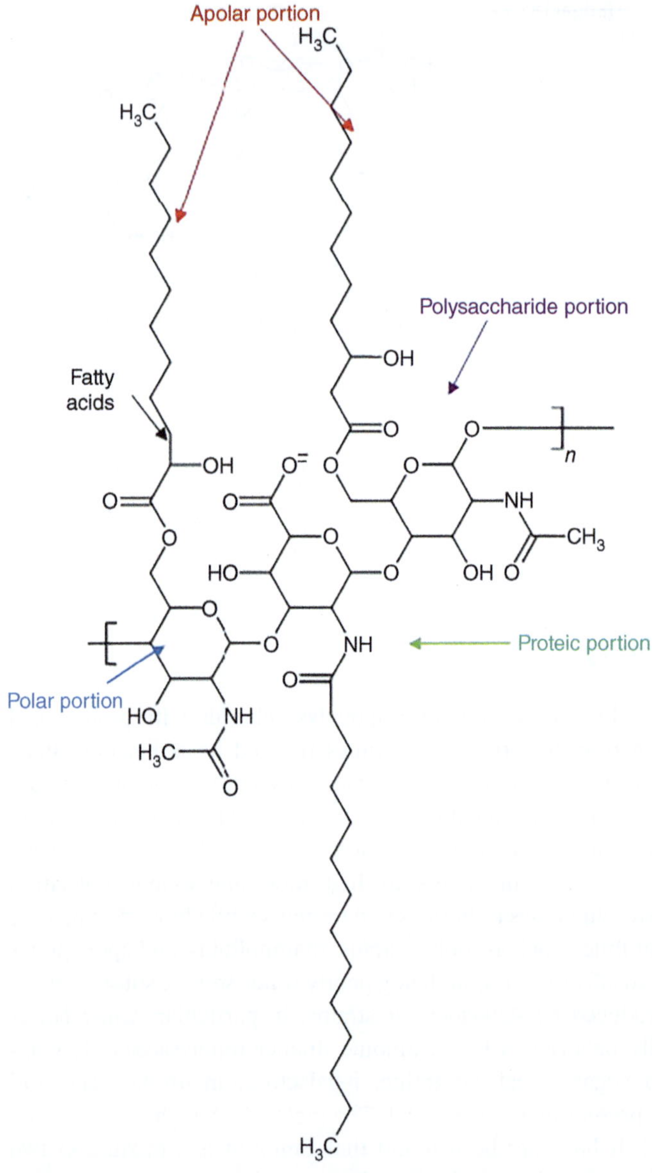

Figure 13.5 Partial structure of the lipopolysaccharide component of Emulsan. (Reprinted from Portal D'Almeida et al. [13] / with permission of Elsevier.)

by the external phase, a crude oil-in-water emulsion was seen as an interesting alternative. The emulsion could subsequently be broken by addition of a demulsifier, or it could be used as such in the burner if the water content was low enough. In the 1970s and 80s, there was considerable interest in Emulsan for the transport of heavy oil, and plans were made for very large Emulsan production.

However, the heavy oil transport concept failed to materialize, at least on a large scale. One contributing factor was that the stability of the bioemulsifier could not be guaranteed and if the emulsion broke during transport in a long pipeline, the consequences would be dire. Emulsan found other, but smaller, applications instead.

Emulsan and other related bioemulsifiers are used for remediation of oil spills. They disperse the hydrocarbons, thus increasing their bioavailability and rate of biodegradation. There are several examples in the literature on the use of bioemulsifiers produced by *Acinetobacter* strains to treat oil spills in marine environments [13]. For instance, Portal D'Almeida et al. report on the use of an *Acinetobacter* strain isolated from an oil spill off the coast of Brazil for bioremediation. The bioemulsifier was found to be robust to variations in temperature, pH, and salt concentration and was found to be efficient both on oil spills in the water and on oil-contaminated sand. It also reduced the stains of oil on stones [14]. The use of bioemulsifiers for remediation of soil polluted with hydrocarbons is reviewed by Calvo et al. [15].

Bioremediation of oil spills can also be viewed as a natural process. The effects of the disaster with the oil rig Deepwater Horizon in 2010 were studied in detail by environmentalists and marine researchers in general. In reports subsequently published about the event, regarded as the largest oil spill in history, it was shown that bacteria inside the oil slick degraded the oil at a rate five times faster than microbes outside the slick. At the time, the rate of the spontaneous disappearance of the oil surprised those who oversaw the disaster. The bacteria evidently produced biosurfactants to disperse the oil and increase the oil–water interfacial area and then started to chow down on the oil. The rate at which the oil disappeared surprised the microbiologist, in particular since the bacteria had a limited supply of nutrients such as nitrogen and phosphorous. Figure 13.6 illustrates bioremediation of an oil spill.

Emulsan has also found use in formulations for cleaning of oil tanks and similar equipment. An efficient dispersion of the oil into small drops is an efficient way to speed up the cleaning process and biological components are viewed favorably because they are regarded as environmentally benign. Bioemulsifiers produced by *Acinetobacter junii* have been tested in microbial enhanced oil recovery (MEOR). In this process, the bioemulsifier-producing bacteria are injected together with the appropriate nutrients into the reservoir and low-molecular-weight biosurfactants together with macromolecular bioemulsifiers are generated in situ [16].

More recently, Emulsan has found applications also outside the petroleum industry. It is used as a stabilizer in the food and beverage industry. It is particularly useful as a stabilizer of emulsion-based products such as sauces, creams, mayonnaise, and baked goods [17]. The low toxicity of Emulsan and the nonpathogenic profile of most *Acinetobacter* species are valuable assets for this application.

An interesting observation made when Emulsan was explored as an emulsifier and stabilizer for oil-in-water emulsions was that it was "substrate specific." It was neither very efficient on aliphatic hydrocarbons nor on aromatics; however, with a mixture of aliphatic and aromatic hydrocarbons, it gave an extremely stable emulsion. This is probably not a coincidence. The Emulsan-producing bacterium uses crude oil as the carbon source and the bioemulsifier is produced to provide a crude oil–water interface that is as large as possible [18].

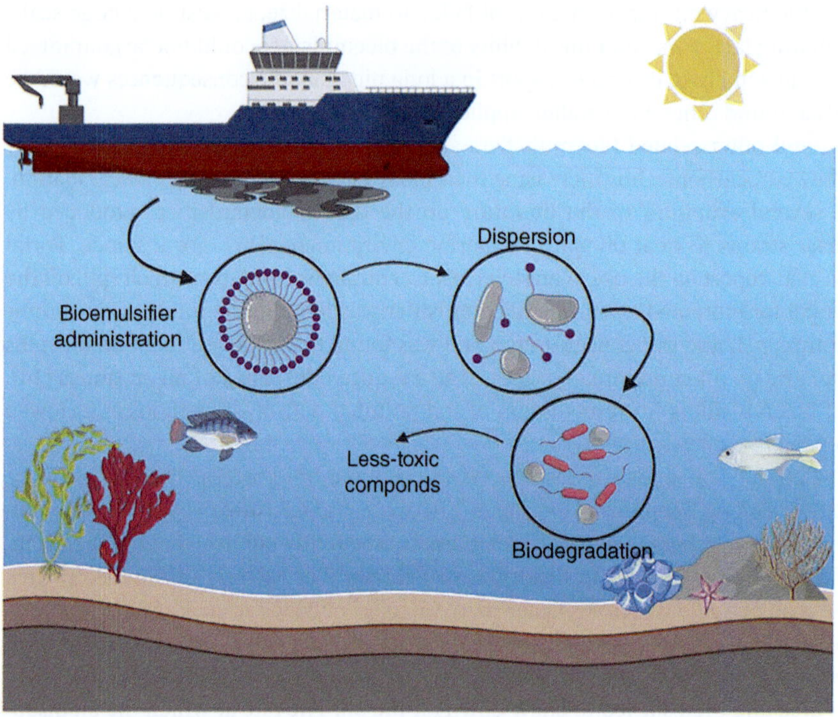

Figure 13.6 An oil spill bioremediation process accelerated by the administration of bioemulsifiers. (Portal D'Almeida et al. [13] / with permission of Elsevier.)

Emulsan has also been produced from other carbon sources. Rocha et al. used natural cashew apple juice as fermentation medium for production of a biosurfactant by the bacterium *Acinetobacter calcoaceticus* [19]. The cashew nut industry is very important in Brazil, so a residue from this industry is an inexpensive substrate. The product obtained was essentially the same as the one obtained using crude oil as carbon source.

The low toxicity of Emulsan together with its ability to create stable emulsions has triggered an interest in the product in the pharmaceutical industry, notably for delivery of poorly water-soluble drugs [20].

13.4 Other Lipopolysaccharides

Other bacteria than *Acinetobacter* strains also produce bioemulsifiers of lipopolysaccharide type, usually together with low-molecular-weight amphiphilic compounds. One example is *Mucor hiemalis* isolated from Caatinga soil [21]. The product obtained from the fermentation using waste soybean oil as carbon source was very surface active, reducing the surface tension of water from 72 to 32 mN/m and gave excellent stability to oil-in-water emulsions. The high surface activity was attributed to a low-molecular-weight cationic biosurfactant and the emulsion stability was believed to be caused by a lipopolysaccharide. The LPS consisted of 48% lipids, 38% carbohydrate, and 8% protein and had a strongly negative net charge. Thus, like many biomacromolecules that are referred to as LPS, it also contained a protein portion.

The alkaliphilic bacterium *Cronobacter sakazakii*, isolated from oil-contaminated wastewater, was used for production of a biomacromolecule that gave stable water continuous emulsions of a variety of oils such as aliphatic hydrocarbons, aromatic hydrocarbons, carbon tetrachloride, and triglyceride oils [22]. The bioemulsifier was found to be a heterolipopolysaccharide. The sugars were glucose, galactose, mannose, and xylose, all in contents between 14 and 24%, as well as a small amount of arabinose (2%). As is often the case with bacterial bioemulsifiers, the LPS also contained a protein fraction.

Alasan is an anionic bioemulsifier produced by *Acinetobacter radioresistens* KA53. It is a complex of polysaccharides and proteins with a high alanine content. Thus, it is not a LPS although its properties are in many respects similar to those of Emulsan and other LPS. The molecular weight is around 1000 kDa, which is similar to that of Emulsan. Unlike Emulsan, it is an efficient emulsifier for a broad range of hydrocarbons, including pure alkanes and pure aromatics. It also emulsifies crude oil efficiently. The protein component, 45 kDa in molecular weight, is essential for the ability to emulsify and create stable emulsions. The important role of the protein component in the emulsifying potential of bioemulsifiers has long been known [23].

Yeasts can also produce bioemulsifiers. Liposan is a product produced by *Candida lipolytica* yeast [24]. It consists of 83% carbohydrate and 17% protein. Thus, it is not an LPS but, like Alasan discussed above, its properties resemble those of LPS such as Emulsan. The presence of the protein fractions in Liposan is essential for its emulsifying properties, as it is for Emulsan and Alasan. The carbohydrate part in Liposan is a heteropolysaccharide, composed of glucose, galactose, galactosamine, and galacturonic acid. Liposan is an efficient stabilizer for a large variety of oil-in-water emulsions. The oils include hydrocarbons and triglycerides such as cottonseed, soybean, sunflower, safflower, and olive oil.

Several yeasts isolated from wastewater from the dairy industry have been found to produce bioemulsifiers, some of which were efficient emulsifiers and stabilizers of emulsions without lowering the surface tension of water to low values [25]. The bioemulsifiers were LPS and mannose, galactose, xylose, and arabinose were the dominant sugars.

References

1 Duval-Terrié, C., Cosette, P., Molle, G. et al. (2003). Amphiphilic biopolymers (amphibiopols) as new surfactants for membrane protein solubilization. *Protein Science* 12: 681–689.

2 Karlberg, M., Thuresson, K., and Lindman, B. (2005). Hydrophobically modified ethyl(hydroxyethyl)cellulose as stabilizer and emulsifying agent in macroemulsions. *Colloids and Surfaces A* 262: 158–167.

3 Charpentier, D., Mocanu, G., Carpov, A. et al. (1997). New hydrophobically modified carboxymethylcellulose derivatives. *Carbohydrate Polymers* 33: 177–186.

4 Sahu, K.M., Patra, S., and Swain, S.K. (2023). Host-guest drug delivery by b-cyclodextrin assisted polysaccharide vehicles: a review. *International Journal of Biological Macromolecules* 240: 124338.

5 Lindman B. (2011). Amphiphilic biopolymers: DNA and cellulose. In: *Colloids in Biotechnology* (ed. M. Fanun), 1–8. Boca Raton, FL: CRC Press.

6 Norgren, M., Costa, C., Alves, L. et al. (2023). Perspectives on the Lindman hypothesis and cellulose interactions. *Molecules* 28: 4216.

7 Kronberg, B., Holmberg, K., and Lindman, B. (2014). *Surface Chemistry of Surfactants and Polymers*, 144–145. Chichester: Wiley.

8 Erridge, C., Bennett-Guerrero, E., and Poxton, R. (2002). Structure and function of lipopolysaccharides. *Microbes and Infection* 4: 837–851.

9 Perfumo, A., Smyth, T.J.P., Marchant, R. et al. (2009). Production and roles of biosurfactant and bioemulsifiers in accessing hydrophobic substrates. In: *Handbook of Hydrocarbon and Lipid Microbiology* (ed. K.N. Timmis), 1502–1512. Heidelberg: Springer-Verlag.

10 Uzoigwe, C., Burgess, J.G., Ennis, C.J., and Rahman, P.K.S.M. (2015). Bioemulsifiers are not biosurfactants and require different screening approaches. *Frontiers in Microbiology* 6: 245.

11 Zuckerberg, A., Diver, A., Peeri, Z., and et al (1979). Emulsifier of Arthobacter RAG-1: chemical and physical properties. *Applied and Environmental Microbiology* 3: 414–420.

12 Mercaldi, M.P., Dams-Kozlowska, H., Panilaitis, B. et al. (2008). Discovery of the dual polysaccharide composition of emulsan and isolation of the emulsion stabilizing component. *Biomacromolecules* 9: 1988–1996.

13 Portal D'Almeida, A., Lima de Albuquerque, T., and Rocha, M.V.P. (2024). Recent advances in Emulsan production, purification, and application: exploring bioemulsifiers unique potentials. *International Journal of Biological Macromolecules* 278: 133672.

14 Portal D'Almeida, A., Lima de Albuquerque, T., Maciel Melo, V.M. et al. (2023). Biosurfactant production by *Acinetobacter venetianus* and its application in bioremediation. *Chemical Engineering and Technology* 46: 1106–1114.

15 Calvo, C., Manzanera, M., Silva-Castro, G.A. et al. (2009). Application of bioemulsifiers in soil oil bioremediation processes. Future prospects. *Science of the Total Environment* 407: 3634–3640.

16 Dong, H., Xia, W., Dong, H., and et al. (2016). Rhamnolipids produced by indigenous *Acinetobacter junii* from petroleum reservoir and its potential in enhanced oil recovery. *Frontiers in Microbiology* 7: 10.33389.

17 Alizadeh-Sani, M., Hamishehkar, H., Khezerlou, A. et al. (2018). Bioemulsifiers derived from microorgansims: applications in the drug and food industry. *Advanced Pharmaceutical Bulletin* 8: 191–199.

18 Ron, E.Z. and Rosenberg, E. (2001). Natural roles of biosurfactants. *Environmental Microbiology* 3: 229–236.

19 Rocha, M.V.P., Oliveira, A.H.S., Souza, M.C.M. et al. (2006). Natural cashew apple juice as fermentation medium for biosurfactant production by *Acinetobacter calcoaceticus*. *World Journal of Microbiology and Biotechnology* 22: 1295–1299.

20 Yi, G., Son, J., Yoo, J. et al. (2019). Emulsan-based nanoparticles for in vivo drug delivery to tumors. *Biochemical and Biophysical Research Communications* 508: 326–331.

21 Ferreira, I.N.S., Rodriguez, D.M., and Campos-Takaki, G.M. (2020). Biosurfactant and bioemulsifier as promising molecules produced by *Mucor hiemalis* isolated from Caatinga soil. *Electronic Journal of Biotechnology* 47: 51–58.

22 Jain, R.M., Mody, K., Mishra, A., and et al (2012). Isolation and structural characterization of biosurfactant produced by an alkaliphilic bacterium *Cronobacter sakazakii* isolated from oil contaminated wastewater. *Carbohydrate Polymers* 87: 2320–2326.

23 Sar, N. and Rosenberg, E. (1983). Emulsifier production by *Acinetobacter calcoaceticus* strains. *Current Microbiology* 9: 309–314.

24 Cirigliano, M.C. and Carman, G.M. (1985). Purification and characterization of liposan, a bioemulsifier from *Candida lipolytica*. *Applied and Environmental Microbiology* 50: 846–850.

25 Monteiro, A.S., Bonfim, M.R.Q., Domingues, V.S. et al. (2010). Identification and characterization of bioemulsifier-producing yeasts isolated from effluents of a dairy industry. *Bioresource Technology* 101: 5186–5193.

14

Characterization of Natural Surfactants

14.1 Introduction

The fermentation process used for the microbial production of biobased chemicals, such as biosurfactants, generally yields complex mixtures of products, often with a wide span in molecular weight and in solubility characteristics. Isolation, identification, and chemical characterization of biosurfactants from such a crude mixture of biomolecules require carefully designed and detailed analytical processes using state-of-the-art methods.

The chemical characterization of the naturally occurring surfactant is obviously of prime importance but understanding the self-assembly characteristics is also essential for a more complete understanding of the behavior of the biomolecule. This chapter therefore deals not only with the analytical methods used to assess the chemical structure of biosurfactants but it also covers methods to investigate their self-assembly into micelles and liquid crystalline phases. The chapter also covers methods to study the producing strains and the biosynthesis pathways.

Microbial glycolipids consist of an oligosaccharide moiety covalently bound to one or more lipid chains. Rhamnolipids, sophorolipids, trehalose lipids, and mannosylerythritol lipids are the most important classes. These differ in the nature of the hydrophilic sugar head group (rhamnose, sophorose, trehalose, and mannopyranosylerythritol, respectively), as well as in the number, the length, and the degree of unsaturation of the lipid chains [1].

Lipopeptides are microbial secondary metabolites isolated from the cell broth of different kinds of microorganisms, notably from the genera *Bacillus*, *Pseudomonas*, and *Streptomyces* [2]. Their hydrophilic part, the "head" in surfactant terms, consists of mainly hydrophilic amino acids, while their hydrophobic part, the "tail," is usually an alkyl chain that may be linear or branched but may also contain nonpolar amino acids. Surfactin and surfactin-like lipopeptides produced by members of the genus *Bacillus* are important examples of lipopeptides.

Antimicrobial peptides are peptides typically consisting of between 12 and 50 amino acids. They are usually positively charged, and the charge originates from arginine, lysine, and/or histidine residues. Negatively charged antimicrobial peptides also exist and the negative charge originates from glutamic acid and/or aspartic acid residues.

Lipopolysaccharides are high-molecular-weight biosurfactants/bioemulsifiers, produced from different bacteria. Formally, they are composed of a linear polysaccharide backbone with lipid chains grafted on the sugar units; however, in reality, they are often more complex

Natural Surfactants: Science, Technology, and Industrial Aspects, First Edition. Guido Bognolo and Krister Holmberg.

than that. The product may contain a protein, which may be either free or covalently attached to the polysaccharide chain. The product may also contain a polysaccharide that does not have lipid chains attached. The polysaccharide is usually a heteropolysaccharide, i.e. it is composed of a mixture of different sugars. These may be either uncharged or negatively charged. It goes without saying that a complete analysis of these products is a very demanding task and the majority of the lipopolysaccharides discussed in the literature are still only partially characterized [3]. The most studied lipopolysaccharide is Emulsan, a commercial product isolated from *Acinetobacter calcoaceticus* with a molecular weight of around 1000 kDa.

The analytical methods used for detecting and characterizing biological surface-active agents vary widely from simple thin-layer chromatography (TLC) and colorimetric assays that provide an approximate indication of the presence and the type of biosurfactant, to advanced methods that give information on the molecular masses and of structural features. However, there is a lack of consistent methodologies and/or regulatory protocols for detection and/or structural characterization of these molecules. Twigg et al. have drawn attention to the reality that many reports on microbially produced surface-active agents lack

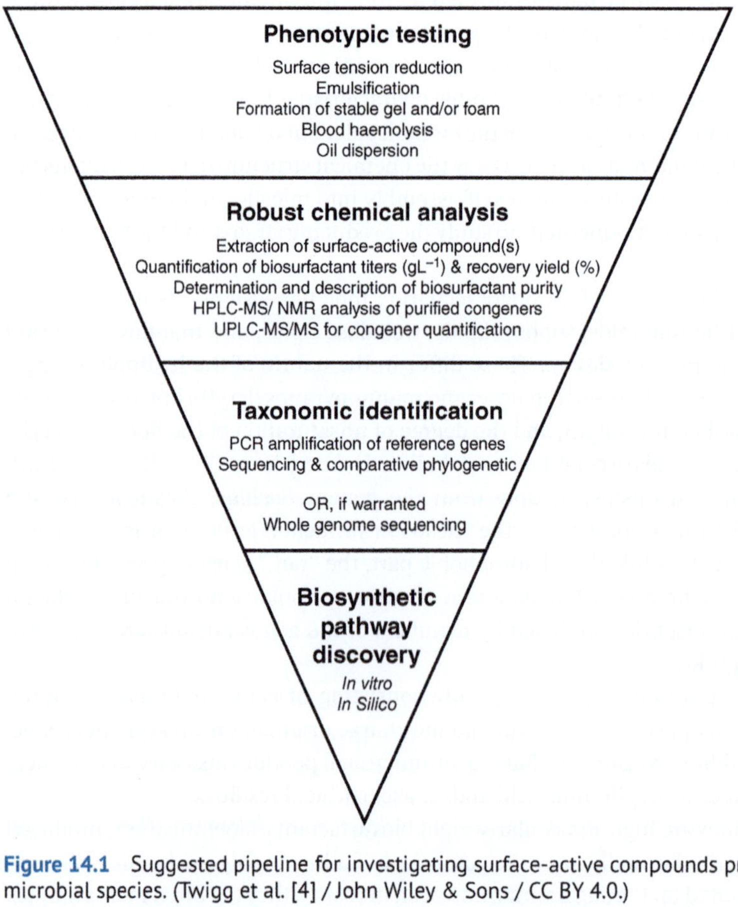

Figure 14.1 Suggested pipeline for investigating surface-active compounds produced by microbial species. (Twigg et al. [4] / John Wiley & Sons / CC BY 4.0.)

robust experimental evidence generated by validated analytical methods [4]. They suggest the methodology shown in Figure 14.1 for a rational approach to the characterization of biological surface-active compounds.

14.2 Methods for the Characterization of Naturally Occurring Surfactants

The simplest characterization techniques to perform are assays for phenotypic traits that indicate the production of surface-active compounds.

Blood agar method: Biosurfactants can cause lysis of erythrocytes. This principle is used for the hemolysis assay, which was developed by Mulligan et al. [5]. Cultures are inoculated on sheep blood agar plates and incubated for two days at 25 °C. Positive strains will cause lysis of the blood cells and exhibit a colorless, transparent ring around the colonies. However, the method has many limitations [6]. First, it is not specific, as lytic enzymes can also lead to clearing zones (zones of inhibition). Second, hydrophobic substrates cannot be included as the sole carbon source in this assay. Third, diffusion restriction of the surfactant can inhibit the formation of clearing zones. In addition, some biosurfactants do not show any hemolytic activity at all. Thus, the blood agar method is no more than a crude preliminary screening that must be supported by other techniques, e.g. those based on surface activity measurements.

Oil spreading method: First, 50 mL of distilled water is put in a petri dish (15 cm diameter), followed by the addition of 100 μL of crude oil to the surface of the water. Then, 10 μL of cell-free culture broth is dropped on the crude oil surface. The diameter of the clear zone on the oil surface is compared to that obtained with 10 μL of distilled water as control [7].

Cetyltrimethylammonium bromide (CTAB) agar plate assay: This is a semiquantitative assay for the detection of extracellular glycolipids or other anionic surfactants. It was developed by Siegmund and Wagner [8]. The isolated bacteria are cultivated on a light blue mineral salt agar plate containing the cationic surfactant CTAB and the basic dye methylene blue. If anionic surfactants are secreted by the microbes growing on the plate, they form a dark blue-purple, insoluble ion pair with CTAB and methylene blue. Accordingly, rhamnolipid-producing strains can be recognized by the dark blue halos formed around the colonies.

Emulsifying ability: The emulsifying ability of a biosurfactant can be measured according to the method of Cooper and Goldenberg [9, 10]. To 2 mL of culture supernatant or biosurfactant crude extract, 2 mL of sunflower oil is added and vortexed at high speed for 2 minutes. The mixture is allowed to stand for 10 minutes prior to measurement. The emulsification activity is defined as the height of the emulsion layer divided by the total height and expressed as percentage.

Two colorimetric assays most often reported for the early detection of rhamnolipids and sophorolipids are the orcinol and the anthrone reagent assays [11–13].

The orcinol assay involves adding a solution of orcinol (5-methylresorcinol) in sulfuric acid to the sample being tested, followed by heating; rhamnose sugars present in rhamnolipids are dehydrated and react with the orcinol to form a green-pigmented compound, which is measured by recording absorbance at 421 nm. Absorbance readings are then compared with a standard curve generated using samples of known rhamnolipid concentrations or simply rhamnose. The orcinol assay is routinely utilized in a wide variety of publications to demonstrate and quantify bacterial production of rhamnolipids [14].

The anthrone reagent assay is used to quantify sophorolipids in culture liquid or after partial purification. The reagent contains a strong acid that first converts carbohydrates to monosaccharides and then catalyzes their dehydration into furfural (from pentoses) and hydroxyfurfural (from hexoses), which condense with anthrone (anthracen-9(10*H*)-one) to produce colored compounds. Absorbance is recorded at 620 nm. Note that in both the orcinol and the anthrone method, the reagent reacts with all carbohydrates including pentoses and sugar alcohols and may give erratic responses.

Kubicki et al. have more recently described a colorimetric method that has the potential to both identify and quantify biosurfactants in samples of culture supernatant. Unlike both the orcinol and the anthrone assays, this assay exploits the surface activity of the biosurfactant and not the chemical nature of the sugar moiety. This is achieved by utilizing Victoria Pure Blue BO dye, which is solubilized by the biosurfactant in the sample. This assay has been adapted onto a microtiter plate allowing high-throughput screening of samples. The exploitation of the phenotypic effect of surface tension reduction to produce a colorimetric response allows this assay to detect different types of biosurfactants [15].

Surface and interfacial activity are an indication of the presence of surface-active agents in the broth or supernatant liquor. A simple tensiometer, for instance one based on the du Noüy ring method, can directly measure both surface and interfacial tension and there are many manufacturers of such tensiometers. The modern instruments are fully automated and easy to handle. The measurement of surface tension results in a so-called surface tension plot, where the surface tension is given as a function of surfactant concentration (or more often the logarithm of surfactant concentration). The values obtained are equilibrium surface tension values and it is important to allow sufficient time at each measuring point for equilibrium to be reached. For small amphiphilic molecules, such as glycolipids or lipopeptides, this is normally not an issue but larger molecules, such as biomacromolecules, travel slowly in solution and once at the surface, they rearrange to attain the energetically most favorable conformation. This may take many minutes and even hours for large molecules. Figure 14.2 shows a typical surface tension plot.

The CMC is the concentration at which the first micelles form in solution. It is an important characteristic of a surfactant and CMC values are given in the literature for most types of biosurfactants. The CMC value is often obtained from the surface tension plot from the concentration where there is a clear break on the curve, as shown in Figure 14.2. For technical surfactants, and most biosurfactants can be regarded as technical surfactants rather than pure species, the determination of the CMC is not so straightforward, however. There is often a pronounced minimum of the curve around the CMC. This is particularly common for anionic surfactants and the minimum in surface tension is due to the presence of a more surface-active impurity than the surfactant itself. This phenomenon is very relevant for biosurfactants, which usually have an anionic polar headgroup. Figure 14.3 illustrates the problem. For such amphiphiles, other methods to assess the CMC than surface tension plots may be preferred. Conductometry is such a technique. Since the conductivity per surfactant unimer is larger when the unimer is free in solution than when it resides

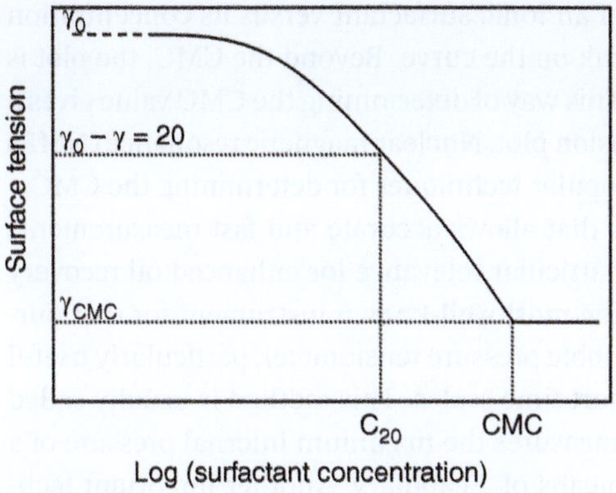

Figure 14.2 A surface tension plot showing the surface tension as a function of the logarithm of the surfactant concentration. The plot usually starts around 72 mN/m, γ_o on the diagram, which is the surface tension of water. The surfactant accumulates at the surface, which is manifested in a decline of the surface tension. Eventually, a fully developed monolayer of surfactant occupies the surface. An ideal plot is linear until the critical micelle concentration (CMC), is reached. At that point, the first micelle is formed and beyond the CMC more and more micelles will form but the unimer concentration remains constant. Since the micelles are not surface active, the surface tension will remain constant above the CMC. The surfactant concentration needed to reduce the surface tension by 20 mN/m, the C_{20} value, indicates the efficiency of the surfactant and the surface tension at the CMC, the γ_{CMC} value, is the effectiveness of the surfactant. (Kronberg et al [16] / John Wiley & Sons.)

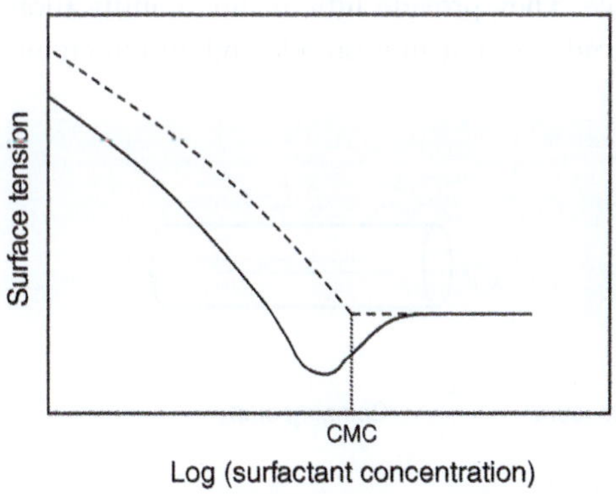

Figure 14.3 Surface tension plots for a pure surfactant (dotted line) and for a technical surfactant (full line). The technical surfactant contains impurities that are more surface active than the pure surfactant. This is a common phenomenon for amphiphiles with a charged head group. The impurities adsorb at the air–water interface along with the surfactant and the mixed surface layer is very surface active. When micelles start to form, the hydrophobic impurities desorb from the surface and enter the micelles instead. This leads to an increase in surface tension up to the value obtained with the pure surfactant. (Kronberg et al. [17] / John Wiley & Sons.)

in a micelle, a plot of specific conductivity of an ionic surfactant versus its concentration is linear up to the CMC, where there is a break on the curve. Beyond the CMC, the plot is again linear but with a decreased slope [18]. This way of determining the CMC value gives a more undisputed answer than the surface tension plot. Nuclear magnetic resonance (NMR) spectroscopy and light scattering are other popular techniques for determining the CMC.

Nowadays, sophisticated equipment exists that allows accurate and fast measurements of ultralow oil–water interfacial tension, of particular relevance for enhanced oil recovery research. The spinning drop tensiometer is the most well-known instrument for that purpose. Another important instrument is the bubble pressure tensiometer, particularly useful for measuring dynamic surface tension at short time scales. This method is usually called the maximum bubble pressure method and measures the maximum internal pressure of a gas bubble, which is formed in a liquid by means of a capillary. Another important technique is the pendant drop method, which is widely used to determine the change in surface or interfacial tension at longer time scales. The image of a drop hanging from a needle into the air (for surface tension measurement) or into another liquid (for interfacial tension measurement) is recorded. Using the Young-Laplace equation, the surface tension or the interfacial tension can be calculated from this image and the change in the drop shape with time is transformed into a change in surface or interfacial tension with time. Figure 14.4 illustrates the spinning drop tensiometer, the maximum bubble pressure instrument, and the pendant drop method, along with several other methods to determine surface and interfacial tension.

The advantage of these phenotypic assays is that they are relatively simple to perform, can be adapted to be high throughput to analyze multiple samples, and are inexpensive. However, the phenotypic assays should under no circumstances be solely utilized to confirm surface-active compounds in a sample. They provide little or no quantification of the concentration of surface-active compounds present in a sample and absolutely no

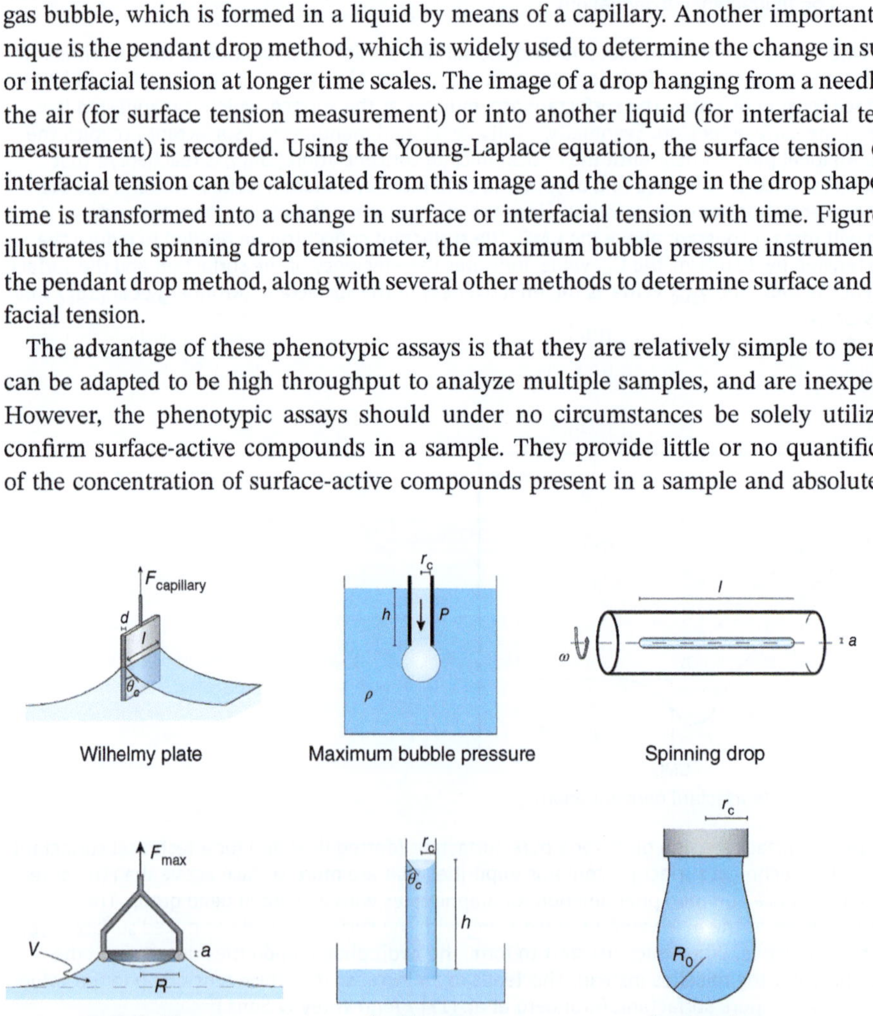

Figure 14.4 Illustrations of common methods to determine surface and interfacial tension. (Berry et al. [19] / with permission of Elsevier.)

structural information about the amphiphile. They can also be very sensitive to impurities. It is therefore suggested that these techniques should only be used as a preliminary screening in studies where large culture collections are investigated for microorganisms with a potential for production of surface-active compounds, or for functional studies with well-defined microbial strains [4].

The use of analytical chemistry techniques to prove and characterize surface-active compound production is significantly more conclusive than the phenotypic and colorimetric tests discussed above. These techniques include thin layer chromatography (TLC), high-performance thin-layer chromatography (HPTLC), Fourier transform infrared spectroscopy (FTIR) and attenuated total reflection Fourier transform infrared spectroscopy (ATR-FTIR), high-performance liquid chromatography (HPLC), mass spectrometry (MS), high-performance liquid chromatography-mass spectrometry (HPLC-MS), tandem mass spectrometry (MS/MS), nuclear magnetic resonance (NMR) spectroscopy, and size exclusion chromatography (SEC). All these techniques require (partial) purification of surface-active compounds from the cell-free supernatant samples.

The methods for purification of surface-active compounds should be chosen depending upon the type of compound predicted to be produced by the microorganism being investigated and are discussed, for example by Heyd et al. [20]. It is important to note that to obtain a meaningful characterization of surface-active compounds, sample purity is fundamental, as different techniques require differing levels of sample purity. Sample purity is also an important consideration when carrying out functional testing of surface-active compounds and in designing scale-up protocols for commercial applications [4].

TLC is one of the most commonly used techniques to detect a component in a crude biosurfactant. It separates nonvolatile substances in a mixture. TLC is relatively simple to execute and an excellent tool to assess qualitatively the purity of a sample. It is performed on a plate of a nonreactive material (usually glass) coated with a thin layer of an adsorbent material (the stationary phase). The sample (e.g. the supernatant liquor from a fermentation flask) is applied at the bottom of the plate, which is then eluted with a solvent, or more often a mixture of solvents (the mobile phase) and moves up the plate by capillary action. As with all chromatography, some compounds are more attracted to the mobile phase, while others are more attracted to the stationary phase and reach different positions on the plate. Therefore, different compounds move up the TLC plate at different speeds and become separated. Separated spots are self-evident if the sample is colored, otherwise, the plate is viewed under UV light or stained.

TLC is quick, simple, and gives high sensitivity at a relatively low cost. It can monitor reaction progress, identify compounds in a mixture, and give a rough indication of purity. TLC may also be used to isolate a small amount of a compound from a mixture (preparative TLC). The disadvantage is that testing different stationary and mobile phases is often necessary to obtain well-defined and separated spots and this can be a time-consuming activity.

High-performance thin-layer chromatography (HPTLC) is an extension of TLC and provides superior resolution and lower limits of detection than TLC. It enhances the resolution of compounds by using higher-quality TLC plates with finer particle sizes as the stationary phase. The separation is achieved by allowing the mobile phase to enter the plate layer precisely along an outer circular path, after which it flows toward the center at a nearly constant speed.

HPLC stands for both **high-pressure liquid chromatography** and **high-performance liquid chromatography.** It is an evolution of the column liquid chromatography (LC) in which a liquid mobile phase passes through a column packed with a chromatographic medium. Molecules in the mobile phase are separated based on differences in their interactions with the stationary and mobile phases. The residence time in the column will depend on these interactions and by optimizing the solvent used as mobile phase, a complex mixture of substances can often be resolved into individual components. It is common to use a mixture of solvents, and it is also common to use a gradient of solvents for this purpose. In HPLC, these operations, including the changes in eluting solvents, are automated and the gravity that governs the elution in LC is replaced by high-pressure pumps. The single components in the elute are detected with IR, UV, refractive index, or some other type of detector, and a computer produces graphical plots of the intensity of the signal (related to the component concentration in the sample) versus time or elution volume, showing peaks corresponding to the different components in the sample. Each component appears with a distinct time (its retention time) and with an area proportional to its amount.

HPLC is an excellent tool to separate complex mixtures and identify single molecular species. Coupled with mass spectrometry or tandem mass spectrometry, it provides highly reliable information for the identification and characterization of biosurfactants [21, 22]. Chiral stationary phases can separate racemic mixtures into the pure enantiomers and can be scaled up to preparative use, i.e. recovery of finite quantities of the separated ingredients. HPLC and ultra-high performance liquid chromatography (UHPLC) systems have also been used for quantitative detection of lipopeptides derived from marine microorganisms [23].

Fourier transform infrared spectroscopy (FTIR) or **attenuated total reflectance-Fourier transform infrared spectroscopy** (ATR-FTIR) enables the determination of the biosurfactants' molecular components and structures by measuring the absorption of the infrared radiation by the chemical bonds of different groups in the molecules. It has been proposed as a rapid technique allowing the identification of rhamnolipids and their quantification [24] and has been utilized in a number of studies reporting glycolipid production by various bacterial strains [25]. However, to allow a reliable identification of the biosurfactant, the sample must be of very high purity, as the chemical bonds are not unique to the biosurfactant but may be present in a large number of other extracellular compounds produced by the microorganisms. If high purity is not guaranteed, it is advisable to use other analytical techniques or to complement FTIR with other methods.

Mass spectrometry (MS) is an analytical tool useful for measuring the mass-to-charge ratio (m/z) of one or more molecules present in a sample. It can be used to assess their exact molecular weight, as well as to provide information for determining their structure. Coupling MS with HPLC or UHPLC gives the advantage of defining retention time(s) of biosurfactants along with their mass spectral fingerprint, permitting the separation and identification of individual components from complex samples [21].

Tandem mass spectrometry (MS/MS or MS^2) is a powerful technique to identify components with similar structures in complex mixtures. It uses two mass spectrometers, often designated MS1 and MS2. The molecules in the sample are first ionized and MS1 separates the ions by their m/z ratio. Ions of a specific m/z ratio are then selected and made to split into smaller fragment ions, for instance by collision-induced dissociation (CID). These fragments are then introduced into MS2, where they become separated by their m/z ratio and subsequently detected. It has, for instance, been utilized to provide confirmatory evidence of rhamnolipid production by a species of *Marinobacter* [1].

Nuclear magnetic resonance (NMR) spectroscopy is an analytical tool based on the re-orientation of so-called NMR-active nuclei in an external magnetic field. The re-orientation involves absorption of radiation in the radio frequency region, typically from 4 to 900 MHz. The method is based on the fact that the resonance frequency of each NMR-active nucleus depends on its chemical environment, which means that the NMR spectrum gives information about the individual functional groups in a molecule. It also gives information about the environment around these groups. The proton, ^{1}H, has a nuclear spin and is therefore NMR-active. The dominant carbon isotope, ^{12}C, does not have a nuclear spin but ^{13}C, which makes up 1.1% of all carbon, does have a spin and is NMR-active. Both proton NMR and carbon NMR are therefore common techniques, but NMR can be applied to many other elements as well. NMR is, together with mass spectrometry, the most important technique to characterize organic substances. It unambiguously resolves the molecular structure of biosurfactants when combined with separation and purification methods such as preparative chromatographic methods or fraction collection with HPLC to obtain pure compounds [26]. The purification needed often requires some prerequisite knowledge of the general structure of the compound(s) being produced (i.e. rhamnolipid, sophorolipid, lipopeptide). Therefore, NMR analysis may not be suitable for high-throughput analysis of multiple strains of interest. Additionally, high-resolution NMR requires the use of a good solvent in perdeuterated form. This may be a problem for high-molecular-weight biomacromolecules. As an example of the use of NMR for characterization of an unusual biosurfactant, it was employed for determining the structure of a 15-membered macrodilactone-containing glycolipid being produced by *Pantoea ananatis* [27].

NMR spectroscopy can be used for other purposes than elucidation of the structure of an organic molecule, for instance a biosurfactant. NMR diffusometry is an excellent technique to determine the size and shape of micelles in solution. It is based on the principle that the rate of self-diffusion in solution is inversely proportional to the hydrodynamic radius of a molecule, an aggregate, or a particle (the Stokes-Einstein equation). Thus, an individual biosurfactant diffuses faster than a spherical micelle of the same amphiphile and the spherical micelle diffuses faster than a rod-like micelle, etc. This technique is widely used in surface chemistry research and is also suitable for determining the CMC [28].

Size exclusion chromatography (SEC) is often used for the analysis of the molecular weight and structural characterization of biological macromolecules such as proteins and polysaccharides. It provides information about the molecular weight distribution since it separates molecules according to their hydrodynamic volumes. Larger molecules elute first because they cannot enter the small pores of the packing material. SEC has a simple operating procedure and high versatility, and a wide range of matrices are commercially available. SEC is normally used for aqueous systems and the columns are typically filled with a hydrophilic gel such as Sephadex, which is a crosslinked dextran gel. Its close relative, gel permeation chromatography (GPC), is used with organic solvents and the columns contain a hydrophobic gel such as crosslinked polystyrene. GPC and SEC are based on the same principle, but GPC is primarily used for research on organic polymers. SEC coupled with multi-angle light scattering detection (SEC-MALS)

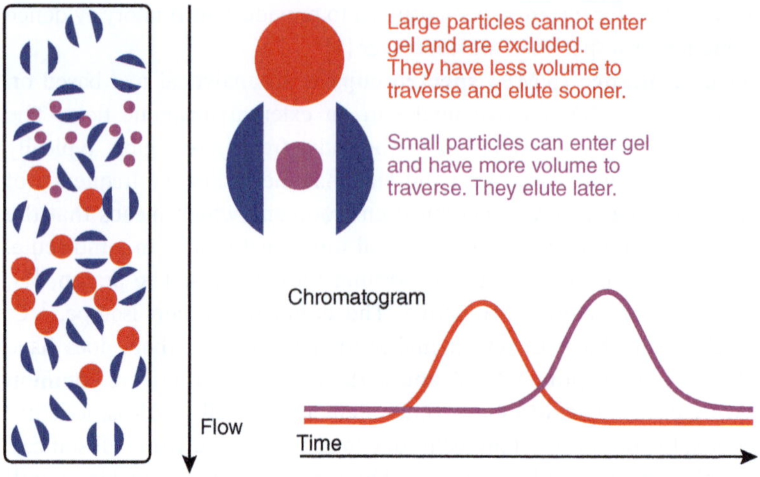

Figure 14.5 In size exclusion chromatography (SEC), small molecules or particles travel a longer distance in a porous medium than large molecules or particles. The large molecules or particles therefore elute earlier than the small ones. (Takometer / Wikimedia Commons / CC BY 2.5.)

has become very popular for determining the molecular weight of biomacromolecules, as discussed in a review by Zompra et al. [3]. Figure 14.5 illustrates the principle for SEC.

Electrospray ionization mass spectrometry (ESI-MS) uses electrical energy to assist the transfer of ions from solution into the gaseous phase before they are subjected to mass spectrometric analysis. The technique is useful for producing ions from biomacromolecules such as proteins and polysaccharides because it overcomes the propensity of such molecules to fragment when ionized. It produces multiple-charged ions, thus extending the range of the analyzer to higher-molecular-weight species [29].

MALDI-TOF stands for matrix-assisted laser desorption ionization time-of-flight. It is an ionization method that uses a laser-enery absorbing matrix to create ions from molecules with minimum fragmentation. These are separated according to their mass-to-charge ratio and measured by determining the time it takes for the ions to travel to a detector at the end of a time-of-flight tube. The resulting spectrum with mass-to-charge values along the x-axis and intensity along the y-axis is compared to a database from known organisms. MALDI-TOF spectrometry is particularly useful for studying macromolecules and is widely used in biochemical research. It is now an established tool within microbiology.

MALDI-TOF MS can identify gram-positive, gram-negative, aerobic, and anaerobic bacteria, as well as mycobacteria, yeast, and molds, typically at the species level, with accuracy as good as and often better than traditional methods with respect to sequencing. The method is also more reliable than traditional and molecular methods for microorganism identification. A small amount of biomass is required, so testing can often be performed from the primary culture, as long as a single well-isolated colony is available. Sample preparation is relatively simple, and analysis of a large number of samples is possible within an hour. Further, a priori knowledge of the type of organism being tested is not required, thus allowing both highly experienced and less experienced microbiologists to perform the testing [30].

Well-known lipopeptides, such as surfactin A, iturin A, and fengycin, all produced by a *Bacillus subtilis* species, were detected by MALDI-TOF MS in a work that clearly illustrates the power of the technique [31]. Although normally MALDI-TOF requires isolated and HPLC-pure samples, Vater et al. developed a method for the rapid sensitive detection and efficient structural characterization of lipopeptide biosurfactants by using whole microbial cells and crude culture extracts [32]. Furthermore, MALDI-TOF MS/MS analysis can define the amino acid sequence of lipopeptides without performing hydrolysis and subsequent amino acid analysis. Unlike Edman degradation, MS/MS analysis does not need the molecule to be completely pure since the mass analyzer can select only the ions of interest [33].

Sato et al. applied MALDI-TOF MS to screen glycolipid-producing species. They demonstrated that *Pseudozyma antarctica* produced essentially mannosylerythritol lipids, whereas *Pseudozyma hubeiensis* produced mannosylerythritol lipids and cellobiose lipids [34].

MALDI-TOF-MS was used by Price et al. to obtain a detailed structural characterization of novel rhamnolipids produced by *Pseudomonas* species. They determined the different glycolipid profiles among the different species and characterized some novel compounds by combining this technique with carbohydrate and lipid analysis utilizing GC–MS and NMR spectroscopy [35].

CID is an MS technique to induce fragmentation of ions in the gas phase. The ions (typically molecular ions or protonated molecules) are accelerated by applying an electrical potential and then allowed to collide with neutral molecules. In the collision, some of the kinetic energy is converted into internal energy, which results in bond breakage and the fragmentation of the molecular ion into smaller fragments. These fragment ions can then be analyzed by tandem mass spectrometry (MS/MS) [36]. CID has been found to be a powerful tool for the detection and identification of structural variations in rhamnolipids produced by various *Pseudomonas* strains or by a specific strain under different culture conditions. CID can go as far as distinguishing subtle structural differences in isomers like RhA-C8-C10 and RhA-C10-C8, which are almost inseparable chromatographically.

Remarkable results in terms of analytical precision and accuracy can be obtained by combining different techniques. For instance, NMR combined with high-resolution MS and tandem mass spectrometry is a very accurate method to determine the chemical structure of unknown biosurfactant molecules. This combination of techniques was used to study marine bacteria not previously known to produce biosurfactants, and to characterize biosurfactant compounds produced by engineered microbial strains [37, 38].

Abdel-Mawgoud et al. have published detailed protocols for HPLC/MS and HPLC/MS/MS analysis of rhamnolipids and their lipid precursor, the 3-(3-hydroxyalkanoyloxy) alkanoic acids (HAA), directly in bacterial culture supernatants. The use of ^{13}C-labeled rhamnolipids as internal standards can further enhance the precision of the quantification [39].

Ekramul Haque et al. used H-NMR and ESI-MS to characterize a biosurfactant produced by *Pseudomonas aeruginosa* and tested for crude oil bioremediation [40]. This combination of techniques turned out to be very powerful.

Hosseini et al. used a combination of H-NMR, liquid chromatography-mass spectrometry (LC-MS), and FTIR for the structural identification of a biosurfactant effective in assisting crude oil biodegradation [41].

Zompra et al. have published a comprehensive review of the analytical methods that can be used for the characterization of biosurfactants and bioemulsifiers [3].

14.3 Methods to Determine the Self-Assembly Behavior of Naturally Occurring Surfactants

Methods to determine the critical micelle concentration (CMC), such as constructing a plot of surface tension versus surfactant concentration or of specific conductivity versus surfactant concentration, or the use of NMR diffusometry or light scattering, have been discussed earlier in the chapter. However, more advanced scattering techniques, such as **small-angle X-ray scattering** (SAXS) and **small-angle neutron scattering** (SANS), as well as **cryogenic transmission electron microscopy** (cryo-TEM), may also be used for the study of size and shape of self-assembled aggregates [42].

The data collected by the small-angle scattering techniques, SAXS and SANS, are extremely reliable when the experiments and data treatment are properly done. Nonetheless, these techniques suffer from a number of drawbacks. First, they are indirect techniques, as they probe the Fourier space, thus needing model-dependent and model-independent analyses to extract quantitative data. For scientists experienced with the techniques, this is usually not an issue; however, complex structures can make modeling tedious, or even impossible. Second, the availability of SAXS and SANS instruments is limited. They require access to large-scale facilities like a synchrotron light source (for SAXS) and a neutron source (for SANS). Fan and Wang have shown the power of combining different numerical approaches to the treatment of small-angle scattering data with the direct Cryo-TEM imaging method [42].

A particularly popular scattering technique employed to study colloids in solution is **dynamic light scattering** (DLS), which has the double advantage of being accessible in many laboratories and easy to use. DLS provides information on the hydrodynamic diameter (diameter of the colloid plus its hydration corona) and size distribution, and the technique is very helpful for studies of colloidal stability. However, DLS gives no information on the morphology and is characterized by a number of possible biases (e.g. scattering from dust or aggregates), which can lead to misinterpretation [43]. Another point of caution, of particular relevance for samples with a broad size distribution, is that DLS tends to overemphasize the larger particles, which means that the average size obtained from the measurement will be too large. Hence, DLS should be used with care and preferably as a complement to other techniques, such as microscopy or NMR diffusometry.

Whenever possible, scattering techniques should be coupled to direct observations. This is not only of importance to avoid misinterpretation but also to correctly attribute the scattering signals, which can often be produced by multiple structures, and which therefore are difficult to unambiguously attribute to a specific structure [37].

Microscopy techniques can be invasive or noninvasive. The former modifies the concentration by drying and the latter can be performed in the parent solution, with a resolution from the nanometer to the millimeter scale. Cryo-TEM has a nanometer resolution. It is a convenient technique, which preserves the hydration and the original concentration of the sample in its parent solvent, thus providing access to the nano-to-microscale morphology of the aggregate. For these reasons, it is generally preferred to conventional TEM, which requires sample drying [44].

More advanced studies on the functionality of surface-active compounds may include analysis of the dynamics and intra/intermolecular interactions in the aggregates formed by these compounds. Probing the elastic constants of lipid membranes, the thermodynamic and kinetic parameters of self-assembly but also the possible existence of raft regions in lipid membranes are hot topics in biophysics and quantification of intermolecular forces are all important aspects of advanced characterization of self-assembled systems [45–47]. Accessing this type of information requires experience, even if the analytical techniques are relatively accessible and easy to use.

14.4 Methods to Determine Phase Diagrams of Naturally Occurring Surfactants

The techniques used to determine the phase diagrams of naturally occurring surfactants are the same as for conventional surfactants. A combination of different experimental techniques is generally necessary.

Rheological measurements rely on the different viscosity ranges of the mesophases. Solutions containing spherical micelles are generally of low viscosity. If the micelles start to become elongated (sometimes called rod-like or worm-like), which they may do at higher concentrations or, for nonionic amphiphiles, at elevated temperature, the viscosity may increase and the increase may be considerable, as can be seen from Figure 14.6 [49, 50].

The thread-like micelles can become very long, up to several 100 nm. Under certain conditions, such as very high concentration, the growth can lead to branched structures. Solutions of long thread-like micelles resemble solutions of linear polymers in many respects. The long micelles can be very flexible but may also be rigid and the degree of flexibility will affect the viscosity. The flexibility of ionic micelles is strongly dependent on electrolytes. Salt addition may induce a change from rigid rods to very flexible structures.

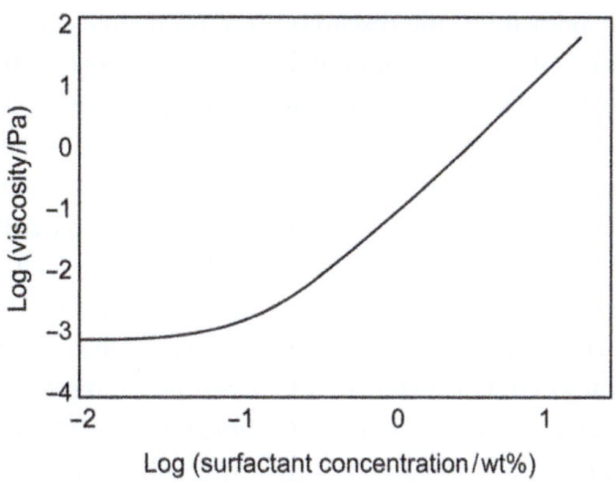

Figure 14.6 Log viscosity versus log surfactant concentration for a nonionic surfactant at constant temperature. At low concentration, the micelles are spherical and at higher concentration, they become increasingly more elongated. (Kronberg et al. [48] / John Wiley & Sons.)

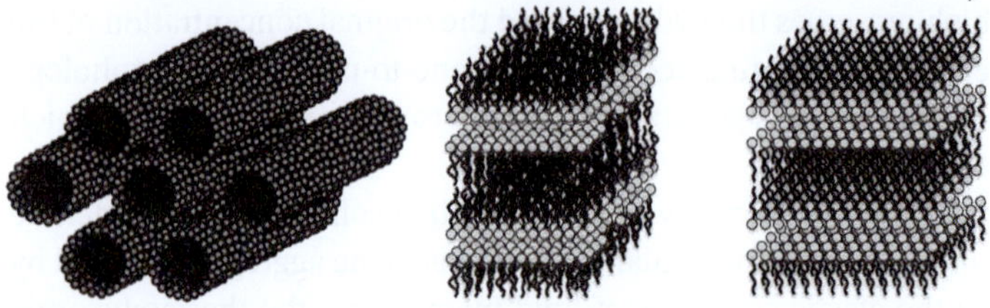

Figure 14.7 Packing of amphiphiles in two common liquid crystalline phases, hexagonal (left) and lamellar (middle), and in a lamellar gel, where the tails are in a solid state. (Kronberg et al. [51] / John Wiley & Sons.)

At low concentration, the micelles behave as independent entities, i.e. they do not overlap. At higher concentration, above the so-called overlap volume fraction, the micelles become entangled, which leads to high viscosity. This is the reason for the steep rise in viscosity for the surfactant in Figure 14.6. The overlap concentration for that system, a nonionic surfactant, is around 0.1 wt% [50].

At even higher concentrations, surfactant liquid crystals may form. The type of liquid crystal that forms will depend not only on the surfactant geometry but also on the concentration. The following order is typical with increasing surfactant concentration:

Micellar solution – hexagonal liquid crystal – cubic liquid crystal – lamellar liquid crystal – cubic liquid crystal – reversed hexagonal liquid crystal – reversed micellar solution.

Figure 14.7 shows representative examples of surfactant packing at high concentrations.

There are many methods to identify the different structures that amphiphiles can give rise to at higher concentrations. **Scattering of normal and polarized light** can be useful for the identification of different liquid crystalline phases. Isotropic phases, such as micellar solutions, sponge phases, and cubic liquid crystalline phases, are clear and transparent, while anisotropic phases scatter light and appear cloudy. Viewing a sample through crossed polarizers gives a bright image for anisotropic phases and a black image for isotropic phases. The pattern in a polarized microscope is also very different for different liquid crystalline phases. A **polarization microscope** is therefore a useful tool to distinguish, for instance between a hexagonal and a lamellar phase, which are both anisotropic [52, 53].

NMR spectroscopy is also useful for characterization of liquid crystalline phases. This is relevant to the research on natural surfactants since many biosurfactants are known to form liquid crystalline phases at high concentration in solution. NMR spectroscopy, especially observations of quadrupole splittings in deuterium NMR, where different patterns are obtained for different liquid-crystalline phases, is very useful for characterization of phase diagrams [54, 55]. For isotropic phases, a narrow singlet is obtained, while a doublet is observed for an anisotropic phase. The magnitude of the splitting is larger for a lamellar phase than for a hexagonal phase so the technique can be used for distinguishing these two anisotropic phases. An advantage of the NMR method is that in a two-phase region, where more than one liquid crystalline phase appears, the spectra of both phases are superimposed. For instance, if in a two-phase region there is one isotropic phase, for instance a cubic phase, and one anisotropic phase, for instance a hexagonal phase, there will be one singlet and one doublet. In a three-phase region consisting of one isotropic phase and two

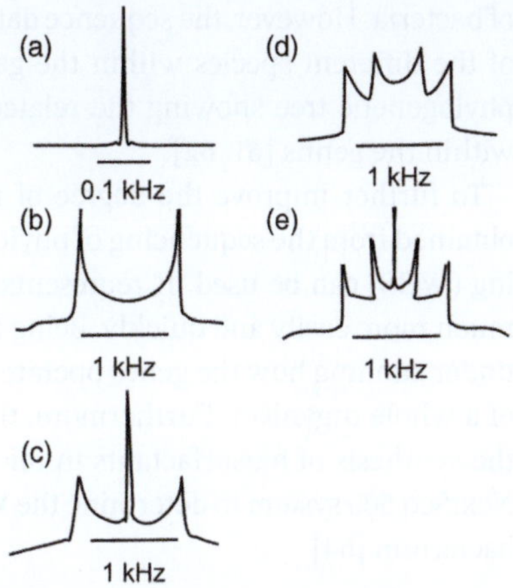

Figure 14.8 Quadrupole splittings in deuterium NMR. The spectra represent (a) an isotropic phase, (b) an anisotropic phase, (c) a two-phase region with one isotropic and one anisotropic phase, (d) a two-phase region consisting of two anisotropic phases, and (e) a three-phase region with one isotropic and two anisotropic phases. (Khan et al. [54] / American Chemical Society.)

anisotropic phases, there will be one singlet and two doublets with different magnitudes of the splitting. This is illustrated in Figure 14.8.

14.5 Methods for Strain Identification

In addition to the characterization of the surface-active compound produced by a microorganism, it is often important to accurately determine the identity of that organism. There are many methodologies for this, and they broadly fall into two general categories:

- Culture-driven
- Molecular biology-driven

Culture-driven methodologies consist of the observation of the colony morphology of bacterial or yeast cultures using microscopy following various staining techniques (i.e. in the case of bacteria, Gram staining) and a combination of biochemical assays such as Analytical Profile Index (API) tests. Studies have been published where culture-driven approaches have been solely utilized for the identification of the strain being described [56, 57]. However, in most cases, these techniques alone are not able to accurately define a microorganism at the species level.

Molecular biology techniques are useful to identify microbial strains. Microbial genomes incorporate elements that consecutively possess domains with conserved and variable sequence, and the variable sequence domains are often strain-dependent and can therefore be utilized to phylogenetically type the organism [37]. Such genetic elements are commonly referred to as reference sequences. In bacteria, the most ubiquitous example is the gene encoding the 16S subunit of ribosomal RNA (16S rRNA) [58]. In eukaryotic microorganisms, the internal transcribed spacer (ITS) region in the 18S subunit of ribosomal RNA is often used [59]. Alternatively, the D1/D2 domains of the large subunit (LSU) rRNA can be used, as demonstrated by Kurtzman et al., to identify several novel sophorolipid-producing yeast species [60].

Although an improvement over culture-based methodologies, reference gene sequencing can only be used to assign an unknown microbial strain to the genus level, for some species of bacteria. However, the sequence data can be compared with similar data from type strains of the different species within the genus via multiple sequence alignment to generate a phylogenetic tree showing the relatedness of the strain of interest to the various species within the genus [61, 62].

To further improve the degree of microbial identification above the level of accuracy obtained from the sequencing of phylogenetic reference genes, the whole-genome sequencing (WGS) can be used. It represents a valuable shortcut, helping scientists to find genes much more easily and quickly. Being able to study the entire genome sequence will help in understanding how the genes operate to direct the maintenance, development, and growth of a whole organism. Furthermore, the WGS can be used to predict the genes involved in the synthesis of biosurfactants in microbes [63]. Nayarisserj and Singh used the Illumina NextSeq 500 system to determine the WGS of *Bacillus tequilensis,* a biosurfactant-producing bacterium [64].

14.6 Methods to Identify the Biosynthesis Pathway of Surface-Active Compounds

The attribution of a microbial strain to the production of specific surface-active compounds is evidence of the biosynthetic pathway/enzymes the organism is utilizing for production.

The biosynthesis of mono- and di-rhamnolipids is due to the action of two, respectively three enzymes, RhlA, RhlB, and RhlC, encoded by three genes, *rhlA*, *rhlB,* and *rhlC*. These genes can be identified by using either *in vitro* or *in silico* methodologies. The *in vitro* approach utilizes *rhlA-C* sequence data from well-characterized rhamnolipid-producing strains to generate sequence alignments for the identification of conserved regions that can then be used to design primers. These primers are used to screen chromosomal DNA extracted from isolates of interest via PCR, a nucleic acid amplification technique that involves denaturing, renaturing, elongating, and amplifying a short segment of DNA or RNA. This methodology was used in the study of rhamnolipid-producing marine bacteria [1, 37]. These conserved regions of either DNA or amino acid sequence can also be utilized to carry out an *in silico* probe of either the sequenced genome of the strain of interest or previously published genomes of closely related strains. This approach was used by Tripathi et al. [1].

PCR-based amplification of biosynthetic genes in glycolipid-producing yeast is less straightforward. However, more WGS has become available, also for ascomycetes and the more complex basidiomycetes species like the mannosylerythritol and cellobiose lipids. For most of the well-described compounds, a reference synthetic gene cluster is available, and recently also the first biosynthetic pathway for liamocin was described. (Liamocin is a biosurfactant with a mannitol headgroup partially *O*-acylated with polyester tails containing three or four 3,5-dihydroxydecanoic ester groups which may or may not be acetylated.) [65].

Like mannosylerythritol lipids, sophorolipids are retrieved in a narrow taxonomic group, the *Starmerella* clade; similarly, putative clusters can be retrieved here.

Many different software tools exist for the retrieval of biosynthetic clusters, with examples used in the field of biosurfactant research [65]. Further information regarding genome mining techniques and their application in natural product discovery can be found in reviews by Machado et al. [66] and Ziemert et al. [67].

References

1 Tripathi, L., Twigg, M.S., Zompra, A. et al. (2019). Biosynthesis of rhamnolipid by a marinobacter species expands the paradigm of biosurfactant synthesis to a new genus of the marine microflora. *Microbial Cell Factories* 18: 164.

2 Raaijmakers, J.M., De Bruijn, I., Nybroe, O. et al. (2010). Natural functions of lipopeptides from Bacillus and Pseudomonas: more than surfactants and antibiotics. *FEMS Microbiology Reviews* 34: 1037–1062.

3 Zompra, A.A., Chasapi, S.A., Twigg, M.S. et al. (2022). Multi-method biophysical analysis in discovery, identification, and in-depth characterization of surface-active compounds. *Frontiers in Marine Science* 9: 1023287.

4 Twigg, M.S., Baccile, N., Banat, I.M. et al. (2021). Microbial biosurfactant research: time to improve the rigour in the reporting of synthesis, functional characterization and process development. *Microbial Biotechnology* 14: 147–170.

5 Mulligan, C., Cooper, D., and Neufeld, R. (1984). Selection of microbes producing biosurfactants in media without hydrocarbons. *Journal of Fermentation Technology* 62: 311–314.

6 Jain, D.K., Collins-Thompson, D.L., Lee, H., and Trevors, J.T. (1991). A drop-collapsing test for screening surfactant-producing microorganisms. *Journal of Microbiological Methods* 13: 271–279.

7 Plaza, G.A., Zjawiony, I., and Banat, I.M. (2006). Use of different methods for detection of thermophilic biosurfactant-producing bacteria from hydrocarbon-contaminated and bioremediated soils. *Journal of Petroleum Science and Engineering* 50: 71–77.

8 Siegmund, I. and Wagner, F. (1991). New method for detecting rhamnolipids excreted by Pseudomonas species during growth on mineral agar. *Biotechnology Techniques* 5: 265–268.

9 Cooper, D.G. and Goldenberg, B.G. (1987). Surface active agents from two Bacillus species. *Applied and Environmental Microbiology* 53: 224–229.

10 Thenmozhi, R. and Nagasathya, A. (2010). Screening for surface-active agent producing bacteria in used engine oil contaminated soil samples. *Research Journal of Agriculture and Biological Sciences* 6: 983–986.

11 Hodge, J.E. and Hofreiter, B.T. (1962). Determination of reducing sugars and carbohydrates. In: *Methods in Carbohydrate Chemistry* (ed. R.L. Whistler and M.L. Wolfrom), 380–394. New York: Academic Press.

12 Gunther, N.W., Nunez, A., Fett, W. et al. (2005). Production of rhamnolipid by *Pseudomonas chlororaphis*, a nonpathogenic bacterium. *Applied and Environmental Microbiology* 71: 2288–2293.

13 Priji, P., Sajith, S., Unni, K.N. et al. (2017). Pseudomonas sp. BUP6, a novel isolate from Malabari goat produces an efficient rhamnolipid type biosurfactant. *Journal of Basic Microbiology* 57: 21–33.

14 Laabei, M., Jamieson, W.D., Lewis, S.E. et al. (2014). A new assay for rhamnolipid detection - important virulence factors of *Pseudomonas aeruginosa*. *Applied Microbiology and Biotechnology* 98: 7199–7209.

15 Kubicki, S., Bator, I., Jankowski, S. et al. (2020). A straightforward assay for screening and quantification of biosurfactants in microbial culture supernatants. *Frontiers in Bioengineering and Biotechnology* 8: 958.

16 Kronberg, B., Holmberg, K., and Lindman, B. (2014). *Surface Chemistry of Surfactants and Polymers*, vol. 236. Chichester, UK: Wiley.

17 Kronberg, B., Holmberg, K., and Lindman, B. (2014). *Surface Chemistry of Surfactants and Polymers*, 242–243. Chichester, UK: Wiley.

18 Rosen, M.J. and Kunjappu, J.T. (2012). *Surfactants and Interfacial Phenomena*, 4the, 160–161. Hoboken, NJ: Wiley.

19 Berry, J.D., Neeson, M.J., Dagastine, R.R. et al. (2015). Measurement of surface and interfacial tension using pendant drop tensiometry. *Journal of Colloid and Interface Science* 454: 226–237.

20 Heyd, M., Kohnert, A., Tan, T.-H. et al. (2008). Development and trends of biosurfactant analysis and purification using rhamnolipids as an example. *Analytical and Bioanalytical Chemistry* 391: 1579–1590.

21 Rudden, M., Tsauosi, K., Marchant, R. et al. (2015). Development and validation of an ultra-performance liquid chromatography tandem mass spectrometry (UPLC-MS/MS) method for the quantitative determination of rhamnolipid congeners. *Applied Microbiology and Biotechnology* 99: 9177–9187.

22 Smyth, T.J., Rudden, M., Tsaousi, K. et al. (2014). Protocols for the isolation and analysis of lipopeptides and bioemulsifiers. In: *Hydrocarbon and Lipid Microbiology Protocols* (ed. T.J. McGenity, B. Nogales, and K.N. Timmis), 3–28. Berlin, Heidelberg: Springer.

23 Biniarz, P. and Łukaszewicz, M. (2017). Direct quantification of lipopeptide biosurfactants in biological samples *via* HPLC and UPLC-MS requires sample modification with an organic solvent. *Applied Microbiology and Biotechnology* 101: 4747–4759.

24 Leitermann, F., Syldatk, C., and Hausmann, R. (2008). Fast quantitative determination of microbial rhamnolipids from cultivation broths by ATR-FTIR spectroscopy. *Journal of Biological Engineering* 2: 13.

25 Nalini, S. and Parthasarathi, R. (2014). Production and characterisation of rhamnolipids produced by *Serratia rubidaea* SNAUo2 under solid-state fermentation and its application as biocontrol agent. *Bioresource Technology* 173: 231–238.

26 Van Bogaert, I.N.A., Buyst, D., and Martins, J.C. (2016). Synthesis of bolaform biosurfactants by an engineered *Starmerella bombicola* yeast. *Biotechnology and Bioengineering* 113: 2644–2651.

27 Gauthier, C., Lavoie, S., Piochon, M. et al. (2019). Structural determination of ananatoside a: an unprecedented 15-membered macrodilactone-containing glycolipid from *Pantoea ananatis*. *Carbohydrate Research* 471: 13–18.

28 Nydén, M. and Holmberg, K. (2002). Measuring micelle size and shape. In: *Handbook of Applied Surface and Colloid Chemistry, Vol. 2* (ed. K. Holmberg), 281–298. Chichester, UK: Wiley.

29 Ho, C.S., Lam, C.W.K., Chan, M.H.M. et al. (2003). Electrospray ionisation mass spectrometry: principles and clinical applications. *The Clinical Biochemist Reviews* 24: 3–12.

30 Rychert, J. (2019). Benefits and limitations of MALDI-TOF mass spectrometry for the identification of microorganisms. *Journal of Infectiology and Epidemiology* 2: 1–5.

31 Kim, P.I., Ryu, J., Kim, Y.H., and Chi, Y.T. (2010). Production of biosurfactant lipopeptides iturin A, fengycin and surfactin A from *Bacillus subtilis* CMB32 for control of *Colletotrichum gloeosporioides*. *Journal of Microbiology and Biotechnology* 20: 138–145.

32 Vater, J., Kablitz, B., Wilde, C. et al. (2002). Matrix-assisted laser desorption ionization-time of flight mass spectrometry of lipopeptides in whole cells and culture filtrates of bacillus subtilis C-1 isolated from petroleum sludge. *Applied and Environmental Microbiology* 68: 6210–6219.

33 Pueyo, M.T., Bloch, C. Jr., Carmona-Ribeiro, A.M., and di Mascio, P. (2009). Lipopeptides produced by a soil *Bacillus megaterium* strain. *Microbial Ecology* 57: 367–378.

34 Sato, S., Fukuoka, T., Saika, A., and et al. (2019). A new screening approach for glycolipid-type biosurfactant producers using MALDI-TOF/MS. *Journal of Oleo Science* 68: 1287–1294.

35 Price, N.P.J., Ray, K.J., Vermillion, K., and Kuo, T.M. (2009). MALDI-TOF mass spectrometry of naturally occurring mixtures of monorhamnolipids and dirhamnolipids. *Carbohydrate Research* 344: 204–209.

36 Johnson, A.R. and Carlson, E.E. (2015). Collision-induced dissociation mass spectrometry: a powerful tool for natural product structural elucidation. *Analytical Chemistry* 87: 10668–10678.

37 Twigg, M.S., Tripathi, L., Zompra, A. et al. (2018). Identification and characterisation of short chain rhamnolipid production in a previously uninvestigated, non-pathogenic marine pseudomonad. *Applied Microbiology and Biotechnology* 102: 8537–8549.

38 Van Renterghem, L., Roelants, S.L.K.W., Baccile, N. et al. (2018). From lab to market: an integrated bioprocess design approach for new-to-nature biosurfactants produced by *Starmerella bombicola*. *Biotechnology and Bioengineering* 115: 1195–1206.

39 Abdel-Mawgoud, A.M., Lépine, F., and Déziel, E. (2014). Liquid chromatography/mass spectrometry for the identification and quantification of rhamnolipids. *Methods in Molecular Biology* 1149: 359–373.

40 Haque, E., Bin Riyaz, M.A., Shankar, S., and Hassan, S. (2021). Compositional characterisation of biosurfactant produced by *Pseudomonas aeruginosa* ENO14-MH271625 and its application in crude oil bioremediation. *International Journal of Pharmaceutical Investigation* 11: 204–207.

41 Hosseini, S., Sharifi, R., Habibi, A., and Ali, Q. (2024). Molecular identification of a rhamnolipid produced by *Pseudomonas oryzihabitans* during biodegradation of crude oil. *Frontiers in Microbiology* 15: 1459112.

42 Fan, Y. and Wang, Y. (2019). Applications of small-angle X-ray scattering/small-angle neutron scattering and cryogenic transmission electron microscopy to understand self-assembly of surfactants. *Current Opinion in Colloid & Interface Science* 42: 1–16.

43 Hassan, P.A., Rana, S., and Verma, G. (2015). Making sense of Brownian motion: colloid characterization by dynamic light scattering. *Langmuir* 31: 3–12.

44 Cui, H., Hodgdon, T.K., Kaler, E.W. et al. (2007). Elucidating the assembled structure of amphiphiles in solution via cryogenic transmission electron microscopy. *Soft Matter* 3: 945–955.

45 Loh, W., Brinatti, C., and Tam, K.C. (2016). Use of isothermal titration calorimetry to study surfactant aggregation in colloidal systems. *Biochimica et Biophysica Acta* 1860: 999–1016.

46 Monzel, C. and Sengupta, K. (2016). Measuring shape fluctuations in biological membranes. *Journal of Physics D: Applied Physics* 49: 243002.

47 Sezgin, E., Levental, I., Mayor, S., and Eggeling, C. (2017). The mystery of membrane organization: composition, regulation and physiological relevance of lipid rafts. *Nature Reviews Molecular Cell Biology* 18: 361–374.

48 Kronberg, B., Holmberg, K., and Lindman, B. (2024). *Surface Chemistry of Surfactants and Polymers. 115.* Chichester, UK: Wiley.

49 Bao, X.-N., Zhang, W.-D., Gang, H.Z. et al. (2021). Formation of viscoelastic micellar solutions by a novel cationic surfactant and anionic salt system. *Colloids and Surfaces A* 611: 125795.

50 Kronberg, B., Holmberg, K., and Lindman, B. (2014). *Surface Chemistry of Surfactants and Polymers*, 113–118. Chichester, UK: Wiley.

51 Kronberg, B., Holmberg, K., and Lindman, B. (2024). *Surface Chemistry of Surfactants and Polymers, 122.* Chichester, UK: Wiley.

52 Mitchell, D.J., Tiddy, G.J.T., Waring, L. et al. (1983). Phase behaviour of polyoxyethylene surfactants with water. Mesophase structures and partial miscibility (cloud points). *Journal of the Chemical Society, Faraday Transactions 1: Physical Chemistry in Condensed Phases* 79: 975–1000.

53 Klaus, A., Tiddy, G.J.T., Touraud, D. et al. (2010). Phase behavior of an extended surfactant in water and a detailed characterization of the concentrated phases. *Langmuir* 26: 16871–16883.

54 Khan, A., Fontell, K., Lindblom, G., and et al. (1982). Liquid crystallinity in a calcium surfactant system: phase equilibria and phase structures in the system calcium octylsulfate/decan-1-ol/water. *Journal of Physical Chemistry* 86: 4266–4271.

55 Kronberg, B., Holmberg, K., and Lindman, B. (2014). *Surface Chemistry of Surfactants and Polymers*, 125–126. Chichester, UK: Wiley.

56 Toribio, J., Escalante, A.E., and Soberón-Chávez, G. (2010). Rhamnolipids: production in bacteria other than *Pseudomonas aeruginosa*. *European Journal of Lipid Science and Technology* 112: 1082–1087.

57 Nordin, N., Zakaria, M.R., Halmi, M.I.E. et al. (2013). Isolation and screening of high efficiency biosurfactant-producing bacteria *Pseudomonas* sp. *Journal of Biochemistry, Microbiology and Biotechnology* 1: 25–31.

58 Langille, M.G.I., Zaneveld, J., Caporaso, J.G. et al. (2013). Predictive functional profiling of microbial communities using 16S rRNA marker gene sequences. *Nature Biotechnology* 31: 814–821.

59 Schoch, C.L., Seifert, K.A., Huhndorf, S. et al. (2012). Nuclear ribosomal internal transcribed spacer (ITS) region as a universal DNA barcode marker for fungi. *Proceedings of the National Academy of Sciences of the USA* 109: 6241–6246.

60 Kurtzman, C.P., Price, N.P.J., Ray, K.J., and Kuo, T.M. (2010). Production of sophorolipid biosurfactants by multiple species of the *Starmerella* (*Candida*) *bombicola* yeast clade. *FEMS Microbiology Letters* 311: 140–146.

61 Pruesse, E., Quast, C., Knittel, K. et al. (2007). SILVA: a comprehensive online resource for quality checked and aligned ribosomal RNA sequence data compatible with ARB. *Nucleic Acids Research* 35: 7188–7196.

62 Quast, C., Pruesse, E., Yilmaz, P. et al. (2013). The SILVA ribosomal RNA gene database project: improved data processing and web-based tools. *Nucleic Acids Research* 41: D590–D596.

63 Stein, L. (2001). Genome annotation: from sequence to biology. *Nature Reviews. Genetics* 2: 493–503.

64 Nayarisserj, A. and Singh, S.K. (2023). Genome analysis of biosurfactants producing bacterium *Bacillus tequilensis*. *PLoS One* 18: e0285994.

65 Xue, S.-J., Liu, G.-L., Chi, Z. et al. (2020). Genetic evidences for the core biosynthesis pathway, regulation, transport and secretion of liamocins in yeast-like fungal cells. *The Biochemical Journal* 477: 887–903.

66 Machado, H., Tuttle, R.N., and Jensen, P.R. (2017). Omics-based natural product discovery and the lexicon of genome mining. *Current Opinion in Microbiology* 39: 136–142.

67 Ziemert, N., Alanjary, M., and Weber, T. (2016). The evolution of genome mining in microbes – a review. *Natural Product Reports* 33: 988–1005.

15

Distinctive Features of Low- and High-Molecular-Weight Amphiphiles

15.1 Introduction

In the context of surface and colloid chemistry, there is a distinction to be made between low- and high-molecular-weight amphiphiles. Both have a pronounced surface activity, but their movement in aqueous solutions differs, which is important. According to the Stokes-Einstein equation, the diffusion rate is inversely proportional to the hydrodynamic radius of a molecule or a particle; thus, a high-molecular-weight molecule moves slowly to surfaces and interfaces compared with low-molecular-weight species. However, once at a surface or an interface, the high-molecular-weight amphiphile can become almost irreversibly attached. This differs from the behavior of low-molecular-weight amphiphiles, the adsorption of which is reversible; adsorption and desorption occur constantly at a short time scale. These differences in physicochemical behavior in solution are the reason why the mode of action of high- and low-molecular-weight amphiphiles differs.

Both are ubiquitous in nature, and both are also common as synthetic compounds. In nature, as well as in technical formulations, they can act independently but it is also common that they act in concert. The man-made low-molecular-weight amphiphiles are normally referred to as surfactants and surfactants are used for a variety of purposes, such as wetting of solid surfaces, emulsification, foaming, solubilization of poorly water-soluble substances (such as soil, dyes, and drugs), etc. The low-molecular-weight amphiphiles that nature provides are often referred to as polar lipids. Well-known types that are covered by different chapters in this book are phospholipids (such as lecithin), saponins, glycolipids (such as sophorolipids, rhamnolipids, mannosylerythritol lipids, trehalose lipids, cellobiose lipids, and xylolipids), and lipopeptides (such as surfactin).

The man-made high-molecular-weight amphiphiles are sometimes called polymeric surfactants, sometimes surface-active polymers. They are often used as stabilizers of dispersed systems, such as emulsions, suspensions (solid dispersions), and foams. They are effective in preventing coalescence of emulsion drops and agglomeration of solid dispersions. To make long-lived (although not "stable" in a thermodynamic sense) emulsions or foams, it is common to use a combination of a low-molecular-weight amphiphile as emulsifier or foaming agent and a surface-active polymer as stabilizer of the emulsion or foam. The stabilizer may be present from the beginning or post-added after the dispersed system is formed. Nature's products of this type belong to the category of biomacromolecules. Lipopolysaccharides and membrane proteins are two well-known examples of

Natural Surfactants: Science, Technology, and Industrial Aspects, First Edition. Guido Bognolo and Krister Holmberg.

biomacromolecules with a pronounced surface activity. Lipopolysaccharides are treated in one chapter of this book.

There are also borderline cases, amphiphiles that have an intermediate molecular weight. EO-PO block copolymers, usually a triblock copolymer composed of a central polyoxypropylene block surrounded by two polyoxyethylene blocks, are a well-known example among man-made surface-active molecules. Antimicrobial peptides, which are the topic of one chapter of this book, are a typical example from nature.

15.2 Emulsions

In the following pages, we will deal mostly with emulsions, but the same considerations apply mutatis mutandis to other dispersed systems, such as foams and solid dispersions.

Ordinary emulsions, or macroemulsions, are thermodynamically unstable and with time the oil and the water phases eventually separate under the effect of gravity. (Microemulsions, although the name implies a similarity with emulsions, are very different, the most fundamental difference being that they are thermodynamically stable).

The emulsion drops may flocculate to form aggregates. These aggregates may rise to the top if the density of the dispersed phase is lower than that of the continuous phase, which is normally the case for oil-in-water emulsions. This is called creaming. If the density of the dispersed phase is higher than that of the continuous phase, then there may be sedimentation instead. The drops of an emulsion that has undergone flocculation or creaming have largely retained their integrity, and such systems can be restored to their original state by mechanical agitation without too much energy input.

The drops may also coalesce into larger drops, and as this process continues, it results in the formation of a separate phase at the top (or at the bottom if the density of the drops is higher than that of the continuous phase). Emulsions that have separated into two distinct phases cannot be restored unless they are reemulsified by strong mechanical action. Figure 15.1 illustrates flocculation, creaming, coalescence, and phase separation.

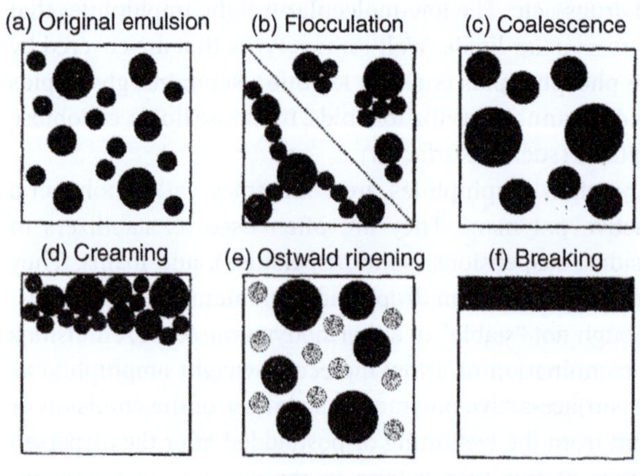

Figure 15.1 Typical mechanisms for destabilization and breaking of emulsions.

The use of an emulsion stabilizer, typically a high-molecular-weight amphiphile or small solid particles, is a way to reduce the rate of this type of phase separation and extend the lifetime of the emulsion. The use of particles to stabilize emulsions has become popular. Such particle-stabilized emulsions are often referred to as Pickering emulsions after one of the pioneers in the field. Also, the particles need to be amphiphilic, however. They should prefer the oil–water interface and not want to reside in neither the oil nor the water phase.

Ostwald ripening is another important destabilization mechanism for emulsions. It is a process where larger drops grow at the expense of small ones. It reflects the fact that the chemical potential in small drops is larger than in big drops. Hence, there will be a transport of molecules from the small drops to the large ones. Ostwald ripening cannot be avoided unless the emulsion is perfectly monodispersed, which is hardly realistic. The rate with which Ostwald ripening occurs depends on the polydispersity of the emulsion – the higher the polydispersity, the more rapid is Ostwald ripening – and on the solubility of the dispersed phase in the continuous phase. Thus, a hexadecane-in-water emulsion exhibits a much lower rate of Ostwald ripening than a hexane-in-water emulsion. A commonly used method to reduce the rate of Ostwald ripening is to add a small amount of a very water-insoluble oil to the hydrocarbon that is emulsified in water. Squalene is a typical oil used for that purpose. This creates a thermodynamic barrier to the diffusion of the hydrocarbon from small to large drops. The choice of emulsifier also plays a role in the rate of Ostwald ripening. Ostwald ripening is also illustrated in Figure 15.1.

Foams behave in a similar manner. Large bubbles grow at the expense of small ones. However, the mechanism is different from Ostwald ripening of emulsions. The transport of gas through to the continuous phase is due to a higher Laplace pressure in the small bubbles. The gas pressure inside a bubble is inversely proportional to the size of the bubble. An Ostwald ripening-related process also operates for many solid dispersions, in particular for dispersions of nanoparticles.

Oil-in-water emulsions can be stabilized by:

- *Electrostatic effects:* Charges at the surface will attract counterions. The surface charges typically originate from the use of ionic surfactants as emulsifiers but may also come from a charged polymer adsorbed at the surface. Charged nanoparticles, such as silica nanoparticles or cellulose nanocrystals, may also be the origin of the charges. When two such drops approach each other in solution, there will be a cloud of counterions in between them. A local high concentration of counterions is entropically unfavorable and leads to an osmotic flow of water, pushing the particles apart. Thus, electrostatic stabilization is related to the entropy penalty of a local high concentration of counterions between approaching drops. Anionic surfactants, in particular sulfonates and sulfates, are commonly used to obtain electrostatic stabilization. Oligomeric or polymeric anionic amphiphiles, such as alkylnaphthalene sulfonates or polycarboxylates, are also used for the purpose.
- *Steric effects:* Water-soluble chains protruding from the surface of the drop into the bulk medium can prevent coalescence of the drops provided that the density of the chains is high enough. Also here the stabilization is related to an entropy penalty. When two drops with protruding chains approach each other, there will be a local high concentration of chains in between the particles. This will lead to an osmotic flow of water to level out

the concentration of the chains and that flow will push the drops apart. Fatty alcohol ethoxylates with long polyoxyethylene chains are commonly used to give steric stabilization. Such chains are highly hydrated, i.e. they are surrounded by structured layers of water molecules. Overlap of these hydration layers is also entropically unfavorable. This is sometimes referred to as hydration stabilization.

- *Electrosteric effects* are a combination of the previous two effects. Electrosteric stabilization is usually obtained by employing a mixture of one surfactant that gives electrostatic stabilization and one that provides steric stabilization. Electrosteric stabilization can also be obtained by a single molecule, typically a macromolecule with both charged (normally anionic) sites at the end of long protruding chains. Electrosteric stabilization is what prevents the cells in the body from agglomeration. The cells have a negative surface charge due to sialic acid residues in glycolipids and glycoproteins. In addition, they have long polysaccharide chains protruding into the body fluid.

Figure 15.2 illustrates electrostatic and steric stabilization. Figure 15.3 shows electrosteric stabilization obtained with a mixture of dodecyl sulfate and ethoxylated (5 EO) dodecanol, a typical surfactant mixture for the preparation of latices.

A good emulsifier should reduce the oil–water interfacial tension to low values to facilitate the breakdown of the oil into small drops. This means that the emulsifier should have a strong driving force for the oil–water interface, i.e. it should have a pronounced interfacial activity. (Interfacial activity is not necessarily the same as surface activity; there are surfactants that are extremely surface active but only moderately interfacially active.) The emulsifier should also diffuse rapidly in the bulk water so that it adsorbs at the surface of the oil drop that has been generated by homogenization or some other mechanical action. As mentioned above, rapid diffusion demands low-molecular-weight species.

As mentioned above, an emulsion stabilizer may be a polymer or solid particles. The stabilizer should have some surface activity, i.e. it should have a driving force to go to the surface of the drop. However, it need not be as surface active as the low-molecular-weight emulsifier, for which one of the tasks is to reduce the oil–water interfacial tension to very low values. The stabilizer will diffuse relatively slowly to the surface of the oil drop. Stirring will increase the rate with which it moves but transport through the stagnant layer around the drop will be by diffusion and not by convection. Then the Stokes-Einstein equation

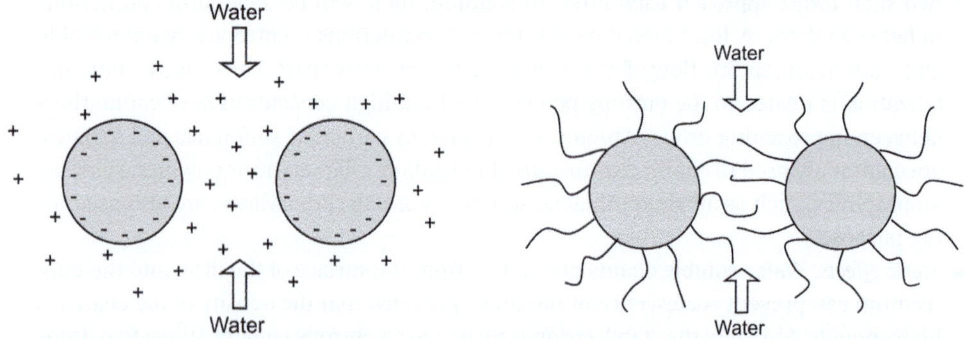

Figure 15.2 Illustration of electropstatic stabilization (left) and steric stabilization (right). (Kronberg et al. [1] / John Wiley & Sons.)

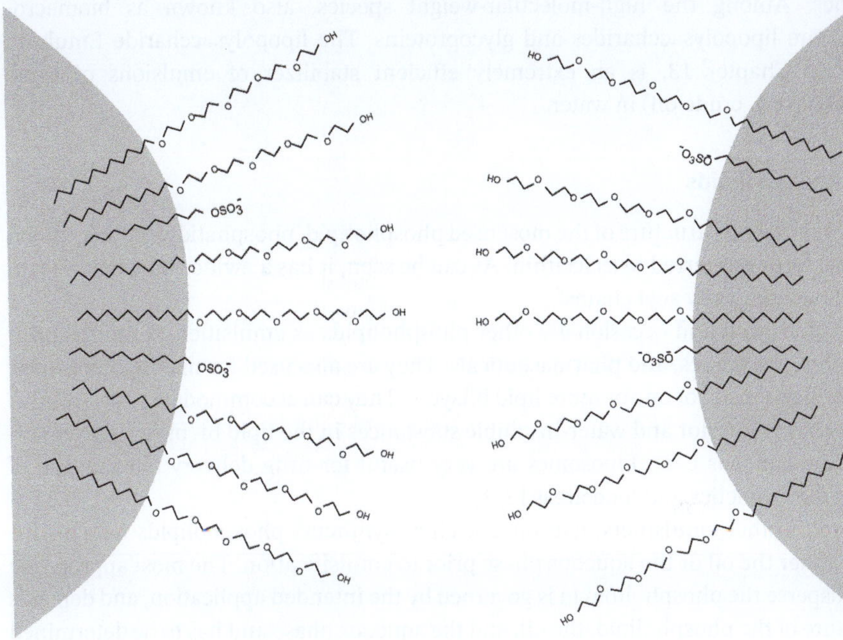

Figure 15.3 Illustration of electrosteric stabilization, keeping two drops or particles apart. (Kronberg et al. [1] / John Wiley & Sons.)

dictates the speed, and a large polymer may need minutes to reach the surface. The polymer will then usually rearrange at the surface to find its energetically most favorable position and this process can take hours. Once anchored at the surface, a macromolecule or a solid particle will remain there. They often become almost irreversibly adsorbed. This is very different from the adsorption behavior of low-molecular-weight amphiphiles, for which the lifetime at the surface is short.

The net result, when a mixture of a low-molecular-weight emulsifier and a high-molecular-weight emulsion stabilizer is used for making an oil-in-water emulsion, is that the oil droplets will eventually be covered by the emulsion stabilizer, which coexists with the low-molecular-weight amphiphile (or amphiphiles). A well-known example from nature where such a complex surface layer stabilizes an emulsion is the fat globules in milk. These triglyceride particles (or drops) are stabilized against coalescence by high-molecular-weight biomacromolecules, such as proteins and glycoproteins coexisting with low-molecular-weight amphiphiles, such as phospholipids and cholesterol.

15.3 Natural Amphiphiles as Emulsifiers and Emulsion Stabilizers

Nature produces a large number of low- and high-molecular-weight amphiphiles that are useful as emulsifiers and emulsion stabilizers, respectively. Among the low-molecular-weight species, also called polar lipids, are phospholipids, saponins, glycolipids, and lipopeptides. Among the high-molecular-weight species, also known as biomacromolecules, are lipopolysaccharides and glycoproteins. The lipopolysaccharide Emulsan, discussed in Chapter 13, is an extremely efficient stabilizer of emulsions of some hydrocarbons (e.g. crude oil) in water.

15.3.1 Phospholipids

Figure 15.4 shows the structure of the most used phospholipid, phosphatidylcholine, which in technical form is referred to as lecithin. As can be seen, it has a zwitterionic polar headgroup and two long fatty acyl chains.

The use of lecithin and occasionally other phospholipids as emulsifiers is limited principally to food, cosmetics, and pharmaceuticals. They are also used for making liposomes, i.e. small vesicles made of one or more lipid bilayers. They can accommodate water-soluble substances in the interior and water-insoluble substances in the lipid bi- or multilayer surrounding the aqueous core. Liposomes are very useful for drug delivery but are also of interest in the cosmetics and food areas [2, 3].

Unlike most other emulsifiers, natural as well as synthetic, phospholipids may be dispersed in either the oil or the aqueous phase prior to emulsification. The most appropriate phase to disperse the phospholipid in is governed by the intended application, and depends on the nature of the phospholipid, the oil, and the aqueous phase and has to be determined empirically.

As mentioned above, lecithin is the most widely used phospholipid and lecithin emulsions are robust. They are stable down to relatively low pH (2–3) and moderate salinity (below 100 mmol NaCl). The emulsion stability is related to the lecithin composition, notably the ratio of phosphatidyl choline to phosphatidyl ethanolamine [4]. Formulation-composition maps have been developed to predict the optimum lecithin–oil–water ratios for producing stable emulsions [5].

Certain types of phospholipids may be effective at retarding the oxidation of emulsified lipids because of their natural free radical scavenging capacity [6, 7]. This can be very important for certain food applications.

Lecithin has been combined with caseins to form antimicrobial emulsions [8], with whey proteins to form lutein-loaded emulsions [9], and with monoacylglycerols to form infant formula emulsions [10].

The functionality of lecithin may be improved by utilizing cosolvents, such as ethanol, which alter the properties of the surfactant monolayer (such as optimum curvature), and this may be advantageous for the formation of the emulsion and for its stability [11].

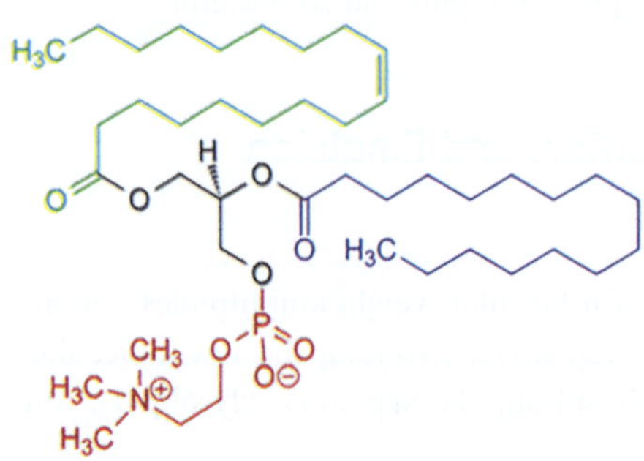

Figure 15.4 Lecithin, the most used phospholipid. (Jü / Public Domain.)

Natural lecithins can be modified by cleavage of one of the fatty acid tails from the glycerol backbone to create more polar surfactants, "lysolecithins," that are suitable for forming oil-in-water emulsions, especially when used in combination with other emulsifiers [12, 13]. The cleavage of the ester linkage can be done through alkaline hydrolysis or by a phospholipase A2-catalyzed process. The enzymatic cleavage is regioselective; only the ester in the 2-position of glycerol is cleaved. It is also reversible. In media with low water activity, such as water-in-oil microemulsions, phospholipase A2 catalyzes the synthesis of lecithin from lysolecithin and a fatty acid. That is a way to synthesize lecithins with an unusual fatty acid, e.g. an ω-3 polyunsaturated acid, in the 2-position [14].

The physical and chemical stability of lecithin-coated lipid droplets can be improved by coating with oppositely charged biopolymers to form multilayer emulsions. As an example, the cationic chitosan has been used to coat anionic lecithin-covered drops [15].

15.3.2 Saponins

Saponins are highly surface active, and because of this, they have been used for centuries for textile cleaning. Figure 15.5 shows the structure of a typical saponin. The polar headgroup is a trisaccharide that is anionic at neutral and alkaline pH and the hydrophobic tail is a fused ring system, a triterpenoid.

Saponins are excellent emulsifiers and can generate oil-in-water emulsions with small drops, below 200 nm in size, that are stable to aggregation over a range of conditions of pH, ionic strength, and temperature, as has been demonstrated by McClements and coworkers [16–18]. The ability of saponins to form stable emulsions has been attributed in part to their ability to form interfacial layers with a high dilatational elasticity, which may inhibit droplet deformation and coalescence [19].

Saponin-stabilized oil-in-water emulsions showed better lipid oxidative stability than those stabilized by synthetic emulsifiers, which was attributed to their free radical scavenging capacity [20]. This can be important for food applications.

A practically important observation made by McClements et al. is that saponins are not suitable for producing emulsions by the phase inversion method because of their insolubility in the oil phase [21].

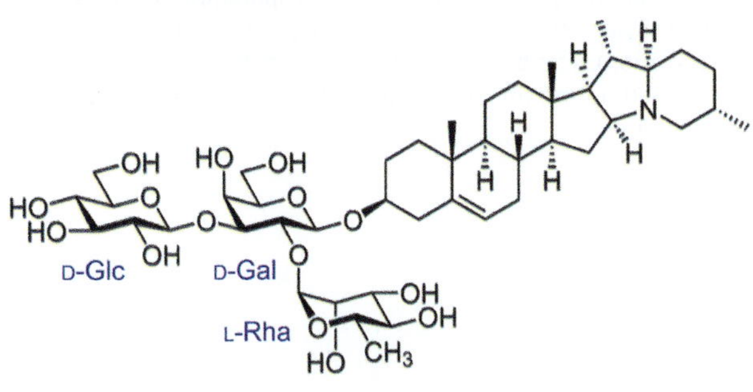

Figure 15.5 A saponin (a steroidal glycoalkaloid) with an oligosaccharide chain attached at carbon atom 3 of the fused ring system. (Calvero / Public Domain.)

15.3.3 Glycolipids

Glycolipids, such as sophorolipids, rhamnolipids, mannosylerythritol lipids, trehalose lipids, cellobiose lipids, and xylolipids, are low-molecular-weight amphiphiles. They are discussed in some detail in separate chapters, but it is appropriate here to highlight again their attractiveness to the cosmetic industry, which increasingly values sulfate-free and ethylene oxide-free emulsifiers. Their antimicrobial properties are also a big asset.

15.3.4 Lipopeptides

Lipopeptides, such as surfactin, fengycin, iturin, and lichenysin, are low-molecular-weight amphiphiles. They are very surface active and have exceptionally low CMC values. They are of interest as emulsifiers for cosmetic applications, but the high cost of production limits their use. Their properties and uses are discussed in a separate chapter.

15.3.5 Surface-Active Proteins

Most surface-active proteins are glycoproteins, i.e. the protein backbone is substituted with oligo- or polysaccharide chains. These chains may be uncharged but are more commonly negatively charged. The glycosylation not only provides amphiphilicity but also affects the folding, enhances the stability of the structure, and protects the protein from proteases. Figure 15.6 shows a schematic of a polymer with hydrophilic side chains grafted on a hydrophobic backbone.

There are also nonglycosylated proteins with pronounced amphiphilicity. Certain transmembrane proteins are prime examples of very surface-active proteins that lack oligo-or polysaccharide chains attached to their backbone. (There are also glycosylated transmembrane proteins.) The surface activity of these biopolymers is due to a structure composed of alternating blocks of hydrophilic and hydrophobic amino acids. Bacteriorhodopsin is a prime example. It penetrates the membrane no less than seven times. Figure 15.7 illustrates transmembrane proteins.

Many critical biological processes take place at the interface between hydrophilic and hydrophobic domains and a wide range of organisms produce surface-active proteins and

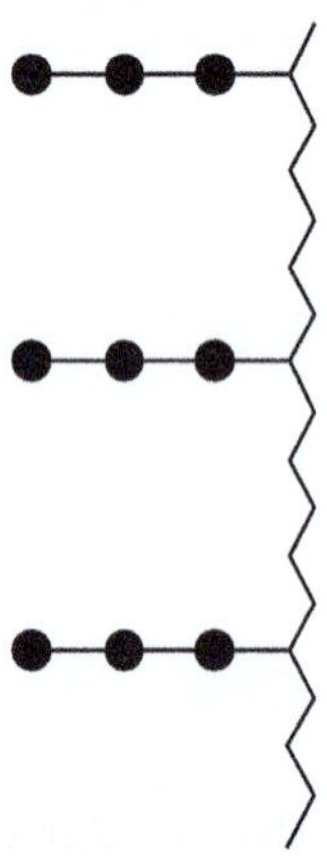

Figure 15.6 An illustration of a glycoprotein. The structure is very simplified. Oligosaccharide chains are grafted on a protein backbone. The grafting is either from asparagine residues (*N*-glycosylation) or from serine or threonine residues (*O*-acylation). The oligosaccharide chains (the connected black dots) need not be linear; on the contrary, they are usually branched.

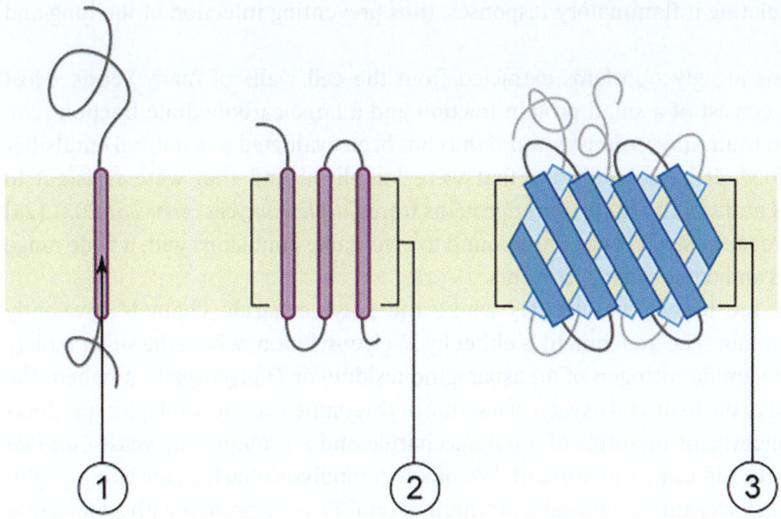

Figure 15.7 Transmembrane proteins: 1) a single-pass membrane protein (α-helix), 2) a multipass membrane protein (α-helix), 3) a multipass membrane protein (β-sheet). The membrane is represented in light yellow. The parts that reside in the lipid bilayer are composed of hydrophobic amino acids and the loops on the inside and the outside consist of hydrophilic amino acids. (Foobar / Wikimedia Commons / CC BY 2.5.)

peptides that reduce surface and interfacial tension and mediate growth and development at these boundaries. Higher-order organisms produce surface-active proteins with a wide variety of functions, including the provision of protective foam environments for vulnerable reproductive stages, evaporative cooling, and gas exchange across airway membranes. In general, the biological functions supported by these diverse biomacromolecules require them to have an amphiphilic nature, and this is achieved by a diverse range of molecular structures, with some proteins undergoing significant conformational change or intermolecular association to generate the structures that are most surface active [22].

Frogs of tropical and subtropical regions of the world build foam nests as biocompatible incubators for the development of fertilized eggs and tadpoles. These foam nests are made of aqueous protein mixtures (ranaspumins), which are very surface active. The proposed model suggests self-assembly of protein aggregates at the air–water interface, in which foam formation is facilitated by certain very surface-active proteins in the mixture. The structure is further stabilized by aggregation and cross-linking into multilayer structures [23]. Latherin is a protein with strong surfactant properties found in the sweat and saliva of horses and other equine species. In sweat, latherin functions to provide effective evaporative cooling by enhancing the wetting of the waterproof horsehair [24].

The mammalian pulmonary surfactant is a lipoprotein complex composed of 90% lipids and 10% proteins. Four surface-active proteins have been identified as important for the functioning of the pulmonary surfactant, which reduces the surface tension of the alveolae at the air–liquid interface, allowing respiration to occur with ease and preventing alveolar collapse. All four surface-active proteins are important in the pulmonary surfactant system; two are also involved in host defense by inhibiting and inactivating foreign pathogens, as well as by modulating inflammatory responses, thus preventing infection of the lung and the airway [25].

Mannoproteins are glycoproteins extracted from the cell walls of many yeasts. Most mannoproteins consist of a small protein fraction and a large carbohydrate fraction [26]. A mannoprotein from *Kluyveromyces marxianus* has been evaluated as a natural emulsifier for processed food. It gave emulsions that were long-lived and that were resistant to variations in pH and salinity [27]. Mannoproteins from *Saccharomyces cerevisiae* 2031 [28] and from *Acinetobacter* sp. [29] were also found to give stable emulsions with a wide range of hydrocarbons and other organic solvents.

Glycoproteins are biomacromolecules where the polysaccharide chain is covalently bonded to the protein. The attachment is either by *N*-glycosylation, where the sugar moiety is coupled to the amide nitrogen of an asparagine residue, or *O*-glycosylation, where the sugar is attached to the hydroxyl oxygen of a serine or threonine residue. Biological products may also be noncovalent mixtures of a polysaccharide and a protein. The yeast *Candida lipolytica* produces Liposan, a mixture of 83% of a heteropolysaccharide consisting of glucose, galactose, galactosamine, and galacturonic acid, and 17% of a protein with a molecular weight of 27 600 [30]. It has emulsifying properties similar to those of lipopolysaccharides and is particularly effective in emulsifying fats. As such, it is used in weight management products, for example, under the registered trademark Liposan Ultra® [31].

15.3.6 Lipopolysaccharides

Lipopolysaccharides are a unique family of macromolecular glycolipids, produced by most Gram-negative bacteria. They are large molecules, and they are essential for the integrity of the outer-membrane permeability barrier and participate extensively in the host-pathogen interplay.

Lipopolysaccharides play a very important role in biology, which is discussed in some detail in Chapter 13. They can also be regarded as a general-purpose amphiphilic macromolecule and as such be useful for stabilization of emulsions and other dispersed systems, as discussed above. A schematic representation of their structure is shown in Figure 15.8. Hydrophobic side chains are grafted on the polysaccharide backbone. The chains are usually C8–C16 fatty acids, and the attachment is by ester bonds. The polysaccharide need not be linear; on the contrary, it is often highly branched. Also, the acyl chains may contain branches.

Emulsan is the trade name of a polyanionic lipopolysaccharide produced by *Acinetobacter calcoaceticus*. It is an 80:20 mixture of a lipopolysaccharide and a high-molecular-weight cationic exopolysaccharide. The composition is complex and is discussed in some detail in the chapter on lipopolysaccharides. Emulsan is not very surface active, but it is extremely efficient as an emulsifier and emulsion stabilizer for crude oil. It is used for remediation of soil contaminated by crude oil, dispersion of crude oil spills, and cleaning of oil tanks. It is also tested for microbial enhanced oil recovery [32, 33].

Emulsan is the most established lipopolysaccharide for technical use, but a wide variety of similar products have been isolated from bacteria and yeasts. In most cases, the product is not a plain lipopolysaccharide; instead, it is a mixture of products, such as a lipopolysaccharide and a protein or a lipopolysaccharide and a heteropolysaccharide.

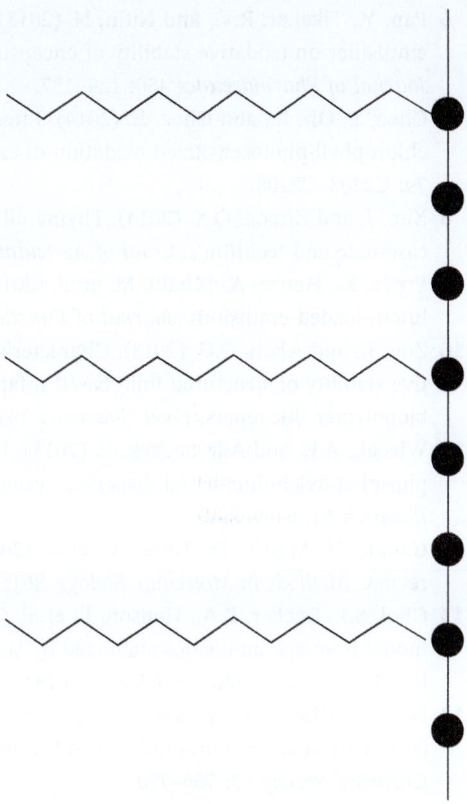

Figure 15.8 An illustration of a lipopolysaccharide. The structure is very simplified. The polysaccharide chain (the connected black dots) need not be linear. The lipophilic side chains may also be branched.

Low-molecular-weight amphiphilic compounds are often also produced. This is a field under rapid development and the analytical challenges involved in determining the structures of the different components are extreme, not least because of the large span in molecular weights of the biomolecules that make up the product.

References

1 Kronberg, B., Holmberg, K., and Lindman, B. (2014). *Surface Chemistry of Surfactants and Polymers*. Chichester, UK: Wiley.

2 Pichot, R., Watson, R.L., and Norton, I.T. (2013). Phospholipids at the interface: current trends and challenges. *International Journal of Molecular Sciences* 14: 11767–11794.

3 Liu, W., Ye, A., Singh, H., and Sagis, L.M.C. (2015). Progress in applications of liposomes in food systems. In: *Microencapsulation and Microspheres for Food Applications* (ed. L.M.C. Sagis), 151–172. Cambridge, MA: Academic Press.

4 Guiotto, E.N., Cabezas, D.M., Diehl, B.W.K., and Tomás, M.C. (2013). Characterization and emulsifying properties of different sunflower phosphatidylcholine enriched fractions. *European Journal of Lipid Science and Technology* 115: 865–873.

5 Thakur, R.K., Villette, C., Aubry, J.M., and Delaplace, G. (2007). Formulation-composition map of a lecithin-based emulsion. *Colloids and Surfaces A* 310: 55–61.

6 Pan, Y., Tikekar, R.V., and Nitin, N. (2013). Effect of antioxidant properties of lecithin emulsifier on oxidative stability of encapsulated bioactive compounds. *International Journal of Pharmaceutics* 450: 129–137.

7 Choe, J., Oh, B., and Choe, E. (2014). Effect of soybean lecithin on iron-catalyzed or chlorophyll-photosensitized oxidation of canola oil emulsion. *Journal of Food Science* 79: C2203–C2208.

8 Xue, J. and Zhong, Q.X. (2014). Thyme oil nanoemulsions coemulsified by sodium caseinate and lecithin. *Journal of Agricultural and Food Chemistry* 62: 9900–9907.

9 Frede, K., Henze, A., Khalil, M. et al. (2014). Stability and cellular uptake of lutein-loaded emulsions. *Journal of Functional Foods* 8: 118–127.

10 Zou, L. and Akoh, C.C. (2013). Characterisation and optimisation of physical and oxidative stability of structured lipid-based infant formula emulsion: effects of emulsifiers and biopolymer thickeners. *Food Chemistry* 141: 2486–2494.

11 Wiacek, A.E. and Adryanczyk, E. (2015). Interfacial properties of phosphatidylcholine-based dispersed systems. *Industrial & Engineering Chemistry Research* 54: 6489–6496.

12 Casado, V., Martin, D., Torres, C. et al. (2012). Phospholipases in food industry: a review. *Methods in Molecular Biology.* 861: 495–523.

13 Choi, S.J., Decker, E.A., Henson, L. et al. (2011). Formulation and properties of model beverage emulsions stabilized by sucrose monopalmitate: influence of pH and lyso-lecithin addition. *Food Research International* 44: 3006–3012.

14 Na, A., Eriksson, C., Eriksson, S.-G. et al. (1990). Synthesis of phosphatidylcholine with (n-3) fatty acids by phospholipase A2 in microemulsion. *Journal of the American Oil Chemists' Society* 67: 766–770.

15 Ogawa, S., Decker, E.A., and McClements, D.J. (2003). Influence of environmental conditions on the stability of oil in water emulsions containing droplets stabilized by lecithin-chitosan membranes. *Journal of Agricultural and Food Chemistry* 51: 5522–5527.

16 Ozturk, B., Argin, S., Ozilgen, M., and McClements, D.J. (2014). Formation and stabilization of nanoemulsion-based vitamin E delivery systems using natural surfactants: Quillaja saponin and lecithin. *Journal of Food Engineering* 142: 57–63.

17 Yang, Y., Leser, M.E., Sher, A.A., and McClements, D.J. (2013). Formation and stability of emulsions using a natural small molecule surfactant: Quillaja saponin (Q-Naturale). *Food Hydrocolloids* 30: 589–596.

18 Yang, Y. and McClements, D.J. (2013). Encapsulation of vitamin E in edible emulsions fabricated using a natural surfactant. *Food Hydrocolloids* 30: 712–720.

19 Stanimirova, R., Marinova, K., Tcholakova, S. et al. (2011). Surface rheology of saponin adsorption layers. *Langmuir* 27: 12486–12498.

20 Uluata, S., McClements, D.J., and Decker, E.A. (2015). Physical stability, autoxidation, and photosensitized oxidation of omega-3 oils in nanoemulsions prepared with natural and synthetic surfactants. *Journal of Agricultural and Food Chemistry* 63: 9333–9340.

21 Mayer, S., Weiss, J., and McClements, D.J. (2013). Vitamin E-enriched nanoemulsions formed by emulsion phase inversion: factors influencing droplet size and stability. *Journal of Colloid and Interface Science* 402: 122–130.

22 Sunde, M., Pham, C.L.L., and Kwan, A.H. (2017). Molecular characteristics and biological functions of surface-active and surfactant proteins. *Annual Review of Biochemistry* 86: 585–608.

23 Cooper, A., Kennedy, M.W., Fleming, R.I. et al. (2005). Adsorption of frog foam nest proteins at the air-water interface. *Biophysical Journal* 88: 2114–2125.

24 McDonald, R.E., Fleming, R.I., Beeley, J.G. et al. (2009). Latherin: a surfactant protein of horse sweat and saliva. *PLoS One* 4: e5726.

25 Wright, J.R. (2005). Immunoregulatory functions of surfactant proteins. *Nature Reviews. Immunology* 5: 58–68.

26 Uzoigwe, C., Burgess, J.G., Ennis, C.J. et al. (2015). Bioemulsifiers are not biosurfactants and require different screening approaches. *Frontiers in Microbiology* 6: 245.

27 Lukondeh, T., Ashbolt, N.J., and Rogers, P.L. (2003). Evaluation of *Kluyveromyces marxianus* FII 510700 grown on a lactose-based medium as a source of natural bioemulsifier. *Journal of Industrial Microbiology & Biotechnology* 30: 715–730.

28 Alcantara, V.A., Pajares, I.G., Simbahan, J.F. et al. (2013). Downstream recovery and purification of a bioemulsifier from *Saccharomyces cerevisiae* 2031. *Philippine Agricultural Scientist* 96: 349–358.

29 Jagtap, S., Yavankar, S., Pardesi, K., and Chopade, B. (2010). Production of bioemulsifier by *Acinetobacter* species isolated from healthy human skin. *Indian Journal of Experimental Biology* 48: 70–76.

30 Cirigliano, M.C. and Carman, G.M. (1985). Purification and characterisation of Liposan, a bioemulsifier from *Candida lipolytica*. *Applied and Environmental Microbiology* 50: 846–850.

31 https://www.liposan.com/products (accessed 23 April 2025).

32 Choi, J.-W., Choi, H.-G., and Lee, W.-H. (1996). Effects of ethanol and phosphate on emulsan production by *Acinetobacter calcoaceticus* RexcellentAG-1. *Journal of Biotechnology* 45: 217–225.

33 Ron, E.Z. and Rosenberg, E. (2001). Natural roles of biosurfactants. *Environmental Microbiology* 3: 229–236.

16 Commercial Exploitation of Natural Surface-Active Agents

Foreword

The reader should not be surprised by the quasi-absence of references in the two sections, Home Care and Cosmetics. This is because of the nature of these markets, where the purchasing decision is entirely with the individual consumers and where, in consequence, their feelings and perceptions prevail. In these markets, nontechnical realities make the success of a product: advertising, marketing campaign, marketing gimmicks, packaging, promotional campaigns, scent, brand images, etc. Individual behaviors dwarf the actual performance of the products. In consequence, an in-depth technical literature is missing as opposed to the other industry segments where a demonstrated and demonstrable effectiveness and cost effectiveness of the products must be provided and confirmed by scientifically supported evidence.

Also, there is a much larger knowledge and familiarity with the products and effects of home care and cosmetic products than for plant protection and oil recovery, such that less background explanations are needed.

16.1 Introduction

The ubiquitous presence of surface-active agents in every aspect of life has been highlighted elsewhere in the book. This chapter will focus on the areas where natural surface-active agents have already established a significant presence. It will discuss the features at the roots of this presence and where/how these features could be exploited to achieve a broader penetration.

In a broad sense, a surface-active agent is expected to provide one or more of the following effects:

- **Wetting,** i.e. covering the surface of a solid by a liquid. (Wetting can also involve covering one liquid with another liquid, immiscible with the first. However, this is less important from a practical point of view.) Wetting is essential in most applications of surface-active agents, but is particularly relevant in laundry cleaning, hard surfaces cleaning, crop protection, paints, inks, coatings, and textile dyeing

Natural Surfactants: Science, Technology, and Industrial Aspects, First Edition. Guido Bognolo and Krister Holmberg.

- **Emulsifying and dispersing,** i.e. compatibilizing nonpolar or solid compounds with polar media or, much less commonly, vice versa. Laundry cleaning, hand dishwashing, crop protection, cosmetics, paints, inks, and coatings are examples. Emulsifying/dispersing is not only a way to make an emulsion or a solid dispersion (suspension), but it also usually provides a reasonable lifetime of the dispersed system. However, it is important to realize that the emulsion or the solid dispersion never becomes thermodynamically stable. Sooner or later, they will always turn into a macroscopic two-phase system.
- **Solubilizing,** i.e. entrapment of nonpolar substances into surfactant self-assemblies in water. The cosmetic and pharmaceutical industries are especially interested in this effect.

Given the very large number of specific applications of surface-active agents, the division above is necessarily a broad brush categorization, but it serves the purpose of guiding the following discussion.

The surface-active products and their usage vary dramatically between regions and countries. Industrialized economies have a complex demand for effects that are covered by an ample range of formulations, rather than individual products, in laundry and home care, personal hygiene, cosmetics, pharmaceutical preparations, crop protection, and a myriad of industrial applications. Millions of tons of anionic, nonionic, cationic, and amphoteric surfactants, fully or partially from nonsustainable sources, are produced and consumed in these countries. Although a minor contributor to the global CO_2 emissions compared to, for example, the steel, cement, and transport industries, the ubiquitous presence of surface-active agents puts them at the center of attention and spurs a mindset and a movement of public opinion that pushes the products and the industry that produces them toward sustainability.

In emerging economies, the needs are much less sophisticated and in line with the socio-economic realities of the countries, ranging from basic to quasi sophisticated. The drive for natural, sustainable, eco-friendly, etc., is much less perceived and demanded.

The majority of the following discussion will reflect the realities of the most industrialized countries, starting with the driving forces shaping the markets for the surface-active agents.

Laundry detergents include the products for the home washing of household textiles, and those for the washing of fabrics from commercial and public institutions like hospitals, nursing homes, hotels, and restaurants. The washing task is different and so are the formulations. In an oversimplified differentiation, household laundry uses sophisticated, "gentle" formulations built around surfactants. Industrial laundry formulations are "harsh" and rely on temperature, alkalinity, and bleaching, with a pinch of surfactant just to provide wetting without foaming. It is difficult to envisage how natural surfactants could be usefully included in these formulations.

16.2 Home Care

16.2.1 Household Laundry Detergents

The washing habits, the washing bundle, the textile types, and the color-versus-white trends led to cleaning not being the principal performance demanded from a laundry detergent formulation. It has become the license to operate, being replaced in consumers' appreciation by stain removal, hygiene, and other benefits once considered as ancillary, like convenience of usage, low temperature wash/energy saving, after-wash scent, and color fidelity.

The traditional workhorses among surfactants for cleaning, alkylbenzene sulfonates and ethoxylated fatty alcohols, are no longer dominant. This is partly because general cleaning performance is no longer perceived by consumers as the most valuable attribute of a cleaning product. Properties such as convenience-in-use, hygiene/antimicrobial effect, and nontoxicity have taken over. Another reason is the rise of nonsurface-active formulation ingredients, like enzymes, soil release polymers, and sustainable chelating agents.

The antimicrobial properties of biosurfactants, in particular of the sophorolipids, might respond to the demand for hygiene and nontoxicity, while at the same time providing acceptable surfactancy. Sophorolipids, besides having useful antimicrobial properties, are usually low foaming, which is a particularly useful property in drum washing machine formulations.

Other natural surface-active agents have the physicochemical properties to enter this market, as total or partial replacements for petrochemically or bio-based surfactants. *In primis* in the line of potential substitutes are the saponins, whose surface-active properties have been known and exploited for centuries. The high foaming characteristics of rhamnolipids would be suitable for the formulation of products for the niche market of the hand washing of delicate fibers.

However, for this to happen, saponins or other prospective natural surfactants must be available at cost-effective prices and in the huge volumes demanded by the laundry cleaning market. It is important to put the accent on both aspects. It is not the absolute price against the benchmark of the surfactants or surfactant combinations presently used that matters. It is the value of the effect provided that must be considered. Unfortunately, there is not much information about such aspects in the open literature. However, it is likely that the large detergent companies (the "big soapers") have carried out these types of comparative studies. Assuming that cost-effectiveness is demonstrated in one or more of the key performance aspects, it is the sheer product availability that will drive the game.

In several countries, there exist laundry products formulated with natural surface-active agents, but these are typically the expression of local producers addressing local markets. The fact that they are on the market since some time and that their number is on the rise can be taken as an indication of their acceptance, but so far none of the big soapers has come forward with concrete propositions, although Unilever in 2020 launched the Clean Future Initiative, aimed at replacing ingredients derived from fossil fuels with alternatives from renewable or recycled sources by 2030. This transition does not necessarily mean a transition to natural surfactants because bio-based surfactants would also qualify as renewable.

In conclusion, it appears that in industrialized countries, there is a distinct possibility that natural surface-active agents may achieve a position in the home laundry detergents segment, but the probability for this to happen on a large scale in the foreseeable future is flimsy.

In emerging economies, biological processes and naturally occurring amphiphiles are an opportunity to rely on domestic resources for the production of surfactants, and to disengage from imported petrochemical feedstock, both for the hydrophobic moieties and for the hydrophilic building block ethylene oxide. In some instances, there might arise the dilemma of using agricultural soil to produce a useful but nonessential good instead of food. It can be argued, however, that saponin-rich species grow on barren land that is not well suited for agriculture. Another aspect is that biosurfactants grown on wastes will not threaten farmland.

16.2.2 Hand Dishwashing Cleaners

A hand dishwashing product is appreciated if it is nonirritating to the skin, keeps dispersed the oils and greases that have been removed mechanically, and retains this capacity across the washing task such that no extra addition of product is needed (economy-in-use). Oil and grease removal/dispersion is associated in consumers' appreciation with the presence of an abundant and rich foam: as long as the washing solution is foaming it cleans. True or not, this is the perception. There are other ancillary attributes that make the selling of a hand dishwashing product, but, mildness and foaming are the condition *sine qua non*, they are a must.

On paper, rhamnolipids offer an excellent combination of these properties, and it is not by accident that they are already used in a limited number of mass consumer products. There are undoubtedly further progresses needed to reduce their cost and confirm a large-scale availability, but it seems that hand dishwashing is a market segment where natural surfactants could establish a sustainable presence.

16.2.3 Hard Surface Cleaners

Products for the cleaning of hard surfaces like tiled walls and ceramic tiles must be low foaming, leave no streaking/filming residues, and require no rinsing. No special cleaning power is demanded, as the soil is mostly inorganic (dust, hard water residues) and the mechanical rubbing does much of the job. However, the wash solution must rapidly and uniformly wet the hard surface. Sophorolipids can provide an adequate performance and given a cost-effective price and availability in relatively large volumes, they are realistic candidates to replace conventional surfactants.

16.3 Cosmetics

The term cosmetics used throughout this section refers to substances or combinations of substances intended for application to the body for cleansing, beautifying, promoting attractiveness, or altering appearance. Personal care is often used instead with specific reference to daily hygiene and grooming products. There is obviously an overlap and there is also a regional difference. The term cosmetics is more common in Europe while personal care is more widely used in the United States.

The cosmetic industry provides a relatively favorable environment for the introduction of novel natural surface-active agents for several reasons.

First, the regulations governing cosmetic products vary from country to country, but generally they involve:

- Lists of substances forbidden in cosmetic and personal care preparations, or subject to restrictions of use, e.g. Annex II and III of Regulation (EC) 1233/2009
- Regulations on the labeling and the claims that can be made. For example, in Europe, a cosmetic product cannot be called "Natural" if it contains less than 50% of naturally derived raw materials (ISO 22715)
- Nontesting on animals, e.g. Regulation (EC) 1233/2009
- Permitted colorants
- Permitted preservatives

Thus, in principle, any substance which is not included in the negative lists, and which is not a colorant or a preservative can be used, the responsibility of the decision being left with the formulator provided a generic reassurance of its safety to humans and the environment is provided.

For surface-active agents, a reassurance of safety to humans and to the environment assessed by acute oral toxicity, eye and skin irritation, aquatic life toxicity, and biodegradability are generally considered adequate and sufficient information.

Second, there are many immaterial attributes that are extensively used to market a formulation not least the real or perceived features and benefits of its ingredients. Mildness, moisturizing, soothing, and emolliency are effects perceived subjectively by the users, and there is a large variety of existing or prospective substances that can be claimed to deliver them, without the need for an objective substantiation.

Last, but not least, the market is huge and is never going to fail: it is a reality that the personal care industry uses products to sell dreams. And people need dreams. Dreams to be beautiful, to please, to attract, to look forever young.

Cost-effectiveness is much less important in cosmetics than in other markets. For the top of the range products, cost is irrelevant. It becomes an issue only for the mass consumer brands, where the claims of effect are minimal and the cost/performance in terms of surfactancy properties against conventional surfactants is more relevant.

Natural surfactants have the emotional and captivating advantage of the image, and the other positive features associated with their natural origin. They are undoubtedly sustainable and do not damage the environment. In this respect, they are the obvious and attractive alternative to the traditional surfactants, which are often wholly or partially petroleum-based and which are reluctantly used by the cosmetic formulator as an unavoidable evil. However, those of microbial origin are sometimes viewed with suspicion and concern because of the potential risk of microbial contamination. The commitment and the reassurance from the biosurfactants producers of strictly controlled production processes and rigorous purification practices cannot exclude the risk of an accident. Even if nonpathogenic microorganisms are involved in the biosynthesis, an emotional barrier may remain with some formulators.

Sophorolipids were the first biosurfactants commercially available, and after the initial failure in the oil industry, the production facilities already put in place, as well as the products, had to find a new use. The attention was directed to cosmetics, the more easily accessible and potentially open market where they eventually received attention. Their antibacterial activity [1] and moisturizing power rather than their surfactancy began to be timidly used in color cosmetics, make-ups, anti-dandruff, anti-acne, and deodorant formulations, extending progressively from there to skin care. They began to be used to formulate moisturizing lotions, mild cleansing milks, creams, and emulsions with a rheological behavior ensuring a smooth and pleasant application on the skin. With time sophorolipids have proven to be prominent ingredients with antimicrobial action in topical cosmetic creams and for oral hygiene [2, 3].

Rhamnolipids possess the performance attributes of cleansing, foaming, and mildness that are also of potential interest for the large market segment of rinse-off products, like hair and body shampoos, and shower gels.

Mannosylerithritol lipids might complement or even outperform ceramides.

There is presently a gap in bio-based high molecular weight emulsifiers, as alkylpolyglucosides higher than C12–C14 are difficult to produce. Trehalose lipids might fill that gap.

The preservation of cosmetics, both in package and after package opening, is a real challenge, as the preservatives allowed and their level of usage are strictly regulated. The antibacterial properties of many biosurfactants are a definite asset as they may reduce or in the best case eliminate the need for preservatives.

The cosmetic industry is constantly seeking sulfate-free and polyethylene glycol-free surfactants and emulsifiers, and although natural surfactants are not unique in this respect, this is an additional bonus to the other effects they provide.

Today natural surfactants have established an ubiquitous presence in cosmetic products with a penetration that is constantly growing in terms of formulations types, species used, and volumes consumed, and this trend will continue and gain momentum as natural surfactants and their features are better known and understood, their safety in use and the absence of untoward effects is confirmed and as they become more competitive vis-à-vis the petrochemical or the bio-based surfactants both in terms of price and availability.

The cosmetic industry is constantly on the search for original effects providing perceived benefits and/or reinforcing existing dreams or creating new ones. Natural surfactants present unique opportunities in this respect. It is conceivable that as formulation and market experience accumulate and more biosurfactants become available, other consumer appealing effects will become accessible, for example hair repair and hair growth, for which mannosylerythritol lipids have already shown promise. It is also conceivable that completely new effects may be discovered, such as skin rejuvenation, skin repair, and wrinkle reduction.

16.4 Plant Protection

Plant protection products are "pesticides" that protect crops or desirable or useful plants. Pesticides prevent, destroy, or control harmful organisms (pests), or protect plants or plant products during production, storage, and transport. They are primarily used in the agricultural sector but also in forestry, horticulture, amenity areas, and in home gardens.

Pesticides contain at least one active substance and have one of the following functions [4]:

- protect plants or plant products against pests/diseases, before or after harvest
- influence the life processes of plants (such as substances influencing their growth, excluding nutrients)

- preserve plant products
- destroy or prevent growth of undesired plants or parts of plants

Depending on their functions, pesticides include several categories, of which herbicides (selective and nonselective, preemergent and postemergent) are the largest, followed by insecticides, fungicides, and seed treatment.

Plant protection actives need to be formulated, and in function of their physical chemical properties and their mode of action, they are commercialized in a variety of presentations, for example emulsifiable concentrates, concentrated emulsions, dispersible powders, suspo-emulsions, water soluble granules. Other than the active, the formulations usually contain surfactants for a variety of purposes.

Surfactants are used in plant protection formulations in order to assist the delivery of the actives to the target sites, be it plants, or pathogens, or as vectors. Surfactant helps in making a good contact with the plant surfaces, increases the retention of the pesticide on the sites for long enough to get the active ingredient absorbed into the plants, and finally enables them to reach the target sites of action. Ideally, a surfactant should provide the following benefits:

- Reduce surface tension and enable surface wetting by the spray
- Improve spreading
- Solubilize spray material
- Enhance spray retention
- Increase active penetration

A single surfactant cannot deliver all those effects and generally combinations of anionic sulfonates and ethoxylated nonionics are used for the liquid formulations. A wetting agent and an anionic polyelectrolyte are often used for the powders.

The remarkable wetting power of certain mannosylerythritol lipids is an example of the distinctive and differentiating effects that some natural surfactants have. In that respect, they can be superior to the traditional surfactants used today.

Phytotoxicity, i.e. the detrimental effect that a chemical can have on a plant, is a major drawback of many conventional surfactants irrespective of whether they are petrochemically based or bio-based. In some instances, effective wetting agents or emulsifiers may have to be excluded in favor of less performing substances because of discovered phytotoxicity, leading to reformulation of the crop protection product. Natural surfactants with their biological origin should be attractive propositions.

In plant protection, surfactants can exert a function beyond the simple formulation. Particular structures when used under specific conditions enhance the biological efficiency of the formulation. In this case, they are referred to as "adjuvants." An adjuvant (sometimes also called synergist or an excipient) is defined as a substance other than water that does not have significant pesticidal properties, but which enhances the effectiveness and efficacy of herbicides, insecticides, and fungicides.

Surface-active adjuvants work via physical or mechanical effects, increasing adhesion, spreading, penetration, and in-plant translocation of the actives. For a number of years, ethoxylated (15–20 EO) tallow amines were used in thousands of tons as formulation or tank mix adjuvants for the nonselective herbicide Round-up® from Monsanto before being phased out when it was shown that they were more toxic to aquatic life than glyphosate alone.

Sophorolipids are used together with polyglycerol esters and biodegradable siloxanes in a commercial, biodegradable wetting adjuvant [5].

Rhamnolipids are efficient bio-fungicides. A large number of studies describe their activity on fungi and oomycetes [6]. A rhamnolipid-based fungicide is sold under the trade name Zonix™ by **Jeneil Biotech** Inc. (Saukville, WI, USA).

The wetting ability of mannosylerythritol lipids on plant leaf surfaces has been evaluated based on contact angle measurements. It was found that contact angles of 22°or lower were achieved, which was below what was obtained with conventional nonionics, such as fatty alcohol ethoxylates or ethoxylated sorbitan esters [7]. Mannosylerythritol lipids are not only efficient wetting and spreading agents on leaves, but they are also effective in suppressing powdery mildew in wheat plants [8].

Plants have developed complex defense mechanisms to enhance resistance to phytopathogens. Rhamnolipids trigger local resistance against plant pathogens [9, 10].

Lipopeptides induce plants' systemic resistance. Surfactin, and to a lower extend fengycin, induced significant disease reduction in bean and tomato infected with *Botrytis cinerea* when applied as pure compounds at micromolar concentrations [11]. A study performed with a large range of natural *Bacillus* isolates strengthened the role of surfactin as systemic resistance inducer since a strong correlation was observed between the defense-inducing activity and the amount of surfactin produced by the different strains [12].

Rhamnolipids, lipopeptides, sophorolipids, mannosylerithritol lipids, and cellobiose lipids have been investigated for their antimicrobial properties toward plant pathogens [8, 13–16].

Plant protection is heavily regulated and the regulations in many regions have become more restrictive over the years. For example, in the European Union, the criteria for safety and environmental impact assessment were switched from risk-based to hazard-based in 2011. Before an active substance, a safener, or a synergist can be used in a plant protection product in the European Union, it must be approved by the European Commission.

In the United States, there is a distinction between formulation and spray additives. Formulation additives (surfactants and adjuvants) added to the pesticide by the manufacturer are inert ingredients on the label of the products as sold. Spray adjuvants are separate products added to the pesticide by the applicator, according to label instructions.

Since surfactants/adjuvants do not have pesticidal properties in their own right, they are not required to be registered by the U.S. Environmental Protection Agency (EPA). Many pesticide labels have specific instructions on what type of spray adjuvant to use and how to use/mix it with the pesticide product. These instructions must be followed to be in compliance with the law. In Brazil, adjuvants do not need to be registered. Surfactants are considered to be adjuvants.

16.5 Food

Surfactants are broadly used in food as "direct food additives" to confer emulsifying, foaming, thickening, texture-improving, and/or preserving properties, or to incorporate fat-soluble substances such as vitamins. They are also used as "indirect food additives" to deliver antiadhesive/antimicrobial protection or surface cleaning, essential for example in food preparation, storage, and packaging.

Traditional surfactants are linked with many health-related issues and drawbacks, such as intestinal dysfunction [17], which might lead to severe intestinal permeability, which in turn may elicit various allergic and autoimmune diseases [18]. There is therefore a general interest in effective surface active and emulsifying agents with low or no untoward effects, *in primis* natural surfactants, such as vegetal and egg yolk, saponins, and milk proteins.

A key differentiating feature of the food industry compared to other industries is the much higher level of regulations.

In the European Union, Regulation EC 1333/2008 sets the rules on food additives, definitions, conditions of use, labeling, and procedures. All food additives in the EU must be authorized and listed with conditions of use in the EU's positive list based on:

- A safety assessment
- The technological need
- Ensuring that the use of the additive will not mislead consumers

In the United States, food ingredients are governed by the Food and Drug Administration (FDA) regulations, Title 21 (21CFR-172.590). The list comprises approved food additives, based on toxicological properties and history of safety-in-use. Microorganism-produced ingredients may be the subject of a GRAS notice. For example, various lactic acid bacteria could be used as food additives as they are all permitted GRAS strains.

In Brazil, the National Sanitary Surveillance Agency (ANVISA in Portuguese) has recently published the Resolution RDC No. 740/2022 that authorizes the use of food additives and technological aids on the basis of positive lists in various food categories.

In Japan, the Ministry of Health, Labour and Welfare regulates food additives under a "positive list" system, meaning that additives must be analyzed, approved, reviewed, and added to the list before they can be used in foods in Japan.

In China, additives are allowed in food if they are:

- Covered by National Food Safety Standards GB 2760 and GB 14880
- Within the list of allowable food additives published by the Ministry of Health
- Within the scope of allowed applications and dosage levels

In general, the regulation favors GRAS or additives specified in positive lists because of their benign toxicology and the proven record of safety in use. The approval of new substances is subject to stringent protocols, which discourages the registration of new substances, unless they hold the promise of exceptional performance improvements.

Further penetration of naturally occurring surfactants in food will, in consequence, come from an extended use of existing approved substances. These substances may also prove beneficial in new areas of application, for example the prevention of formation of bacterial films or formation of liposomes that will enable the encapsulation and delivery of vitamins.

The field of food regulatory is complex and requires attentive examination for both assessing the authorization to use additives, among others naturally occurring surface-active agents, the level of use allowed, as well as the categories of foods in which they are admitted. The discussion below just provides an example of this complexity.

In the European Union, the only approved naturally occurring surfactants in Annex II of EC 1333/2008 (list of food additives approved for use and conditions of use) are:

- Lecithins, E-322, identified as mixtures or fractions of phosphatides obtained by physical procedures from animal or vegetable foodstuffs. Their specifications have been defined in Commission Regulation (EU) No. 231/2012. Lecithins may be used on the basis of the principle of *quantum satis*.
- Quillaja extracts, E-999, identified as Quillaja extract CAS 68990-67-0, may be used on the basis of an acceptable daily intake (ADI) of 3 mg/kg of body weight (the acceptable daily intake is defined as the daily intake of a substance which, over the entire lifetime of a human, appears to be without adverse effects or harm to the health of that human).

The authorizations for both lecithins and Quillaja extracts have been reviewed in response to requests to extend their use to other food categories than those for which they are currently approved. The extension of the use of E-999 in solid and liquid food supplements and as a carrier in botanical nutrients was rejected on the basis of the ADI principle [19]. Conversely, the request for the extension of the use of lecithins as additive in foods for infants below 16 weeks of age was granted [20].

16.6 Oil Industry

16.6.1 General

Through the years, the oil industry has consumed surfactants to assist in a variety of tasks from equipment cleaning, oil transportation, soil and oil spill remediation, crude oil demulsification, corrosion inhibition, scale inhibition, defoaming/antifoaming, etc. Considering the number and the fragmentation of the uses, it is not possible to assess with any degree of accuracy the penetration of natural surfactants. There is however a certain consumption, driven principally by their benign toxicology to humans and the marine fauna and by their biodegradability.

It is ironic that this latter aspect turned into a drawback at least in one case. As discussed in the chapter on lipopolysaccharides, the failure of the potent bioemulsifier Emulsan for the pipeline transport of low viscosity emulsions of highly viscous crude oil was due to its biodegradability.

It is fair to say that, to date and in the foreseeable future, the most promising outlet of natural surfactants in relation to the oil industry will be in microbial enhanced oil recovery (MEOR).

16.6.2 Microbial Enhanced Oil Recovery

Several enhanced oil recovery processes are currently performed worldwide using thermal or various types of chemical processes, such as polymer flooding, surfactant flooding, alkali flooding, and combinations of these. Because of their costs and their environmental impact, the use of surfactants is limited. However, there are examples where conventional surfactants are used with good results. China is in the forefront of this development and the largest surfactant flooding operation worldwide is currently at the Daqing oil field where a 0.3% solution of a mixture of petrochemically based surfactants gives a considerable increase in the oil recovery. The use of naturally occurring surfactants would be the answer to the environmental issue, but at a prohibitive cost, such that they have remained the subject of laboratory experience, thereby becoming a favorite subject for the production of an abundant (yet absolutely unrealistic) literature on their potential.

The production of biosurfactants in situ, i.e. in the reservoir, by microorganisms feeding on the oil itself, is conceptually appealing as it might combine environmental aspects and cost-effectiveness, despite numerous hurdles, as we shall see. Tests have been made around the world with generally positive results, although not yet conclusive. Also here, China is currently in the lead with Shengli and Daqing oil fields running many large trials. The large variability of the oilfields' morphology and environment, the crude oil composition, the temperature in the reservoir, and the salinity of the reservoir water are factors that demand different strains and operational methodologies for different oil fields. This causes major difficulties, along with the issue of identifying, farming, and bringing in place the suitable strain.

MEOR is a process in which microorganisms are injected into a mature oil reservoir for the recovery of residual crude oil that was not extracted during the initial and secondary extraction processes. The idea behind MEOR is that when favorable conditions are present in the reservoir, the introduced microbes grow exponentially, and their metabolic products would mobilize the residual oil cost-effectively.

Figure 16.1 shows a model of a microbial enhanced oil recovery operation.

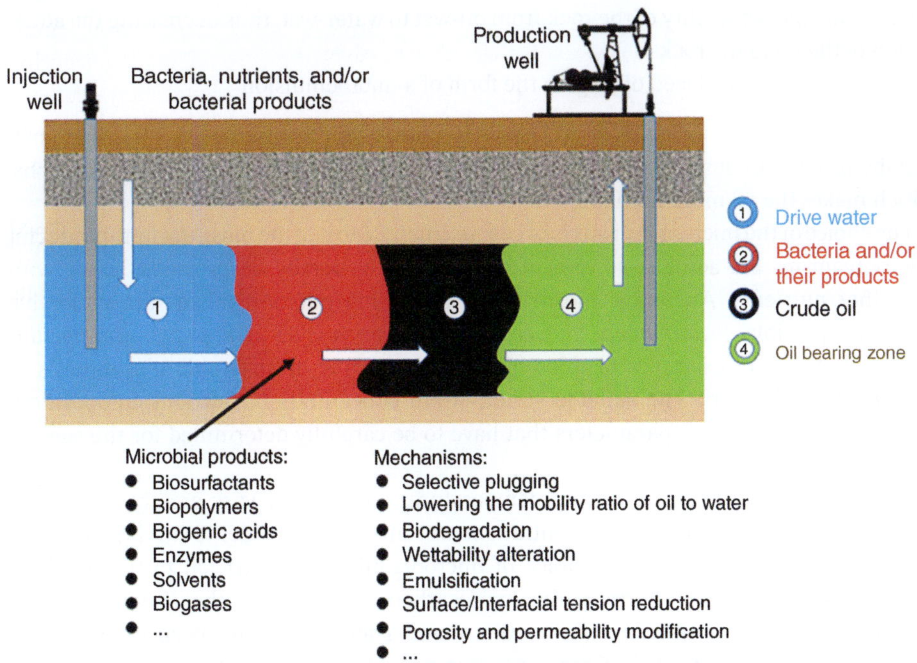

Figure 16.1 Model of a microbial enhanced oil recovery operation and some of the processes that are taking place. (Niu et al. [21] / with permission of Elsevier.)

In the model shown in Figure 16.1, the microorganisms together with nutrients are injected through the water injection well. A common nutrient is molasses, a cheap byproduct from the sugar industry. Then follows a shut-in phase of months, in which the bacteria produce a series of metabolites that contribute to the displacement and mobilization of the crude oil. The oil is recovered via the production well. The process is usually repeated several times. The cycles of injection, shut-in, and production are sometimes referred to as "microbial huff and puff," although "huff and puff" in enhanced oil recovery originally relates to cycles of a steam injection phase, followed by a pause phase, and finally a production phase. Alternatively, the injection may take place via the production well. Also in this case a shut-in phase of several months is necessary. The advantages/disadvantages of the two methods have been discussed by Gao and Zekri [22].

The idea that microbes could remove oil from porous media is not recent. The concept of MEOR was proposed a century ago [23]. In 1946, for the first time, Zobell proposed a model for MEOR, which included reduction of oil–water interfacial tension, increase in rock permeability, and reduction of viscosity of crude oil by microbial strains [24]. The first field test of MEOR was conducted by Yarbrough in 1954 [25]. In the 1970s, MEOR began to be regarded as a promising, economically feasible, and scientifically confirmed tertiary oil recovery process. Since then, many oil-producing nations have conducted field trials with generally encouraging results [26–28].

The injected microorganisms feed on the crude, producing biosurfactants and these promote oil displacement and movement in oil-bearing rocks by [27, 29]:

- Reducing the interfacial tension between oil-brine and oil-rock
- Changing the wettability of the rock from oil-wet to water-wet, thus decreasing the adhesion of the oil to the rock
- Solubilizing the displaced oil, e.g. in the form of a microemulsion

Furthermore, the production of biosurfactants can contribute to the microbial metabolism of viscous oils, in turn resulting in the formation of lighter hydrocarbons, which makes the oil more fluid [30].

The choice of the microorganisms is of course crucial. Most of the biosurfactant-producing microorganisms are aerobic, so oxygen supply must generally be provided, along with the other nutrients. Anaerobic biosurfactant-producing microorganisms are preferable whenever possible. *Pseudomonas stutzeri* Rhl, *Pseudomonas aeruginosa* SG, and *Bacillus subtilis* AnPL-1 are examples of microbial strains capable of anaerobic biosynthesis of biosurfactants [31–33]. The suitable strain, strain concentration, nutrient composition, and treatment period are parameters that have to be carefully determined for the success of the recovery operation.

In developed petroleum reservoirs, temperatures and salinity vary greatly but can be higher than 70 °C and have salt concentrations well above 5%. In order to survive such high temperatures and salinities, the choice of microorganisms is restricted to thermophilic and halophilic species. Thermophilic hydrocarbon degraders of the genera *Bacillus*, *Thermus*, *Thermoanaerobacter*, *Thermococcus*, and *Thermotoga* have been isolated from high-temperature reservoirs in China, resisting up to 90 °C. [34]. Ollivier et al. reported that methanogenic *Methanocalculus halotolerans* isolated from an oil well grows at the highest reported salt concentration – up to 20% NaCl and at temperatures up to 45 °C [35].

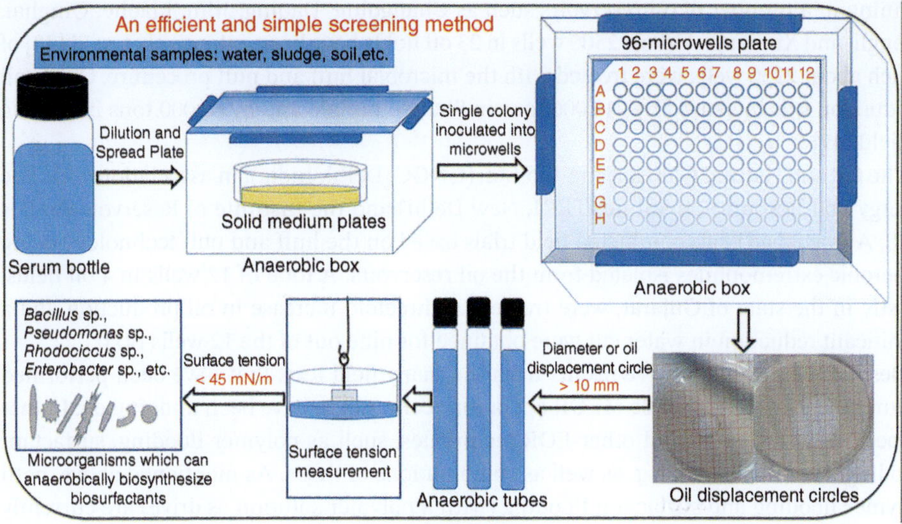

Figure 16.2 Method for screening microbial strains capable of anaerobic biosynthesis of biosurfactants. (Zhao et al. [36] / Frontiers Media S.A. / CC BY 4.0.)

A simple method for screening microbial strains capable of anaerobic biosynthesis of biosurfactants has been proposed by Zhao et al. and is shown in Figure 16.2 [36].

Nikolova and Gutierrez report that more than 400 MEOR field tests have been conducted in the United States alone [37]. However, the majority were single-well stimulation treatments on low-productivity wells; hence, reliable data are sparse. Although single-well stimulations can give benefits under certain circumstances, some major operators, such as Shell, consider that the MEOR process needs to have a field-wide impact [37]. On the other hand, smaller oil and gas companies will consider single-well stimulation if this is an important method by which they produce oil. MEOR may be attractive to independent oil producers, who mostly produce on average less than half a tonne of oil per day from "stripper wells," i.e. oil or gas wells that are nearing the end of their economically useful life [38].

In Europe, MEOR field projects are mostly confined to Norway and conducted by the company Equinor [39], Poland (RAM Biomedicals, Polish and Gas Institute) [40], and Germany (BASF, Wintershall) [41].

In several oil fields in Romania, the average oil production was improved between 100 and 200% after MEOR application [42].

In the United States, field trials have been conducted in the Lisbon oil fields (Arkansas) with injection of *Clostridium acetobutylicum* [42] and in the Bebee field (Oklahoma) using *Bacillus* strains [43].

The crude oil production was increased from 3.6 to 5.9 tons/day by endogenous MEOR technology in the oil reservoirs in Saskatchewan, Canada [44]. The injection of anaerobic fermentation and hydrocarbon-degrading microbes in the Piedras Coloradas oil field of Argentina for 12 months increased oil production by 66% in the six production wells [45].

In China, field trials only started in the last 20 years and show some promising, although mixed, results [46, 47]. There is currently a strong interest in MEOR and there are field tests running in a number of oil reservoirs such as Changqing, Daqing, Jilin, Liaohe, Qinghai, Shengli, and Xinjiang. Almost 2500 wells in 23 oil fields have been extensively tested [48] of which about 2150 wells were treated with the microbial huff and puff procedure. Crude oil production was increased by 219 000 tons in Shengli oilfield and by 64 000 tons in Daqing oilfield [49].

The Oil and Natural Gas Corporation Ltd. (ONGC, Delhi, India), in association with The Energy and Resources Institute (TERI, New Delhi) and the Institute of Reservoir Studies (IRS, Ahmedabad), has conducted field trials based on the huff and puff technology using anaerobic extremophiles isolated from the oil reservoirs. A total of 12 wells in 4 oil fields, mostly in the state of Gujarat, were treated. A threefold increase in oil production and a significant reduction in water cut were obtained for nine out of the 12 wells [27].

Despite the mostly positive results from the many field tests that have been performed around the world, a full-scale MEOR operation seems not to have been undertaken. In that respect, MEOR lags behind other EOR techniques, such as polymer flooding, surfactant flooding, and alkali flooding, as well as combinations of those. As mentioned above, both polymer flooding and surfactant flooding (with a polymer solution as drive) are currently conducted at the Daqing field. A general consensus about the economic feasibility of MEOR has, therefore, not been reached among oil operating companies [37].

Nikolova and Gutierrez have summarized the advantages and disadvantages of MEOR [37]:

- The injected bacteria, nutrients, and/or other natural products can be produced using inexpensive and easy-to-obtain raw substrates or even waste materials, and they are not affected by crude oil price as is the case with conventional EOR processes.
- It is an economically attractive alternative for use in mature oil fields prior to their abandonment.
- No major alteration of the existing field facilities and infrastructure is required to apply the process, making it cheaper and easier to implement than many other EOR methods.
- Microbial processes consume less energy than thermal EOR processes.
- The process is primarily suited for carbonate oil reservoirs where some EOR technologies would not achieve desired results.
- The benefits of bacterial activity within the reservoir are amplified with time, whereas the opposite is true for other EOR technologies.
- It largely involves fully biodegradable products/additives and, hence, is considered a more environmentally compatible process than most other EOR processes.
- Microbial processes can be stimulated in situ within the reservoir, thus minimizing or eliminating the need to accommodate large storage facilities onsite/offshore.

These advantages are based on the assumption that MEOR is successful in the field – i.e. the bacteria, nutrients, and/or bacterial products are uniformly distributed in the swept zone of the reservoir.

Most of the published data on MEOR is based on laboratory tests where MEOR is shown to work. However, these results are rarely replicated in the field due to the general unpredictability of reservoir conditions, such as variable high temperature and pressure, salinity, multiphase flow of fluids in the reservoir, and uncertainty of the rock formation. The real value of MEOR can only be fully assessed on field scale and this has not yet been demonstrated due to the apparent deficiency of both consistency in data collection and processing, and generation of quantitative data.

Below is a summary of the limitations of MEOR processes:

- It is a complex process because the desired bacterial activities depend on the physical and chemical characteristics of the reservoir.
- The majority of MEOR field projects are conducted on stripper wells, which renders MEOR a low incremental oil recovery process.
- It is a slower process than chemical or thermal EOR, and it usually takes weeks or even months before any benefits are observed.
- MEOR is hard to control once implemented in the field, and its success is difficult to predict due to high heterogeneity of reservoirs.
- The cultivation of microorganisms in the laboratory that can grow and/or produce the desired metabolic products (e.g. biosurfactants) under reservoir conditions has proven difficult.

References

1 Van Bogaert, I.N.A., Zhang, J., and Soetaert, W. (2011). Microbial synthesis of sophorolipids. *Process Biochemistry* 46: 821–833.

2 Lydon, H.L., Baccile, N., Callaghan, B. et al. (2017). Adjuvant antibiotic activity of acidic sophorolipids with potential for facilitating wound healing. *Antimicrobial Agents and Chemotherapy* 61: e02547–e02516.

3 Elshikh, M., Moya-Ramírez, I., Moens, H. et al. (2017). Rhamnolipids and lactonic sophorolipids: natural antimicrobial surfactants for oral hygiene. *Journal of Applied Microbiology* 123: 1111–1123.

4 https//http://food.ec.europa.eu/plants/pesticides_en (accessed 13 April 2025).

5 Krause, D. (2022). Evonik Communication Leaflet: Sustainable Surfactants for Crop Protection solutions. Evonik Brochure 12.2022.

6 Crouzet, J., Arguelles-Arias, A., Dhondt-Cordelier, S. et al. (2020). Biosurfactants in plant protection against diseases: Rhamnolipids and lipopeptides case study. *Frontiers in Biotechnology and Bioengineering* 8: 1014.

7 Fukuoka, T., Yoshida, S., Nakamura, J., Koitabashi M., Sakai H., Abe M., Kitamoto D., Kitamoto H. (2015). Application of yeast glycolipid biosurfactant, mannosylerythritol lipid, as agrospreaders. *Journal of Oleo Science* 64: 689–695.

8 Yoshida, S., Koitabashi, M., Nakamura, J. et al. (2015). Effects of biosurfactants, Imannosylerythritol lipids, on the hydrophobicity of solid surfaces and //nfection behaviors of plant pathogenic fungi. *Journal of Applied Microbiology* 119: 215–224.

9 Vatsa, P., Sanchez, L., Clément, C. et al. (2010). Rhamnolipid biosurfactants as new players in animal and plant defense against microbes. *International Journal of Molecular Sciences* 11: 5095–5108.

10 Monnier, N., Cordier, M., Dahi, A. et al. (2020). Semi-purified rhamnolipid mixes protect *Brassica napus* against *Leptosphaeria maculans* early infections. *Phytopathology* 110: 834–842.

11 Ongena, M., Jourdan, E., Adam, A. et al. (2007). Surfactin and fengycin lipopeptides of *Bacillus subtilis* as elicitors of induced systemic resistance in plants. *Environmental Microbiology* 9: 1084–1090.

12 Cawoy, H., Mariutto, M., Henry, G. et al. (2014). Plant defense stimulation by natural isolates of *Bacillus* depends on efficient surfactin production. *Molecular Plant-Microbe Interactions* 27: 87–100.

13 Mnif, I. and Ghribi, D. (2016). Glycolipid biosurfactants: Main properties and potential applications in agriculture and food industry. *Journal of the Science of Food and Agriculture* 96: 4310–4320.

14 Sen, S., Borah, S.N., Bora, A., and Deka, S. (2017). Production, characterization, and antifungal activity of a biosurfactant produced by *Rhodotorula babjevae* YS3. *Microbial Cell Factories* 16: 95.

15 Chen, J., Liu, X., Fu, S. et al. (2020). Effects of sophorolipids on fungal and oomycete pathogens in relation to pH solubility. *Journal of Applied Microbiology* 128: 1754–1763.

16 Penha, R.O., Vandenberghe, L.P.S., Faulds, C. et al. (2020). *Bacillus* lipopeptides as powerful pest control agents for a more sustainable and healthy agriculture: recent studies and innovations. *Planta* 251: 70.

17 Glynn, A., Igra, A.M., Sand, S. et al. (2017). Are additive effects of dietary surfactants on intestinal tight junction integrity an overlooked human health risk?—a mixture study on Caco-2 monolayers. *Food and Chemical Toxicology* 106: 314–323.

18 Csáki, K.F. (2011). Synthetic surfactant food additives can cause intestinal barrier dysfunction. *Medical Hypotheses* 76: 676–681.

19 Follow-up of the re-evaluation of quillaia extract (E 999) as a food additive and safety of the proposed extension of uses. EFSA Journal, 6 February 2024.

20 Opinion on the re-evaluation of lecithins (E 322) as a food additive in foods for infants below 16 weeks of age and follow-up of its re-evaluation as food additive for uses in foods for all population groups. EFSA Journal, 10 November 2020.

21 Niu, J., Liu, Q., Lv, J. et al. (2020). Review on microbial enhanced oil recovery: Mechanism, modelling and field trials. *Journal of Petroleum Science and Engineering* 192: 107350.

22 Gao, C.H. and Zekri, A. (2011). Applications of microbial-enhanced oil recovery technology in the past decade. *Energy Sources, Part A: Recovery, Utilization, and Environmental Effects* 33: 972–989.

23 Samanta, A. (2021). Biosurfactants and its application in oil recovery. In: *Surfactants in Upstream E&P* (ed. T. Solling, M.S. Kamal, and S.M.S. Hussain), 181–205. Cham, Switzerland: Springer.

24 Zobell, C.E. (1946). Action of microorganisms on hydrocarbons. *Bacteriology Reviews* 10: 1–49.

25 Bryant, R.S. and Douglas, J. (1988). Evaluation of microbial systems in porous media for EOR. *SPE Reservoir Engineering* 3: 489–495.

26 Ke, C.-Y., Lu, G.-M., Li, Y.-B. et al. (2018). A pilot study on large-scale microbial enhanced oil recovery (MEOR) in Baolige Oilfield. *International Biodeterioration & Biodegradation* 127: 247–253.

27 Sen, R. (2008). Biotechnology in petroleum recovery: the microbial EOR. *Progress in Energy and Combustion Science* 34: 714–724.

28 She, H., Kong, D., Li, Y., and et al (2019). Recent advance of microbial enhanced oil recovery (MEOR) in China. *Geofluids* 2019: 871392.

29 Sen, R. (2010). *Biosurfactants*. New York, N.J.: Landes Bioscience and Springer Science+Business Media, LLC.

30 Shibulal, B., Al-Bahry, S.N., Al-Wahaibi, Y.M. et al. (2014). Microbial enhanced heavy oil recovery by the aid of inhabitant spore-forming bacteria: an insight review. *Scientific World Journal* 2014: 309159.

31 Zhao, F., Zhang, J., Shi, R. et al. (2015). Production of biosurfactant by a *Pseudomonas aeruginosa* isolate and its applicability to *in situ* microbial enhanced oil recovery under anoxic conditions. *RSC Advances* 5: 36044–36050.

32 Zhao, F., Shi, R., Zhao, J. et al. (2015). Heterologous production of Pseudomonas aeruginosa rhamnolipid under anaerobic conditions for microbial enhanced oil recovery. *Journal of Applied Microbiology* 118: 379–389.

33 Zhao, F., Zhu, H., Cui, Q. et al. (2021). Anaerobic production of surfactin by a new *Bacillus subtilis* isolate and the *in situ* emulsification and viscosity reduction effect towards enhanced oil recovery applications. *Journal of Petroleum Science and Engineering* 201: 108508.

34 Lin, J., Hao, B., Cao, G. et al. (2014). A study on the microbial community structure in oil reservoirs developed by water flooding. *Journal of Petroleum Science and Engineering* 122: 354–359.

35 Ollivier, B., Fardeau, M.-L., Cayol, J.-L. et al. (1998). *Methanocalculus halotolerans* gen. Nov., sp. nov., isolated from an oil-producing well. *International Journal of Systematic Bacteriology* 48: 821–828.

36 Zhao, F., Wang, Y., Hu, X. et al. (2022). How to simply and efficiently screen microbial strains capable of anaerobic biosynthesis of biosurfactants. *Frontiers in Microbiology* 12: 989998.

37 Nikolova, C. and Gutierrez, T. (2020). Use of microorganisms in the recovery of oil from recalcitrant oil reservoirs: current state of knowledge, technological advances and future perspectives. *Frontiers in Microbiology* 10: 2996.

38 Van Hamme, J.D., Singh, A., and Ward, O.P. (2003). Recent advances in petroleum microbiology. *Microbiology and Molecular Biology Reviews* 67: 503–549.

39 Bødtker, G., Lysnes, K., Torsvik, T. et al. (2009). Microbial analysis of backflowed injection water from a nitrate-treated North Sea oil reservoir. *Journal of Industrial Microbiology and Biotechnology* 36: 439–450.

40 Falkowicz, S., Cicha-Szot, R., Dubiel, S. et al. (2015). Microbial flooding increases recovery factor of depleted Pławowice oil field - from lab to the field. *AGH Drilling, Oil, Gas* 32: 345–356.

41 Alkan, H., Biegel, E., Krüger, M., et al. (2014). An integrated MEOR project; workflow to develop a pilot in a German field. In: *Proceedings of the SPE Improved Oil Recovery Symposium*, Tulsa, OK.

42 Safdel, M., Anbaz, M.A., Daryasafar, A. et al. (2017). Microbial enhanced oil recovery, a critical review on worldwide implemented field trials in different countries. *Renewable and Sustainable Energy Reviews* 74: 159–172.

43 Youssef, N., Simpson, D.R., Duncan, K.E. et al. (2007). *In situ* biosurfactant production by *Bacillus* strains injected into a limestone petroleum reservoir. *Applied and Environmental Microbiology* 73: 1239–1247.

44 Town, K., Sheehy, A.J., and Govreau, B.R. (2009). MEOR success in Southern Saskatchewan. *SPE Reservoir Evaluation & Engineering* 13: 773–781.

45 Maure, M.A, Dietrich F.L., Diaz, V.A., et al. (1999). Microbial enhanced oil recovery pilot test in Piedras Coloradas field, Argentina. Paper presented at the Latin American and Caribbean Petroleum Engineering Conference, Caracas, Venezuela, April 1999. Paper Number: SPE-53715-MS.

46 Weidong, W., Junzhang, L., Xueli, G. et al. (2014). MEOR field test at block Luo801 of Shengli oil field in China. *Petroleum Science and Technology* 32: 673–679.

47 Li, C.-F., Li, Y., Li, X.-M. et al. (2015). The application of microbial enhanced oil recovery technology in Shengli oil field. *Petroleum Science and Technology* 33: 556–560.

48 Wang, W. (2012). Laboratory research and field trials of microbial oil recovery technique. *Oil Drilling & Production Technology* 34: 107–113.

49 Gao, C. (2018). Experiences of microbial enhanced oil recovery in Chinese oil fields. *Journal of Petroleum Science and Engineering* 166: 55–62.

Index

Entries given in *italic* refer to figures and figure captions, entries in **bold** refer to tables.

d

q

r